MARINE ALGAE

Kruger Brentt
Publishers

MARINE ALGAE
Taxonomy and Ecology

Christopher Clark
Flavia Soledad Darqui
Joseph Cook

Kruger Brentt
Publishers
2024

Kruger Brentt Publishers UK. LTD.
Company Number 9728962

Regd. Office: 68 St Margarets Road, Edgware, Middlesex HA8 9UU

ISBN: 978-1-78715-259-5

For information on all our publications visit our website at http://krugerbrentt.com/

PREFACE

The study of marine algae, known as phycology, plays a pivotal role in understanding the intricate dynamics of marine ecosystems. Algae, ranging from microscopic phytoplankton to macroscopic seaweeds, form the foundation of aquatic food webs, contributing significantly to global primary production. The taxonomy of marine algae delves into the classification and identification of these diverse organisms, providing a roadmap to comprehend their ecological roles and relationships. This field not only unveils the evolutionary relationships among different algal groups but also sheds light on their adaptation strategies in response to environmental changes. As marine ecosystems face unprecedented challenges such as climate change and human activities, a comprehensive understanding of the taxonomy and ecology of marine algae becomes essential for informed conservation and management efforts. This introductory exploration into the taxonomy of the ecology of marine algae sets the stage for unraveling the complexities of these fascinating organisms and their vital contributions to the health and sustainability of our oceans.

The taxonomy and ecology of marine algae constitute a captivating realm within the field of phycology, offering profound insights into the intricate web of life that thrives beneath the ocean's surface. In the vast expanse of marine environments, algae, ranging from microscopic diatoms to towering kelps, assume roles of paramount importance. The taxonomy aspect involves the systematic classification of these diverse organisms, unraveling their evolutionary relationships and providing a structured framework for understanding their ecological functions. Simultaneously, the ecological dimension explores the dynamic interplay between marine algae and their surroundings. From their fundamental role as primary producers, contributing significantly to global oxygen production, to their interactions with marine organisms and response to environmental fluctuations, the taxonomy of marine algae serves as a gateway to comprehending the nuanced relationships that shape the health and resilience of marine ecosystems. This intricate interplay holds crucial implications for biodiversity conservation, climate change mitigation, and sustainable management practices in our ever-evolving oceans.

The exploration of marine algae encompasses a fascinating journey into the intricate world of taxonomy and ecology, where the microscopic organisms play pivotal roles in the balance and sustainability of marine ecosystems. This book, Taxonomy and Ecology of Marine Algae, delves into a comprehensive understanding of various aspects related to marine algae, ranging from their taxonomy, adaptability, and diversity to their ecological roles and interactions within the marine environment. The book provides a comprehensive exploration of the taxonomy and ecology of marine algae, covering various aspects such as adaptability, diversity, allelopathic interactions, biochemical traits, morphology, phylogeny, and environmental impacts. The book delves into the trophic modes of dinoflagellates, suppression by bacteria, isolation and identification of algal strains, examination of specific algae species, and the impact of growth phases on marine green algae. The content integrates various scientific disciplines, including microbiology, biochemistry, molecular biology, ecology, and taxonomy, providing readers with a holistic understanding of marine algae from different perspectives. Each chapter presents specific research methodologies, including laboratory culture-based characterization, isolation and identification techniques, numerical investigations, and molecular and morphological identification, offering insights into the scientific methods employed in the field. This book serves as a comprehensive resource for researchers, students, and enthusiasts interested in the taxonomy and ecology of marine algae. Each chapter contributes valuable insights into the diverse and dynamic world of marine algae, fostering a deeper appreciation for their ecological significance and the intricate relationships that shape marine ecosystems.

We are grateful to all those persons as well as various books, manuals, periodicals, newsletters, magazines, journals etc. that helped in the preparation of this book. In spite of the best efforts, it is possible that some errors may have occurred into the compilation and editing of the book. Further queries, constructive suggestions and criticisms for the further improvement of the book are always welcome and shall be thankfully acknowledged.

Christopher Clark

Flavia Soledad Darqui

Joseph Cook

CONTENTS

PREFACE V

1. **TAXONOMY AND ECOLOGY OF MARINE ALGAE** 1

1. Introduction 1

2. Papers Details 2

References 5

2. **PLASTICITY AND MULTIPLICITY OF TROPHIC MODES IN THE DINOFLAGELLATE KARLODINIUM AND THEIR PERTINENCE TO POPULATION MAINTENANCE AND BLOOM DYNAMICS** 7

1. Introduction 7

2. Trophic Modes of Dinoflagellates 11

3. Autotrophy of Karlodinium **13**

4. Osmotrophy of Karlodinium **14**

5. Phagotrophy of Karlodinium **15**

6. Evolution of the Feeding Mechanisms in Karlodinium **20**

7. Mixotrophy in Regulating Population Dynamics and HABs Formation of Karlodinium **21**

8. Perspectives for Future Investigations on the Mixotrophy in Karlodinium **23**

9. Conclusions 25

References 26

3. **ALLELOPATHIC INHIBITION BY THE BACTERIA *BACILLUS CEREUS* BE23 ON GROWTH AND PHOTOSYNTHESIS OF THE MACROALGA ULVA PROLIFERA** 37

1. Introduction 37

2. Materials and Methods 38

3. Results 45

4. Discussion 50

5. Conclusions 54

References 54

4. ISOLATION, IDENTIFICATION, AND BIOCHEMICAL CHARACTERISTICS OF A COLD-TOLERANT *CHLORELLA VULGARIS* KNUA007 ISOLATED FROM KING GEORGE ISLAND, ANTARCTICA... 63

1. Introduction 63

2. Materials and Methods 65

3. Results 67

4. Discussion 70

References 72

5. MULTIDISCIPLINARY ANALYSIS OF CYSTOSEIRA SENSU LATO (SE SPAIN) SUGGEST A COMPLEX COLONIZATION OF THE MEDITERRANEAN 77

1. Introduction 77

2. Materials and Methods 79

3. Results 81

4. Discussion 89

5. Conclusions 91

References 92

6. LABORATORY CULTURE-BASED CHARACTERIZATION OF THE RESTING STAGE CELLS OF THE BROWN-TIDE-CAUSING PELAGOPHYTE, *AUREOCOCCUS ANOPHAGEFFERENS*.. 99

1. Introduction 99

2. Materials and Methods 101

3. Results 103

4. Discussion 107

5. Conclusions 109

References 109

7. MORPHOLOGY AND PHYLOGENY OF *SCRIPPSIELLA PRECARIA MONTRESOR* & ZINGONE (THORACOSPHAERALES, DINOPHYCEAE) FROM KOREAN COASTAL WATERS 115

1. Introduction 115

2. Materials and Methods 116

3. Results 119

4. Discussion 126

References 129

8. CHANGES IN FREE-LIVING AND PARTICLE-ASSOCIATED BACTERIAL COMMUNITIES DEPENDING ON THE GROWTH PHASES OF MARINE GREEN ALGAE, *TETRASELMIS SUECICA* 135

1. Introduction 135

2. Materials and Methods 136

3. Results and Discussion 139

4. Conclusions and Remarks 144

References 144

9. COAGULANT PLUS *BACILLUS NITRATIREDUCENS* FERMENTATION BROTH TECHNIQUE PROVIDES A RAPID ALGICIDAL EFFECT OF TOXIC RED TIDE DINOFLAGELLATE 149

1. Introduction 149

2. Material and Methods 151

3. Results 154

4. Discussion 165

5. Conclusions 168

References 168

10. IDENTIFICATION OF NEW SUB-FOSSIL DIATOMS FLORA IN THE SEDIMENTS OF SUNCHEONMAN BAY, KOREA 173

1. Introduction 173

2. Materials and Methods 174

3. Results 176

4. Discussion 187

References 189

11. FIRST REPORT OF THE MARINE BENTHIC DINOFLAGELLATE BYSMATRUM SUBSALSUM FROM KOREAN TIDAL POOLS..........201

1. Introduction 201

2. Materials and Methods 202

3. Results 204

4. Discussion 208

References 212

12. NEW RECORDS OF THE DIATOMS (*BACILLARIOPHYCEAE*) FROM THE COASTAL LAGOONS IN KOREA 215

1. Introduction 215

2. Materials and Methods 216

3. Results 217

4. Discussion 225

References 227

13. MOLECULAR AND MORPHOLOGICAL CHARACTERIZATION OF *COLACONEMA FORMOSANUM* SP. NOV. (COLACONEMATACEAE, RHODOPHYTA) 233

1. Introduction 233

2. Materials and Methods 235

3. Results 236

4. Discussion 241

References 244

14. DETECTION OF THE BENTHIC DINOFLAGELLATES, *OSTREOPSIS* CF. *OVATE* AND *AMPHIDINIUM MASSARTII* (DINOPHYCEAE), USING LOOP-MEDIATED ISOTHERMAL AMPLIFICATION 249

1. Introduction 249

2. Materials and Methods 251

3. Results 257

4. Discussion 261

5. Conclusions 262

References 262

15. NUMERICAL STUDY ON THE MASSIVE OUTBREAK OF THE *ULVA PROLIFERA* GREEN TIDES IN THE SOUTH-WESTERN YELLOW SEA IN 2021 269

1. Introduction 269

2. Data and Methodology 271

3. Results 273

4. Concluding Remarks 279

References 279

INDEX 283

CHAPTER-1

TAXONOMY AND ECOLOGY OF MARINE ALGAE

1. INTRODUCTION

The term "algae" refers to a large diversity of unrelated phylogenetic entities, ranging from picoplanktonic cells to macroalgal kelps. Marine algae are an important primary producer in the marine food chain, responsible for the high primary production of coastal areas, providing food resources in situ for many grazing species of gastropods, peracarid crustaceans, sea urchins or fish. The biodiversity of marine algae is enormous, and they represent an almost untapped resource. This Special Issue shows the current worldwide interest in the field of taxonomy and ecology of marine algae, and it will be a very useful tool to support and inspire future studies about this exciting topic. Recent findings indicate that marine environments have rapidly changed due to global warming over the past several decades. This change leads to significant variations in marine algal ecology. For example, a long-term increase in ocean temperatures due to global warming has facilitated the intensification of harmful algal blooms, which adversely impact public health, aquatic organisms, and aquaculture industries. Thus, extensive studies have been conducted, but there is still a knowledge gap in our understanding of the variation of their ecology in accordance with future marine environmental changes. To fill this gap, studies on the taxonomy and ecology of marine algae are highly necessary.

We have invited algologists to submit research articles that enable us to advance our understanding of the taxonomy and ecology of marine algae. The Issue collected fourteen papers that cover different aspects of the taxonomy and ecology of marine algae, including understudied species, interspecific comparisons, and new techniques.

2. PAPERS DETAILS

Wang and Wu [1] suggested that the significantly enhanced discharge of the Changjiang River since the winter of 2020–2021 was crucial for the outbreak of the *Ulva prolifera* green tides in south-western Yellow Sea, which could have contributed significantly to the nutrient enrichment of the Subei coast. They demonstrated that these large amounts of nutrient inputs, as an effective supplement, were the reason the green tides emerged sharply as an extensive outbreak in 2021. The easterly wind anomaly during spring 2021 contributed to the landing of *Ulva prolifera* off the Lunan coast.

Lee et al. [2] established an integrated loop-mediated isothermal amplification (LAMP) assay based on *Ostreopsis* cf. *ovata* and *Amphidinium massartii*. The detection sensitivity of the LAMP assay for *O.* cf. *ovata* or *A. massartii* was comparable to other molecular assays (PCR and quantitative PCR (qPCR)) and microscopy examination. The detection limit of LAMP was 0.1 cell of *O.* cf. *ovata* and 1 cell of *A. massartii*. The optimized LAMP assay was successfully applied to detect *O.* cf. *ovata* and *A. massartii* in field samples. They provided an effective method of detecting target benthic dinoflagellate species and could be further implemented to monitor phytoplankton in field surveys as an alternative.

Lee and Yeh [3] proposed a new endophytic filamentous red algal species from Taiwan based on morphological observations and molecular analysis, named as *Colaconema formosanumJ. Mar. Sci. Eng.* **2022**, *10*, 105. https://doi.org/10.3390/jmse10010105 https://www.mdpi.com/journal/jmsesp. nov. The new species was confirmed based on morphological observations and molecular analysis. Both the large subunit of ribulose-1,5-bisphosphate carboxylase/oxygenase (*rbc*L) and cytochrome c oxidase subunit I (COI-5P) genes showed high genetic variation between our sample and related species. Anatomical observations indicated that the new species present asexual reproduction by monospores, cylindrical cells, irregularly branched filaments, a single pyrenoid, and single parietal plastids.

Kwon et al. [4] reported 11 unrecorded diatom species (Diploneis didyma, Mastogloia elliptica, Cosmioneis citriformis, Haslea crucigera, Pinnularia bertrandii, Pinnularia nodosa var. percapitata, Gyrosigma sinense, Gomphonema guaraniarum, Gomphonema italicum, Navicula freesei, Trybionella littoralis var. tergestina) among samples collected from the Hwajinpo, Hyangho, Maeho, and Gapyeongri wetlands, and Cheonjinho and Gyeongpoho lagoons in Korea, during a survey from 2018–2020. They presented the taxonomic characteristics, ecological information, habitat environmental conditions, and references for these 11 species.

Park et al. [5] reported the marine benthic dinoflagellate *Bysmatrum subsalsum* from Korean Tidal Pools. They identified two ribotypes of *B. subsalsum* based on molecular phylogeny. The Korean isolates were nested within the ribotype B consisting of the isolates from China, Malaysia, and the French Atlantic, whereas the ribotype A included only the isolates from the Mediterranean Sea. These findings support the idea that there is cryptic diversity within *B. subsalsum*.

Park et al. [6] investigated fossil diatoms in Suncheonman Bay and introduced sub- fossil diatoms recorded in Korea. One sedimentary core has been extracted in 2018. They identified 87 diatom taxa from 52 genera in the sediment core sample. Of these, six species represent new records in Korea: *Cymatonitzschia marina*, *Fallacia hodgeana*, *Navicula mannii*, *Metascolioneis tumida*, *Surirella recedens*, and *Thalassionema synedriforme*. These six newly recorded diatom species were examined by light microscopy and scanning electron microscopy. The ecological habitats for all the investigated taxa are presented.

Prasath et al. [7] studied these four coagulants combined with algae suppression bacteria for their effect on *G. catenatum* to promote a more efficient and environmentally friendly algae suppression method. The results showed that polyaluminum chloride (PAC) is more efficient than other coagulants when used alone because it had a more substantial inhibitory effect. The PAC and Ba3 broth produced a pronounced algae inhibition effect that effectively hindered the germination of algae cysts. They conclude that this combination provides a scientific reference for the prevention and control of marine red tide. They suggested that designing environmentally friendly methods for the management of harmful algae is quite feasible.

Park et al. [8] investigated the community composition of free-living (FLB) and particle- associated (PAB) bacteria in each growth phase (lag, exponential, stationary, and death) of *Tetraselmis suecica* P039 culture using pyro-sequencing. They suggested that the PAB community may have a stronger association with the algal growth than the FLB community. Irrespective of the growth phase, *Roseobacter* clade and genus *Muricauda* were predominant in both FLB and PAB communities, indicating that bacterial communities in *T. suecica* culture may positively affect algae growth, and that they are potentially capable of enhancing the *T. suecica* growth.

Kim et al. [9] investigated the diversity of dinoflagellate genus *Scrippsiella*, a common member of phytoplankton, and their cysts using light and scanning electron microscopy. They observed the Korean isolates of *S. precaria* had two types of the $5''$ plate that either contacted the 2a plate or not. Molecular phylogeny based on internal transcribed spacer(ITS) and large subunit (LSU) rDNA sequences revealed that the Korean isolates were nested within the subclade of PRE (*S. precaria* and related species) in the clade of *Scrippsiella* sensu lato and that the PRE subclade had two ribotypes: ribotype 1 consisting of the isolates from Korea, China, and Australia, and ribotype 2 consisting of the isolates from Italy and Greece. They suggested that lineages between isolates of ribotype 1 were likely related to the dispersal by ocean currents and ballast waters from international shipping, andthe two types of spine shapes and locations of the 5" plates may be a distinct feature of ribotype

Ma et al. [10] proved that the well-known brown-tide-causing pelagophyte *Aureococcus anophagefferens* has a resting stage. They conducted a follow-up study to characterize the resting stage cells (RSCs) of *A. anophagefferens* using the culture CCMP1984. The results indicated that the RSCs of *A. Anophagefferens* are a dormant state that differs from vegetative cells morphologically and physiologically, and that RSCs likely enable the species to survive unfavourable conditions, seed annual blooms, and facilitate its cosmopolitan distribution.

Jódar-Pérez et al. [11] studied seven species of *Cystoseira sl* spp. in Cabo de las Huertas (Alicante, SE Spain) and analyzed their distribution using Permutational Analysis of Variance (PERMANOVA) and Principal Component Ordination plots (PCO). The phylogenetic tree supported actual classification, including, for the first-time, *Treptacantha sauvageauana* and *Treptacantha algeriensis* species. These data supported a complex distribution and speciation of *Cystoseira sl* spp. in the Mediterranean, perhaps involving Atlantic clades.

Jo et al. [12] investigated the growth and lipid composition of the algal strain *C. vulgaris* KNUA007. They found that the strain was able to thrive in a wide range of temperatures, from 5 to 30 °C; however, it did not survive at 35°C. The microalga was tolerant of low temperatures, making it an attractive candidate for the production of biochemical under cold weather conditions. Therefore, this Antarctic microalga may have potential as an alternative to fish and/or plant oils as a source of omega-3 PUFA. The temperature tolerance and composition of *C. vulgaris* KNUA007 also make the isolate desirable for commercial applications in the pharmaceutical industry.

Li et al. [13] investigated the impacts of different concentrations of cell-free filtrate of the bacteria *Bacillus cereus* BE23 on *Ulva prolifera*. The evidence indicates that the alteration of energy dissipation caused excess cellular ROS accumulation that further induced oxidative damage on the photosynthesis apparatus of the D1 protein. The potential allele-chemicals were further isolated by five steps of extraction and insolation (solid phase– liquid phase–open column–UPLC–preHPLC) and identified as N-phenethylacetamide, cyclo (L-Pro-L-Val), and cyclo (L-Pro-L-Pro) by HR-ESI-MS and NMR spectra. The dike-topiperazines derivative, cyclo (L-Pro-L-Pro), exhibited the highest inhibition on *U. prolifera* and may be a good candidate as an algicidal product for green algae bloom control.

Yang et al. [14] reviewed *Karlodinium* due to their representative features as mixoplankton and harmful algal blooms (HABs)-causing dinoflagellates. Their phagotrophy exhibits multiple characteristics: (1) omnivority, i.e., they can ingest a variety of preys in many forms; (2) flexibility in phagotrophic mechanisms, i.e., they can ingest small preys by direct engulfment and much bigger preys by myzocytosis using a peduncle; (3) cannibalism, i.e., species including at least *K. veneficum* can ingest the dead cells of their own species. They suggested that mixotrophy of *Karlodinium* plays a significant role in the population dynamics and the formation of HABs in many ways, which thus deserves further investigation in the aspects of physiological ecology, environmental triggers (e.g., levels of inorganic nutrients and/or presence of preys), energetics, molecular (genes and gene expression regulations) and biochemical (e.g., relevant enzymes and signal molecules) bases, origins, and evaluation of the advantages of being a phagotroph.

REFERENCES

1. Wang, B.; Wu, L. Numerical Study on the Massive Outbreak of the Ulva prolifera Green Tides in the South-western Yellow Sea in 2021. *J. Mar. Sci. Eng.* **2021**, *9*, 1167. [CrossRef]
2. Lee, E.S.; Hwang, J.; Hyung, J.-H.; Park, J. Detection of the Benthic Dinoflagellates, Ostreopsis cf. ovata and *Amphidinium massartii*
3. (Dinophyceae), Using Loop-Mediated Isothermal Amplification. *J. Mar. Sci. Eng.* **2021**, *9*, 885. [CrossRef]
4. Lee, M.-C.; Yeh, H.-Y. Molecular and Morphological Characterization of *Colaconema formosanum* sp. nov. (Colaconemataceae, Rhodophyta)—A New Endophytic Filamentous Red Algal Species from Taiwan. *J. Mar. Sci. Eng.* **2021**, *9*, 809. [CrossRef]
5. Kwon, D.; Park, M.; Lee, C.S.; Park, C.; Lee, S.D. New Records of the Diatoms (Bacillariophyceae) from the Coastal Lagoons in Korea. *J. Mar. Sci. Eng.* **2021**, *9*, 694. [CrossRef]
6. Park, J.S.; Li, Z.; Kim, H.J.; Kim, K.H.; Lee, K.W.; Youn, J.Y.; Kwak, K.Y.; Shin, H.-H. First Report of the Marine Benthic Dinoflagellate Bysmatrum subsalsum from Korean Tidal Pools. *J. Mar. Sci. Eng.* **2021**, *9*, 649. [CrossRef]
7. Park, M.; Lee, S.D.; Lee, H.; Lee, J.-Y.; Kwon, D.; Choi, J.-M. Identification of New Sub-Fossil Diatoms Flora in the Sediments of Suncheonman Bay, Korea. *J. Mar. Sci. Eng.* **2021**, *9*, 591. [CrossRef]
8. Balaji Prasath, B.; Wang, Y.; Su, Y.; Zheng, W.; Lin, H.; Yang, H. Coagulant Plus *Bacillus nitratireducens* Fermentation Broth Technique Provides a Rapid Algicidal Effect of Toxic Red Tide Dinoflagellate. *J. Mar. Sci. Eng.* **2021**, *9*, 395. [CrossRef]

9. Park, B.S.; Choi, W.-J.; Guo, R.; Kim, H.; Ki, J.-S. Changes in Free-Living and Particle-Associated Bacterial Communities Depending on the Growth Phases of Marine Green Algae, *Tetraselmis suecica*. *J. Mar. Sci. Eng.* **2021**, *9*, 171. [CrossRef]

10. Kim, H.J.; Li, Z.; Kang, N.S.; Gu, H.; Kim, D.; Seo, M.H.; Lee, S.D.; Yun, S.M.; Oh, S.-J.; Shin, H.H. Morphology and Phylogeny of Scrippsiella precaria Montresor & Zingone (Thoracosphaerales, Dinophyceae) from Korean Coastal Waters. *J. Mar. Sci. Eng.* **2021**, *9*, 154.

11. Ma, Z.; Hu, Z.; Deng, Y.; Shang, L.; Gobler, C.J.; Tang, Y.Z. Laboratory Culture-Based Characterization of the Resting Stage Cells of the Brown-Tide-Causing Pelagophyte, Aureococcus anophagefferens. *J. Mar. Sci. Eng.* **2020**, *8*, 1027. [CrossRef]

12. Jódar-Pérez, A.B.; Terradas-Fernández, M.; López-Moya, F.; Asensio-Berbegal, L.; López-Llorca, L.V. Multidisciplinary Analysis of Cystoseira sensu lato (SE Spain) Suggest a Complex Colonization of the Mediterranean. *J. Mar. Sci. Eng.* **2020**, *8*, 961. [CrossRef]

13. Jo, S.-W.; Do, J.-M.; Kang, N.S.; Park, J.M.; Lee, J.H.; Kim, H.S.; Hong, J.W.; Yoon, H.-S. Isolation, Identification, and Biochemical Characteristics of a Cold-Tolerant Chlorella vulgaris KNUA007 Isolated from King George Island, Antarctica. *J. Mar. Sci. Eng.* **2020**, *8*, 935. [CrossRef]

14. Li, N.; Zhang, J.; Zhao, X.; Wang, P.; Tong, M.; Glibert, P.M. Allelopathic Inhibition by the Bacteria Bacillus cereus BE23 on Growth and Photosynthesis of the Macroalga Ulva prolifera. *J. Mar. Sci. Eng.* **2020**, *8*, 718. [CrossRef]

15. Yang, H.; Hu, Z.; Tang, Y.Z. Plasticity and Multiplicity of Trophic Modes in the Dinoflagellate Karlodinium and Their Pertinence to Population Maintenance and Bloom Dynamics. *J. Mar. Sci. Eng.* **2021**, *9*, 51. [CrossRef]

CHAPTER-2

PLASTICITY AND MULTIPLICITY OF TROPHIC MODES IN THE DINO-FLAGELLATE KARLODINIUM AND THEIR PERTINENCE TO POPULATION MAINTENANCE AND BLOOM DYNAMICS

1. INTRODUCTION

Microalgae are an important group in terms of global primary productivity. Those microalgae that spend their time on vegetative growth in the water column are categorized as phytoplankton, a counterpart of zooplankton in aquatic ecology [1]. As the terms imply, autotrophy or phototrophy is the most important trophic mode in microalgae or phytoplankton and thus the focus of research on microalgae [1–3], which is reasonable and fair in terms of their major function in aquatic ecosystems as primary producers. However, other trophic modes have been found in many groups or species of microalgae and have attracted increasingly more attention from the scientific community during the last several decades because these non-autotrophic modes have been, or will be, proven to be vital strategies for the population survival and development (e.g., blooms) of phytoplankton [4]. Among 2400 valid species of dinoflagellates, about 50% are strictly heterotrophic, while the other half of species obtained and maintained the ability of photosynthesis [5].

Independent of the number of species, these photosynthetic dinoflagellates occupy an essential place in primary production, particularly in coastal and estuarine ecosystems [6]. As the dark facet of primary producers, dinoflagellates are also the crucial perpetrators of harmful algal blooms (HABs) forming species, given that they are responsible for 75% of documented HABs [7]. However, intriguingly, photosynthetic dinoflagellates are generally of relatively lower photosynthetic capacity per unit of biomass and exhibit lower growth rates in comparison to many of their competitors, such as diatoms [8,9]. Thus, dinoflagellates must have other strategies to balance this competitive disadvantage. Mixotrophy, a nutritional strategy by which organisms are able to obtain nutrients and/or energy by both phototrophic autotrophy and heterotrophy [10–12], is one of these strategies that enhance growth rates via obtaining energy from either dissolved organic compounds [13] or particulate preys [14]. The mixotrophic protists that play roles of both primary producer and consumer have been widely investigated from different aspects [11,12].

The genus *Karlodinium* J. Larsen was erecteded from the genus *Gymnodinium* in 2000 because of the characteristics of their apical groove, ultrastructure, and partial large subunit rDNA sequences [15]. Species of *Karlodinium* are well known for forming HABs and thus causing the consequent fish-killing events [16–18]. The genus includes at least 15 species to date (Table 1). The distribution of the genus *Karlodinium* spreads over four oceans [16] (also see Ocean Biogeographic Information System, https://obis.org/taxon/23 1789). *Karlodinium veneficum* (original name: *Gymnodinium veneficum*; synonym: *Karlodinium micrum*, *Gymnodinium micrum*, *Gyrodinium galatheanum*, *Woloszynskia micra*, and *Gyrodinium estuariale*) is the type species and also the most intensively and extensively investigated one in the field and the laboratory [19–23]. Species in the genus of *Karlodinium*, such as *K. veneficum*, *K. armiger*, *K. corsicum* and *K. aculat*, have been reported to be associated with many toxic events and caused mortality of fishes, mussels and zooplanktons [18,24–30]. Multiple types of toxins have also been detected from these species. The toxins produced by *K. veneficum* are termed as karlotoxins [25] and at least 12 natural analogs of karlotoxins have been identified to date [25,31–33]. *Karlodinium conicum* was also proved to produce karlotoxin [34,35]. From *K. armiger*, however, a different species of toxin, karmitoxin, has been chemically characterized [36,37]. Besides, the presence of some types of NSP toxins in *K. corsicum* has also been testified by mouse tests [27].

Table 1. A collection of *Karlodinium* species regarding their bloom threat, toxicity, trophic modes, and associated mechanisms.

Species	Former names and/or taxonomic synonyms	Distribution	Blooms	Toxicity	Autotrophy	Osmotrophy	Phagotrophy	Peduncle-like Structure
K. armiger	\	Alfacs Bay, Ebro Delta, NW Mediterranean [38]	Yes [29,39]	Yes (Karmitoxin) [37]	Yes	?	Yes [40,41]	Have a peduncle [38]
K. austral	\	North-eastern Tasmania, Port Phillip Bay (Victoria), South Australia and TuggerahLakes [28] and Singapore [42,43]	Yes [18]	Yes [18,28]	Yes [28]	?	Yes [28]	Have a thick, tubular peduncle-like structure [28]
K. antarcticum	\	Southern Ocean [43]	?	No [34]	Yes [43]	?	?	Have a tube-shaped structure [43]
K. azanzae	\	Manila Bay, Philippines [44]	?	Yes [44]	Yes [44]	?	Yes [44]	Have a peduncle [44]
K. ballantinum	\	Mercury Passage, Tasmania, Australia, and Tyrrhenian coastal waters [43] and the Mexican Pacific [45]	?	?	Yes	?	?	Have a tube-shaped structure [43] K.
K. conicum	\	Southern Ocean [43]	?	Yes (KmTx) [34]	Yes	?	?	Have a tube-shaped structure [43]
K. corrugatum	\	Southern Ocean [43]	?	No [34]	Yes	?	?	?
K. corsicum	Gyrodinium Corsicum	Corsica (France), Tyrrhenian Sea and the Spanish Alfacs Bay of Mediterranean Sea [29,46]	Yes [24]	Yes [26]	Yes	?	?	Have a ventral plate [47]
K. decipiens	Karenia digitate	from coastal Tasmania southward to the north polar front, and western European Atlantic waters (Bilbao, Spain) [43]	?	No [34]	Yes	?	?	Have a tube-shaped structure [43]

Species	Former names and/or taxonomic synonyms	Distribution	Blooms	Toxicity	Autotrophy	Osmotrophy	Phagotrophy	Peduncle-like Structure
K. digitatum1	Karenia digitata	Japan coastal waters of Hong Kong, Fujian and Guangdong's Southern, China [48,49]	Yes [49]	Yes [49]	Yes	?	?	Have small finger-like extensions [49]
K. elegans	\	Pingtan coastal water, East China Sea [50]	?	No [50]	Yes [50]	?	?	Have a tube-like structure [50]
K. gentienii	\	The Atlantic coast of Brittany [51]	Yes [51]	Yes [51]	Yes	?	?	Have a tube-shaped structure [51]
K. veneficum	Gymnodinium veneficum; Karlodinium micrum; Gymnodinium micrum; Gyrodinium galatheanum; Woloszynskia micra; Gyrodinium estuariale [38,52]	Cosmopolitan [16]	Yes [29,39,53]	Yes [54]	Yes [20]	Yes	Yes [16,20,21]	Have a peduncle [38]
K. vitiligo	Gymnodinium vitiligo ?K. veneficum [38]	?	?	?	Yes	?	?	?
K. zhouanum	K. jejuense	Widely spread over the coastal waters of China [55,56]	Possible [57]	?	Yes [56]	?	?	Tube-like structure in intercingular region [56]

Note: "\" indicates none. "?" indicates that there is no explicit record in the literature.

The trophic modes of *Karlodinium* have been studied for several decades, and were found to be diverse and typical. However, new findings have been made recently and improved our understanding of this genus, which comprises a group of mixoplankton. At least four *Karlodinium* species have been confirmed to be mixotrophic, namely *K. veneficum* [20,58], *K. armiger* [40,41], *K. aculat* [28] and *K. azanzae* [44], by now. Among these proven mixotrophic species, some exhibit plastic and multiple trophic modes, as well as wide spectrum of prey size and varieties, i.e., *K. armiger* and *K. veneficum* [16,59]. Other species, while direct evidence about their mixotrophy is lacking at present, were also reported to possess peduncle-like structures (Table 1), namely the instrument for phagotrophy, such as *K. gentienii* [51] and *K. zhouanum* [56].

Because possibly all *Karlodinium* species have potential for mixotrophy and the highly flexible trophic modes may be a vital trait of this and other similar groups of dinoflagellates in their ecology and evolution, here we review the knowledge advancement in understanding the trophic modes of dinoflagellates in general and *Karlodinium* in particular, with the hope of inspiring further investigations on the genetic, cellular, physio-chemical, and eco- logical mechanisms of mixotrophy in dinoflagellates (HABs-forming groups particularly) by putting forward our insights and suggestions about the interesting topic.

2. TROPHIC MODES OF DINOFLAGELLATES

In addition to autotrophy or phototrophy, many free-living dinoflagellates live as either heterotrophs or mixotrophs [5,60]. Mixotrophic modes can be further categorized as amphitrophic (heterotrophy or autotrophy alone is sufficient for nutrition) and mixotrophic *sensu stricto* (both forms of nutrition are required) [61].

Dinoflagellates have evolved multiple heterotrophic nutritional strategies [61]: (1) osmotrophy (or resorption), by which the organic macronutrients are taken up by direct passage through the plasma membrane, (2) saprotrophy, a chemoheterotrophic process of digesting organic matter extra-cellularly and (3) endocytosis, which includes pinocytosis (cell drink, a mode of endocytosis by which liquid organic matter is taken up into the cell by invaginating of the cell membrane, and forming a small vesicle inside the cell) and phagocytosis or phagotrophy (cell eating, the endocytosis of particulate food).

Generally, there are three types of feeding mechanisms of phagotrophy that dinoflagel- lates use to uptake food particles (including intact cells): (1) direct engulfment (phagotrophy *sensu stricto*), i.e., a cell phagocytizes an entire food particle, including the prey cell membrane [61,62], (2) tube feeding, i.e., the feeding cells use an feeding appendage to suck food particles (e.g., *Peridiniopsis berolinensis*) [63,64], and (3) pallium feeding, i.e., some species use a feeding veil, namely pallium, to surround and digest the prey outside the cell body of the predator, then the liquefied cytoplasmic content of prey is taken up by the predator, leaving only an empty wall or frustule (e.g., *Zygabikodinium lenticulatum*, *Oblea rotunda* and *Protoperidinium conicum* [64–66]).

Direct engulfment, or phagocytosis *sensu stricto*, of dinoflagellates seems to be mainly found in athecate dinoflagellates, like *Blastodiniales*, *Gymnodiniales*, *Noctilucales*, and *Oxyrrhinales* [61,67]. However, it remains unclear whether this apparent "preference" is due to the higher flexibility or elasticity of the cells of naked species than armored species. In addition, special feeding organelles, such as tentacles, lobopodia and peduncles, were usually found in the sulcal region near the flagellar groove [61].

Tube feeding has been observed to use two types of feeding tubes in dinoflagellates: the peduncle (a protoplasmic strand protruding from the mid-ventral area of the sulcus to connect predator and prey, e.g., *Paulsenella*) [61,68,69], and the phagopod (a non- cytoplasmic feeding tube, e.g., *Amphidinium cryophilum*) [70]. Myzocytosis is a kind of tube feeding by which the feeding cells suck out the contents of prey cells by leaving the plasma membrane outside the predator. The prey plasmalemma is not taken up, and thus the prey cytoplasm is bounded only by the vacuolar membrane in the food vacuole. This mode of nutrition was first described in the naked dinoflagellate *Gyrodinium vorax* Biecheler [62]. The terminology for the uptake organelle of myzocytosis has not been uniformed, and it has been variously referred to as feeding tubes or peduncles [61], and in this review, we use the term "peduncle" to refer to the uptake organelle of myzocytosis. The peduncle was reported to be formed by the emergence of a preformed "microtubular basket" which consists of plates of microtubules [69,71]. Based the light and electron microscopic observations on *Pausenella sp.*, Schnepf et al. (1985) suggested that the food uptake was driven by a hydrostatic gradient which might be attributed to rhythmical ion pumping and based on the existence of a common cavity and the sphincters [71]. In contrast to most suctorian tentacles, peduncles are generally not permanently protruded [61,69], and are usually invisible in predators not feeding on food [69]. The length of feeding tubes also differs in species and even varies in a single cell with feeding status [61]. No prey size spectra are confirmed for tube feeders in the literature, but some researchers have pointed out that the prey size seems not to have an upper limit, as studies reported the ingestion ability of *K. veneficum* and *K. aculat* on rotifer, copepod eggs, and even tissues of fish [16,72,73]. The strictly heterotrophic species *Pfiesteria shumwayae* was found to exhibit lethal effect on fish by myzocytosis, also named "micropredation" [74], a trophic strategy in which a predator feeds on a rather large prey and one feeding individual attacks more than one prey during its life span and attacks the prey intermittently without necessarily eliminating its fitness (e.g., mosquito) [75]. However, both direct engulfers and pallium feeders have prey size spectra restricted to the volume capacity of the predator cell [64].

The feeding processes of phagocytosis were described by several steps including pre-capture behaviour, capture, and prey manipulation [64]. While in pre-capturing, dinoflagellates swimming faster than their prey are referred as the "searching type" and those being able to catch the faster-moving prey are described as the "trapping type" [64]. Search type is induced by chemical substances released from the injured prey and is independent of prey size [63,76]. It is demonstrated that dinoflagellates of similar size but with different speed in comparison to praise, swimming characteristics, and feeding strategies (peduncle vs. tow line) have substantially different responses to the introduction of preys [77]. The feeding dinoflagellates usually capture preys using some specialized appendages named "capture filament" or "tow filament". The capture filament of *Peridiniopsis berolinensis* is a thin filament that originates from the ventral region of the cell near the sulcus [63]. Once the filament anchors to the prey, it retracts, brings the prey closer to the predator (e.g., *Protoperidinium* and Diplosalis group) or contracts entirely, and thereby drags the prey to the sulcal region of the predator (e.g., *Gyrodinium*) [78]. After capturing the prey, most dinoflagellates consume the prey immediately, but the manipulation of prey may differ with other feeding mechanisms [14,78].

Certain dinoflagellates may utilize cleptochloroplasts (transiently alien chloroplasts) obtained from preys [79]; this nutrient strategy is termed kleptochloroplastidy [80]. Myzocytosis is the proven method to acquire kleptochloroplasts from preys [61,81]. Hansen (1998) reviewed a few species that lack chloroplasts but are capable of sequestering chloroplasts from other phytoplankters and then using the "stolen" chloroplasts for photosynthesis [82]. This kind of mixotrophy has been reported among some species belonging to the naked genera *Amphidinium* and *Gymnodinium* [82–84]. It is noteworthy that the latter may contain species from *Karlodinium*, as this genus had not been separated from *Gymnodinium* until 2000 [15]. Li et al. found fragmental pigments from cryptophycean prey in *K. veneficum* that had been ingested with the prey for 41 h, suggesting some chloroplasts of prey could be retained by the dinoflagellate [20].

3. AUTOTROPHY OF *KARLODINIUM*

All *Karlodinium* species have the ability of photosynthesis (Table 1). *Karlodinium veneficum* and *K. armiger* are the best-studied species that have haptophyte origin chloroplasts [20,21,40,59,85,86]. Phototrophic growth rates of *K. armiger* are quite low (a maximum of 0.01 and 0.10 d^{-1}), even at high irradiances [40,59]. In comparison, *K. veneficum* grows faster than *K. armiger* photosynthetically without prey, with growth rates ranging from0.17 to 0.36 d^{-1} [87]. In some cases, the growth rate of *K. veneficum* may even elevate up to0.55 d^{-1} in the light without prey [85]. The photosynthetic growth rate of *K. veneficum* is significantly affected by temperature and salinity [23]. At least some strains of *K. veneficum* were better adapted to "low-light" conditions than were *K. armiger*, whereas characteristics of *K. armiger* were more suitable to cope with "high-light" [54].

Karlodinium aculat grows poorly in the normal conditions without providing food. The monoculture of *K. australe* grown in laboratory and Gse medium stabilized at low concentrations (10^2–10^3 cells mL^{-1}) and failed to reach higher cell concentrations [28]. Lim et al. obtained similar results from a *K. austral* bloom in the cage-farming region of the West Johor Strait of Malaysia (0.31–2.34 × 10^3 cells mL^{-1}). However, *K. veneficum* could reach extremely high cell densities (2–3 × 10^5 cells mL^{-1}) in laboratory cultures [88]. These observations suggest that different species of *Karlodinium* may also differ in their phototrophic growth potential. In contrast to the relatively poor autotrophic ability, the genus is successful in forming harmful blooms. Thus, other nutritional strategies may play a key role in population competition and deserve further investigation.

4. OSMOTROPHY OF *KARLODINIUM*

Osmotrophy, i.e., the uptake of dissolved organic compounds, has been shown to be an efficient nutritional strategy for algae. *Karlodinium veneficum* is the most studied species in *Karlodinium* on osmotrophy. Cell-surface proteolytic activity (leucine aminopeptidase) was detected in *K. veneficum* and suggested to play a role in obtaining nutrition by obtaining amino acids for assimilation, while, alternatively, released amino acids may be degraded by cell-surface amino acid oxidases to provide ammonium, which can be assimilated as a source of nitrogen [89]. Solomon and Glibert found that urease activity in *K. veneficum* was significantly higher than that in other species (including *Heterocapsa triquetra*, the cryptophyte *Storeatula major*, and the haptophyte *Isochrysis* sp) on both a per cell basis and a per cell volume basis [90]. Harmful dinoflagellates like *K. veneficum* may be better suited to utilize urea than other species do according to their high urease activity and large intracellular urea pools, which may explain why these harmful dinoflagellates proliferate rapidly in the water bodies with plenty of urea [90].

Osmotrophy may be an important and ubiquitous trophic strategy for all species in *Karlodinium*, because almost all phytoplankton are osmotrophs in some parts, not least by virtue of being auxotrophic; many need external sources, e.g., vitamins [91]. Phytoplankton exhibit non-auxotrophic osmotrophy to a significant level, mostly in relation to the uptake of primary metabolite compounds, especially amino acids [54]. It has been reported that many dissolved organic compounds, such as amino acids (e.g., glutamine, leucine, thymidine, aspartic acid), carbohydrates (e.g., glucose) and other organic compounds (e.g., acetic acid, coumaric acid, glycerol), can be used as carbon and nitrogen sources, which are commonly released by the algae themselves or bacteria [92–96].

5. PHAGOTROPHY OF *KARLODINIUM*

5.1 Karlodinium veneficum

Karlodinium veneficum exhibited increased ingestion rate on eubacteria when phosphate was limited, which may be an important nutrient-acquiring strategy when inorganic nutrient is limited [97]. *Karlodinium veneficum* was also reported to ingest various kind of small algae by phagocytosis, including *Chroomonas salina*, *Cryptomonas appendiculata*, *C. calceiformis*, *C. maculata*, *Hemiselmis brunnescens*, *H. rufescens*, *Hemiselmis* sp., *Rhinomonas reticulata*, *Rhodomonas salina*, *Rhodomonas* sp., *Storeatula major*, and *Isochrysis galbana*, and most of them are cryptophytes [20–22,85,86,98,99]. The direct engulfment of whole cells of *Storeatula major*, a species of crypto- phyte, by *K. veneficum* and the associated feeding processes were initially documented via video recording under light microscope [20]. This phagocytosis process was described to have three typical steps [20]: (1) Pre-capture behaviour. After adding cryptophyte as prey, most *K. veneficum* cells increased the swimming speed, and some began to swim around the prey; (2) Capture. Generally, *K. veneficum* cells formed a protrusion, which was near the flagellar pores in the sulcal region, and attached to the prey. Once the protrusion contacted the prey cell, phagocytosis began.

In some cases, a thin capture filament projected from the extending sulcal region in the epicone was observed, and then the filament captured and drew the prey cell to the surface of the dinoflagellate in the sulcal region (Figure 1a, SEM micrographs were adopted from Place et al. (2012) [72]; (3) Prey manipulation. After capturing prey firmly, *K. veneficum* usually stopped swimming to draw the whole cell of prey into the dinoflagellate cell through the protrusion. During the process of feeding, a "feeding gap" appeared to form along the cingulum near the flagellar pores and a pair of "lip-like" protrusions (i.e., peduncle) was observed (Figure 1b) [20]. The engulfment behaviour usually took 2 to 3 min at room temperature (20 °C) and, when the ingestion was completed, *K. veneficum* cells resumed swimming and were able to find and phagocytize another prey cell.

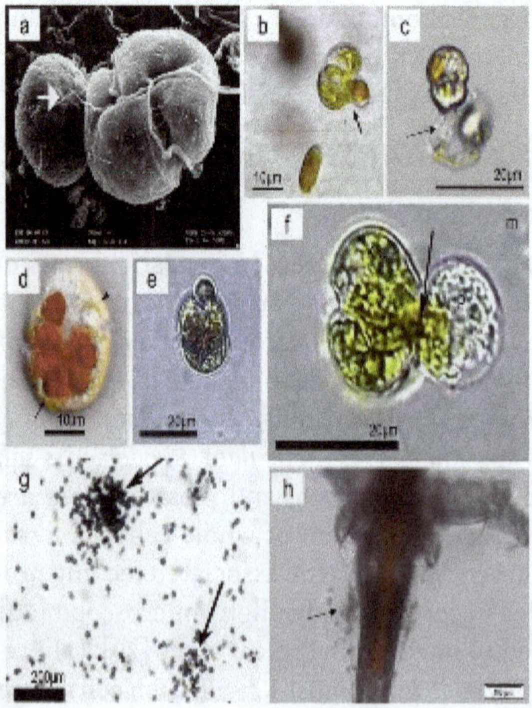

Figure 1. The phagotrophic behavior of *Karlodinium*. (a) SEM of *K. veneficum* feeding on *Rhodomonas* sp. The thin filament is marked with an arrow (the photo was modified from Place et al. [72]). (b) A cell of *K. veneficum* was engulfing whole cells of *Storeatula major*. The "lip-like" protrusion was observed to gradually move laterally along the prey surface, which causes further engulfment of the prey cell (the photo was modified from Li et al. [20]). (c) *Karlodinium veneficum* ingest a dead cell of con-species by myzocytosis using peduncles. *Karlodinium veneficum* was searching for cytoplasm by opening the peduncle (arrow) widely in the dead cell (the photo was modified from Yang et al. [16]).(d) Subsurface ventral view of *K. australe* after feeding overnight on *Rhodomonas salina*. Note light yellow-green chloroplasts (arrowhead) and red food vacuoles (arrow, the photo was modified from de Salas et al. [28]). (e) *Karlodinium armiger* was ingesting the cryptophyte *Rhodomonas salina* by direct engulfment (the photo was modified from Berge et al. [59]). (f) *Karlodinium armiger* was ingesting the raphidophyte *Fibrocapsa japonica* (p) by myzocytosis (the photo was modified from Berge et al. [59]). The peduncle was shown by the arrow. (g) Aggregations (arrows) of *K. armiger* cells in cultures fed the thecate dinoflagellate *Prorocentrum minimum* (the photo was modified from Berge et al. [59]).(h) Aggregations (arrows) of *K. veneficum* cells fed an injured brine shrimp *Artemia salina* (the photo was modified from Yang et al. [16]).

Sheng et al. also represented the phagocytosis process of *K. veneficum* on *Storeatula major* and paid more attention to its pre-capture behaviour. When presented with *S. major*, the velocity, radius, and pitch of *K. veneficum* reduced, but its angular velocity increased [58,77]. The feeding cells of *K. veneficum* significantly reduced their usual vertical migration, probably to remain in the vicinity of their preys [58,100].

A peduncular microtubular strand was observed in *K. veneficum* cells and believed to be a tube feeder; however, the small sized preys such as eubacteria and cryptophytes were obviously ingested by direct engulfment [20,58], indicating an alternative function of the microtubular strand. Although *K. veneficum* was also observed to have the potential to feed on the diatom *Melosira* and copepod *Acartia tonsa* [72], direct evidence of feeding behaviour using peduncles has not been captured. Recently, we observed *K. veneficum* ingested preys via myzocytosis using the peduncle [16], in which the entire feeding process was much the same as that observed in direct engulfment except that only the cytoplasm of prey (cells or larger multicellular individuals) was sucked into *K. veneficum* cells through the peduncle in myzocytosis (Figure 1c) and the time expenditure of myzocytosis, which varied from several seconds to a few minutes, was relatively shorter than that of the direct engulfment on small-sized prey [16]. As the ingestion proceeded, the cell volume of *K. veneficum* gradually increased [16]. Profiting by this mechanism, the prey size spectra would have no upper limit. We demonstrated that *K. veneficum* is virtually an omnivorous feeder, as it could feed on both live and dead bodies/cells of phytoplankton (the dinoflagellates *Margalefidinium polykrikoides*, *Akashiwo sanguinea*, and *Alexandrium leei*, the cryptophyte *Rhodomonas salina*, and the haptophyte *Isochrysis galbana*) and animals (the finfish *Oryzias melastigma*, brine shrimp *Artemia salina*, and rotifer *Brachionus plicatilis*). Importantly, *K. veneficum* also exhibited cannibalism (i.e., feeding on dead cells of its own species), which implies that the dead and weak cells of *K. veneficum* can be ingested by the live cells to recycle nutrients contained in the eaten cells [16]. Cannibalism is one of the simplest trophic interactions [101]. The advantageous aspect of this particular type of phagotrophy is that it allows an efficient nutrient transfer because of the well-matched nutritional value between the food and consumer [102]. We also observed that *K. veneficum* could survive at a lower cell density, without inorganic nutrients supplementing the culture medium, for a year, which was obviously attributable to the cannibalistic phagotrophy [16]. Cannibalism was also observed in *Protoperidinium* when cell abundances were high, and in *Oxyrrhis* when "victim" and "cannibal" differed in sufficient cell size-classes [66,102]. These observations suggest that cannibalism may be a mechanism of withstanding prolonged starvation.

The ingesting ability of *K. veneficum* is affected by environmental factors such as irradiance. It was observed that *K. veneficum* did not exhibit phagocytosis without light and the ingestion rate increased drastically when irradiance rose up to ~ 50 μmol photons $m^{-2} \cdot s^{-1}$ [20].

5.2 Karlodinium australe

Karlodinium australe has also been known to phagocytize particulate foods in food vacuoles since the species was initially described [28]. A thick and tubular peduncle-like structure of this organism was observed in the sulcal region [28]. Phagotrophy in*K. australe* was captured when autotrophically grown cultures were provided with live *R. salina* cells as food (Figure 1d) [28]. However, the intact feeding behaviour has not been captured. A recent study further investigated the feeding mechanism and ecological implication of the phagotrophic mixotrophy of *K. australe. Karlodinium australe* is a phagotroph that can ingest preys via direct engulfment or tube feeding. In accord with its flexible phagotrophic modes, *K. australe* is also an omnivorous mixotroph. Except *R. salina*, a diverse range of organisms could be ingested, such as microalgae (*Isochrysis galbana, Margalefidinium polykrikoides, Karenia mikimotoi* and *Gymnodinium catenatum*) and zooplankton (*Artemia salina* and *Brachionus plicatilis*) [73].

5.3 Karlodinium armiger

Karlodinium armiger is an omnivorous and obligate mixotroph. It seems that *K. armiger* obtain essential growth factor or substance through phagotrophy [59]. This species can ingest many types of preys (except for almost all Bacillariophyceae tested), but yield the highest growth rates when offered cryptophytes as prey [40,59]. However, *K. armiger* cannot grow and survive by feeding in complete darkness or at dim light, even by feeding adequate amounts of preys regularly [59], indicating that *K. armiger* cannot grow on as complete phagotrophy.

Under light microscope, the feeding mechanism of *K. armiger* was occasionally assumed to be direct engulfment (i.e., phagocytosis *sensu stricto*) while it was feeding on small-sized cells of prey; however, it would use tube feeding (myzocytosis) when the preys were larger or thecate [59]. Rigid cell coverings seem to set a barrier to grazing; thus, species like diatoms and thecate dinoflagellates may not be appropriate food [59].

Some details of the phagotrophy process in *K. armiger* were also documented by Berge et al. [59]: *K. armiger* also displayed distinct and intense pre-capture behaviour by increasing swimming speed and frequently changing swimming direction before ingestion. This pre-capture swimming behaviour has also been documented in other phagotrophic dinoflagellates [63,76,103,104]. The predator cell usually encountered a prey cell with its apical part. After contacting with a prey cell, the predator slowed down the swimming speed. During this stage, less than half of the prey cells (*R. salina*) were captured. Occasionally, a capture filament, an up to 10 μm long structure, which has also been reported in other phagocytosing dinoflagellates, such as *Peridiniopsis berolinensis* [63] and *K. veneficum* [19,20], was observed to attach the prey. When the capture succeeding, the predator placed its sulcal area, where the phagocytosis took place, facing the prey and revolved around its anterior-posterior axis. During this feeding stage, a small protrusion sometimes appeared. However, most preys and predators often established close contact immediately without any signs of protrusion. Often, the whole *Rhodomonas* cell was apparently engulfed or sucked into a food vacuole (Figure 1e) [59]. Occasionally, the cytoplasm was separated from the periplast of cryptophyte and taken up through the sulcus, leaving the periplast behind [59].

However, it differed somewhat from the feeding sequence of ingesting intact cells of *R. salina* when the predator cells fed on relatively large preys (> 10 μm). During feeding on large preys like the raphidophyte *Fibrocapsa japonica*, only a small part was sucked into a food vacuole. The cytoplasm separated from the cell membrane of the prey and flowed into food vacuoles of the predator through a narrow part (3–4 μm thickness) of the sulcal area (Figure 1f) [11]. This behaviour resembled myzocytosis or tube feeding [59].

Karlodinium armiger feeds on preys using an unnoticeable feeding tube (peduncle) which allows for ingestion of larger food particles [59]. However, Bergholtz et al. reported the presence of a peduncular microtubular strand in *K. armiger* [38]. The optimal prey size for *K. armiger* was about 13 μm, a size class which is close to the predator and contributesighed ingestion rates. Smaller preys (< 8 μm) resulted in lower ingestion rates (20–24 pg C cell^{-1}·d^{-1}), but still contributed to fairly high growth rates (0.35–0.45 d^{-1}). Although *K. armiger* can feed on preys in a large size spectrum [40], maximum growth rates relied more on prey taxa (cryptophytes) rather than on prey size when the food was saturated [40].

Several cells of *K. armiger* often attacked and fed on prey cells simultaneously [59]. When *K. armiger* reached higher cell densities, aggregates of predator cells swarming intensely around prey cells were easily recognized (Figure 1g) [71]. Aggregates led to fairly high swimming speeds of other *K. armiger* cells in the culture as these cells were obviously attracted to the preys. Such aggregation of predator cells around prey indicated a chemical attraction. Both mobile and immobile cells were observed to be captured and ingested [59]. We also observed the same aggregation of predator cells around a prey in *K. veneficum* (Figure 1h) [16].

The feeding mechanism of *K. veneficum* and *K. armiger* indicates that mixotrophic species of *Karlodinium* may be omnivorous phagotrophs with a relatively wide range of prey species and prey size spectrum than previously recognized. A newly identified species, *K. azanzae*, was also demonstrated to be phagotrophic and able to feed on invertebrates by micropredation [44]. However, direct evidence of myzocytosis feeding, for most other species of *Karlodinium*, has been absent.

Whether the presence or absence of a trophic mode-relevant trait in one, but not in another, species of *Karlodinium* was really caused by interspecific genetic differences, or was due to imbalanced investigations, definitely deserves more intensive study.

6. EVOLUTION OF THE FEEDING MECHANISMS IN *KARLODINIUM*

The feeding mechanism of *Karlodinium* seems to be plastic and of more than one type (e.g., *K. veneficum* feeding by direct engulfment and myzocytosis). According to most studies, only one feeding mechanism was found in a given dinoflagellate species [64]. The flexible feeding mechanisms of *Karlodinium* may lead a new discovery and provide a novel view of the evolution of feeding mechanisms in dinoflagellates, but it is clear that this aspect cannot be adequately summarized due to the current status of knowledge.

Cannibalism of other dinoflagellates has been reported, such as *Fragilidium*, *Peridiniopsis*, *Protoperidinium*, *Pfiesteria*, and *Oxyrrhis* [63,66,102,105–107] and may be widespread in more dinoflagellates, particularly those that are strictly heterotrophic. Cannibalism has been speculated to have particular implications during the evolution of sex because self-ingestion without self-digestion may have led to the evolution of diploidy [108].

Phagotrophy, the internalization of photosynthetic organisms by a eukaryote in a general sense, is essential for the occurrence of present-day endosymbiotic algae and kleptoplastid-containing protists, and even for the origin of plastids themselves [109]. Analysis of field data revealed that up to 40–60% of plankton which have been traditionally labelled as microzooplankton (non-) are actually non-constitutive mixotrophs. They are mixotrophs lacking a constitutive ability of photosynthesis, and thus, can employ acquired chloroplastids for phototrophy other than phagocytose for nutrients [110]. It is interesting that the evolutionary histories of chrysophytes and dinoflagellates, two groups containing the largest amounts of phagotrophic species, can be traced back to the early Paleozoic [111]. This suggests that mixotrophy, or multiple trophic modes, may be a primitive state, and also be indispensable for long term evolutionary success [112]. However, this aspect largely continues to be an unexplored area.

There may be close relationships between phagotrophy and toxicity/allelopathy of *K. veneficum*. Phagotrophy could not be an isolated aspect of the physiological ecology of phytoplankton. It may have coevolved with other physiological capabilities in many taxa, such as the ability to use dissolved organic material and allelopathic tendencies [113]. Many toxic algae have been proved to be phagotrophic or closely related to the known phagotrophs. The toxicity of *K. veneficum* in different strains exhibited a decreasing order that perfectly coincided with the increasing order of laboratorial culturing time [88]. It seems that the toxicity of *K. veneficum* may have receded because of the lack of prey.

7. MIXOTROPHY IN REGULATING POPULATION DYNAMICS AND HABS FORMATION OF *KARLODINIUM*

The significance of mixotrophy in phytoplankton has been increasingly emphasized in recent years. In 2016, a new functional grouping of planktonic protists in an ecophysiological context was proposed to recognize the value of mixotrophy in euphotic aquatic systems and to align with the traditional dichotomy of phytoplankton and zooplankton: (1) phago- heterotrophs as protists lacking photosynthetic autotrophic capacity, (2) photoautotrophs as protists lacking phagotrophic capacity, (3) constitutive mixotrophs (CMs) as phagotrophs with an inherent capacity for phototrophy, and (4) non-constitutive mixotrophs (NCMs) as phagotrophs acquiring their phototrophic capacity by ingesting specific (SNCM) or general non-specific (GNCM) preys [114]. Given that mixotrophs differ widely in their biology, it is apparent that they are also different in their ecological niche and their implications on ecosystem processes [115]. CMs, combining functions of both phagotrophy and phototrophy, are supposed to have the capability to hold the high ground in an ecosystem, ultimately triggering a large area of blooms. Indeed, constitutive mixotrophy has been considered as a major trophic mode for harmful dinoflagellate species in eutrophic coastal waters [4]. Moreover, mixotrophic species tend to dominate in more-mature systems, such as established eutrophic systems and oligotrophic systems in temperate summer, with their flexible nutritional supplies [116]. The mixotroph-dominated ecological structure differs radically in energy flow and material cycling, which is reflected in the shortened and more efficient transformation from nutrient regeneration to primary production. In severe eutrophic water bodies, bloom-forming phytoplankton with mixotrophic mode may sometimes decrease energy flowing to higher trophic levels and thus simplify the food web [115]. Moreover, mixotrophic protists can also take advantage of bacterial production to support primary production [116]. In view of the important role of mixotrophic protists in the marine ecosystem, "mixoplankton" was proposed and emphasized as a new paradigm for marine ecology and is believed to offer a better understanding on the microbial trophic dynamics and the biological pump, along with "phytoplankton" and "zooplankton" [116,117]. This conception may become a new research hotspot.

Mixotrophy is supposed to be a major contributor to the population dynamics of the *Karlodinium* species. Dinoflagellates with different trophic modes may indicate that they employ different survival strategies and occupy different ecological niches, and the phagotrophic tendencies of *Karlodinium* may partially explain some aspects of their bloom dynamics and population ecology. On one hand, phagotrophy may play an important role for phagotrophs in maintaining their population in environments of low light intensity and low nutrient availability [118] via acquiring limiting elements from prey. On the other hand, even in eutrophic habitats, phagotrophic mixotrophs may attain growth higher than that which they could reach in a strict phototrophic mode [4]. Phagotrophy can also contribute to a better budget of essential and major nutrients (C, N and P) in these species. It was documented that the prey-ingestion of *K. armiger* helped to acquire essential inorganic nutrients to stimulate the photo-synthetic capability under nutrient limitation, as it grew very slowly in standard growth medium (f/2) and light without prey, but grew dramatically faster ($\mu = 0.65\ d^{-}1$) when fed preys [119]. *Karlodinium veneficum* also grew much faster with prey than it did strictly autotrophically [20,85]. Adolf et al. studied the balance of autotrophy and heterotrophy of mixotrophic growth of *K. veneficum* [85]. It turned out that the mixotrophic growth of *K. veneficum* was dominated by heterotrophic metabolism, and photosynthesis continued at a lower rate, suggesting a shift toward heterotrophy during grazing. It is confirmed that photosynthesis contributed 27–69% of the gross C uptake with an irradiance at 200 μmol photons $m^{-}2{\cdot}s^{-}1$ and a daily supply of prey cells [85].

Multiple studies have pointed out that the predation of phagotrophic bloom-forming species on their competitors or potential grazers may contribute to the success in monopolizing resources and forming dense, mono-specific blooms [113,118]. A recent bioassay suggests that phagotrophy or micropredation of *K. australe* might play a key role in the lethal effects on the marine animals rather than exotoxicity, especially at lower cell densities [73]. This may explain why many groups of autotrophic phytoplankton can grow rapidly and densely under a combination of light and nutrients in the laboratory but most of them cannot form monospecific blooms in the field [113]. Mixotrophic *Karlodinium* species also show a growth advantage in size [20,59]. For example, in *K. armiger* cultures with sufficient food, it was easy to reach a cell size of up to 9000 $\mu m^{3}{\cdot}cell^{-}1$, and the mean biovolume was approximately twice the size (2500–3000 $\mu m^{3}{\cdot}cell^{-}1$) of non-fed cultures (1200–1500 $\mu m^{3}{\cdot}cell^{-}1$) [11]. Magnifying cell size may help to avoid part of predators specializing in smaller preys. More importantly, the large range of prey types, wide spectrum of prey size, and flexible nutritional modes of *Karlodinium*, such as *K. armiger* and *K. veneficum*, seems to make it a powerful competitor in marine plankton [4].

HABs of *Karlodinium* have been demonstrated to be highly related to mixotrophic predation. Adolf et al. suggested that prey abundance, especially the abundance of nano-planktonic cryptophytes, was a key factor stimulating the formation of toxic *K. veneficum* blooms in eutrophic waters [86]. They also stated the key elements resulting in toxic *K. veneficum* blooms, include (1) eutrophic environments, (2) co-occurrence of cryptophytes and *K. veneficum*, (3) a rapid response of cryptophytes to environmental opportunities (e.g., nutrient input) to bloom, and (4) mixotrophic predation of *K. veneficum* on cryptophytes, aided by allele-chemicals (e.g., karlotoxins) produced by *K. veneficum* that improve prey capture and reduce grazing mortality of toxic strains [1].

Toxins and/or allelochemicals are involved in prey capture in this genus. Both *K. veneficum* and *K. armiger* were observed to immobilize preys by toxins, and then an ingestion process followed [41,58]. HABs of *K. veneficum* were assisted by karlotoxins and contributed to accumulations of toxic *K. veneficum* based on their relatively higher phagotrophic capacity compared to non-toxic cells. High densities of *K. veneficum*, when harmful blooms occurred, exhibited allelopathy to other co-occurred algae by suppressing their physiological activity and growth rates [19,88,120], which induced other microalgae species more favourable to being captured.

It was assumed that once the mixotrophic harmful algal population has reached bloom density, mixotrophic feeding may not play a key role because preys were significantly reduced [121]. However, cannibalism, the recently found nutrient mode in *K. veneficum*, may help in maintaining population levels after the bloom is formed by consuming the dead cells of their own species [49]. This may explain the unusual phenomenon that certain harmful algal blooms maintain high cell densities even when nutrients are exhausted [122]. Based on the significant role of mixotrophy in bloom formation and dynamics in general, many new factors should be taken into consideration when we attempt to prevent and control HABs caused by mixoplankton. For instance, elimination of inorganic nutrient loading may not work well for this type of bloom. Other than inorganic nutrients and hydrological conditions, factors such as dissolved organic matter and even co-occurring plankton species could contribute to the formation of these blooms. In addition, the elimination of HABs desiderates a healthy ecosystem and complex food web, because the more energy flows to higher trophic levels, the less energy mixotrophs can detain.

8. PERSPECTIVES FOR FUTURE INVESTIGATIONS ON THE MIXOTROPHY IN *KARLODINIUM*

8.1 The Ecophysiology of Karlodinium Under Global Changes

Mixotrophy constitutes an energy-saving and a compensatory mechanism to meet the cellular C demands, thereby gaining the necessary energy to cope with the abiotic stress such as cooling and warming under the ultraviolet portion of the spectrum [123]. This metabolic flexibility implies a competitive advantage under multi-driver conditions compared with strict phototrophic or heterotrophic metabolisms as it would allow them to acquire energy and nutrition from both sun and prey depending on the environmental conditions [123]. Thus, more studies ought to be carried out to evaluate the influence of global change on the ecophysiology of mixoplankton, such as *Karlodinium* species.

8.2 Molecular Basis of Phagotrophy-Relevant Genes in Karlodinium and Other Species

At present, we know few details about the molecular or genetic mechanisms involved in mixotrophs in modulating their photoauto- vs. phagohetero-trophic capabilities [114]. The environmental changes may play an important role in impacting the metabolic regulation of mixotrophs under stressful conditions, which need to be taken into account. It was demonstrated that the phagotrophy intensity of *Karlodinium* species increases under nutrient limitation [16,22,119]. Other factors such as prey density, prey species, nutrient concentration, water depth, and salinity were also observed to affect the switch and intensity of phagotrophy [16,21]. We have recently documented that the intensity of phagotrophy in *K. veneficum*, including cannibalism, changed with the growth stage [16]. However, how the change in phagotrophy intensity and the switch among feeding mechanisms are regulated at subcellular and genetic levels continues to be a "Blackbox". Previous studies on the molecular and genomic mechanisms of phagocytosis were based on and limited to a small part of organisms from other groups, like the specialized phagocytotic cells of insects and mammals (e.g., macrophages), the amoebozoans *Dictyostelium discoideum*and *Entamoeba histolytica*, and the ciliate *Tetrahymena thermophila* [124–128]. The molecular studies of phagocytosis in marine microalgae are rare and focus on non-dinoflagellates like the chlorophyte *Cymbomonas tetramitiformis* and chrysophyte *Ochromonas* sp. [129,130]. Considering the possibly early origin of phagotrophy and the relatively close evolution- ary distances within protists, the knowledge obtained from these molecular studies on protists, ciliates in particular, should be a solid basis for generating testable hypotheses about the molecular mechanisms of phagotrophy in *Karlodinium*. Nevertheless, it is now the time to start investigations on the phagotrophy-relevant genes and their expression regulations, and the biochemical (e.g., enzymes, proteins, and signal chemicals) and cellular mechanisms in *Karlodinium*.

8.3 Energetics and Pathways Relevant to the Energy Metabolisms of Phagocytosis of Karlodinium

Once organic particles, as above mentioned, are ingested as foods into *Karlidinium* cells, these "particles" should be subsequently degraded and utilized via a series of energy metabolism-related pathways. In addition, ingestion of organic particles may exert an influence on other metabolic pathways. A transcriptomic analysis about the effects of light and prey availability on the global gene expression of a mixotrophic chrysophyte *Ochromonas* sp. demonstrated that the ingestion of bacterial prey resulted in prominent changes in major metabolic pathways of carbon and nitrogen [130]. With the very limited knowledge regarding to the molecular processes involved in phagotrophy of *Karlodinium*, we postulate that studies focusing on the energetics and energy metabolism pathways involved in phagocytosis may be a key step to comprehensively understand the molecular processes and ecological significance of phago-mixotrophy in *Karlodinium*.

9. CONCLUSIONS

Although *Karlodinium* as a group of small, unarmored dinoflagellates has been long overlooked, owing to the difficulty in identification, and the nutritional modes have been far less studied for most species of the genus, our current knowledge about the trophic modes of *Karlodinium* is worthy of a synthesis, as has been done in this review, to promote forward studies. *Karlodinium* species exhibit plastic and multiple trophic modes and switching between these modes allows *Karlodinium* species to use inorganic and organic, dissolved and particulate nutrients, and live and dead organisms, as nutrients, and even those contained in other individuals of the same species, via multiple instruments (e.g., peduncle and capture filament) and processes (e.g., engulfment, myzocytosis, etc.). *Karlodinium* species may not be able to survive well in any single mode, but the mixotrophic strategy certainly provides competitive advantages over other strictly autotrophic or heterotrophic competitors, by obtaining nutrients from multiple sources and killing competitors and even predators. In addition, the synergism among toxicity and allelopathy found at least in *K. veneficum* may also help to capture preys and avoid predation (Figure 2).

Mixotrophy, particularly phagotrophy, may have been a major contributor to the formation of harmful algal blooms and the achievement of a cosmopolitan distribution in species of *Karlodinium*, *K. veneficum* in particular, which thus deserves more in-depth investigations regarding the knowledge gaps that we have at least partly identified above.

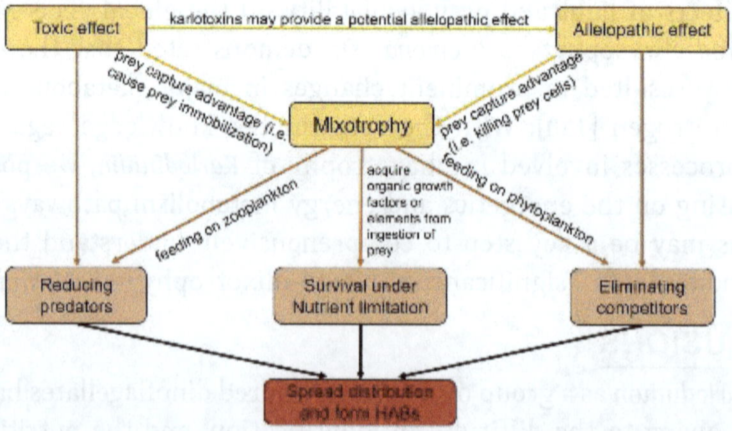

Figure 2. The relationship of phago-mixotrophy with toxicity and allelopathy of *Karlodinium*, and the implications for their global distribution and harmful algal blooms (HABs).

REFERENCES

1. Charles, B.M.; Patricia, A.W. *Biological Oceanography*, 2nd ed.; Wiley-Blackwell: Chichester, UK, 2012; pp. 1–464.
2. Long, S.P.; Humphries, S.; Falkowski, P.G. Photoinhibition of photosynthesis in nature. *Annu. Rev. Plant Physiol. Plant Mol. Biol.* **1994**, *45*, 633–662. [CrossRef]
3. Falkowski, P.G.; Barber, R.T.; Smetacek, V. Biogeochemical controls and feedbacks on ocean primary production. *Science* **1998**, *281*, 200–206. [CrossRef] [PubMed]
4. Burkholder, J.A.M.; Glibert, P.M.; Skelton, H.M. Mixotrophy, a major mode of nutrition for harmful algal species in eutrophic waters. *Harmful Algae* **2008**, *8*, 77–93. [CrossRef]
5. Gomez, F. A quantitative review of the lifestyle, habitat and trophic diversity of dinoflagellates (Dinoflagellata, Alveolata). *Syst. Biodivers.* **2012**, *10*, 267–275. [CrossRef]
6. Lin, S.; Cheng, S.; Song, B.; Zhong, X.; Lin, X.; Li, W.; Li, L.; Zhang, Y.; Zhang, H.; Ji, Z.; et al. The *Symbiodinium kawagutii* genome illuminates dinoflagellate gene expression and coral symbiosis. *Science* **2015**, *350*, 691–694. [CrossRef]
7. Smayda, T.J. Harmful algal blooms: Their ecophysiology and general relevance to phytoplankton blooms in the sea. *Limnol. Oceanog.* **1997**, *42*, 1137–1153. [CrossRef]
8. Cullen, J.J.; Yang, X.; Macintyre, H.L. Nutrient limitation of marine photosynthesis. In *Primary Productivity and Biogeochemical Cycles in the Sea*; Falkowski, P.G., Ed.; Plenum Press: New York, NY, USA, 1992; Volume 43, pp. 69–88.

9. Tang, E.P.Y. Why do dinoflagellates have lower growth rates? *J. Phycol.* **1996**, *32*, 80–84. [CrossRef]
10. Jones, R.I. Mixotrophy in Planktonic protists as a spectrum of nutritional strategies. *Mar. Microbial Food Webs* **1994**, *8*, 87–96.
11. Jones, R.I. Mixotrophy in planktonic protists: An overview. *Freshw. Biol.* **2000**, *45*, 219–226. [CrossRef]
12. Hammer, A.C.; Pitchford, J.W. The role of mixotrophy in plankton bloom dynamics, and the consequences for productivity. *ICES J. Mar. Sci.* **2005**, *62*, 833–840. [CrossRef]
13. Gaines, G. Heterotrophic nutrition. *Biology of Dinoflagellates.* In *The biology of dinoflagellates*; Taylor, F.J.R., Ed.; Blackwell: Oxford, UK, 1987; pp. 224–267.
14. Skovgaard, A. Mixotrophy in *Fragilidium subglobosum* (Dinophyceae): Growth and grazing responses as functions of light intensity. *Mar. Ecol. Prog.* **1996**, *143*, 247–253. [CrossRef]
15. Daugbjerg, N.; Hansen, G.; Larsen, J.; Moestrup, O. Phylogeny of some of the major genera of dinoflagellates based on ultrastructure and partial LSU rDNA sequence data, including the erection of three new genera of unarmoured dinoflagellates. *Phycologia* **2000**, *39*, 302–317. [CrossRef]
16. Yang, H.; Hu, Z.; Shang, L.; Deng, Y.; Tang, Y.Z. A strain of the toxic dinoflagellate *Karlodinium veneficum* isolated from the East China Sea is an omnivorous phagotroph. *Harmful Algae* **2020**, *93*, 101775. [CrossRef] [PubMed]
17. Hallegraeff, G.; Mooney, B.; Evans, K.; Hosja, W. *What Triggers Fish-Killing Karlodinium veneficum Dinoflagellate Blooms in the Swan Canning River System*; Swan Canning Research and Innovation Program Final Report; Swan River Trust: Perth, Australia, 2010; p. 31.
18. Lim, H.C.; Leaw, C.P.; Tan, T.H.; Kon, N.F.; Yek, L.H.; Hii, K.S.; Teng, S.T.; Razali, R.M.; Usup, G.; Iwataki, M.; et al. A bloom of *Karlodinium australe* (Gymnodiniales, Dinophyceae) associated with mass mortality of cage-cultured fishes in West Johor Strait, Malaysia. *Harmful Algae* **2014**, *40*, 51–62. [CrossRef]
19. Adolf, J.E.; Bachvaroff, T.R.; Krupatkina, D.N.; Nonogaki, H.; Brown, P.J.P.; Lewitus, A.J.; Harvey, H.R.; Place, A.R. Species specificity and potential roles of *Karlodinium micrum* toxin. *Afr. J. Mar. Sci.* **2006**, *28*, 415–419. [CrossRef]
20. Li, A.S.; Stoecker, D.K.; Adolf, J.E. Feeding, pigmentation, photosynthesis and growth of the mixotrophic dinoflagellate *Gyrodinium galatheanum*. *Aquat. Microb. Ecol.* **1999**, *19*, 163–176. [CrossRef]
21. Li, A.S.; Stoecker, D.K.; Coats, D.W. Mixotrophy in *Gyrodinium galatheanum* (Dinophyceae): Grazing responses to light intensity and inorganic nutrients. *J. Phycol.* **2000**, *36*, 33–45. [CrossRef]

22. Li, A.S.; Stoecker, D.K.; Coats, D.W. Spatial and temporal aspects of *Gyrodinium galatheanum* in Chesapeake Bay: Distribution and mixotrophy. *J. Plankton Res.* **2000**, *22*, 2105–2124. [CrossRef]

23. Nielsen, M.V. Growth and chemical composition of the toxic dinoflagellate *Gymnodinium galatheanum* in relation to irradiance, temperature and salinity. *Mar. Ecol. Prog.* **1996**, *136*, 205–211. [CrossRef]

24. Delgado, M.; Alcaraz, M. Interactions between red tide microalgae and herbivorous zooplankton: The noxious effects of *Gyrodinium corsicum* (Dinophyceae) on *Acartia grani* (Copepoda: Calanoida). *J. Plankton Res.* **1999**, *21*, 2361–2371. [CrossRef]

25. Deeds, J.R.; Terlizzi, D.E.; Adolf, J.E.; Stoecker, D.K.; Place, A.R. Toxic activity from cultures of *Karlodinium micrum* (=*Gyrodinium galatheanum*) (Dinophyceae)—a dinoflagellate associated with fish mortalities in an estuarine aquaculture facility. *Harmful Algae* **2002**, *1*, 169–189. [CrossRef]

26. Fernández-Tejedor, M.; Soubrier-Pedreño, M.Á.; Ma, D.F. Acute LD of a *Gyrodinium corsicum* natural population for *Sparus aurata*and *Dicentrarchus labrax*. *Harmful Algae* **2004**, *3*, 1–9. [CrossRef]

27. Da Costa, M.R.; Franco, J.; Cacho, E.; Fernández, F. Toxin content and toxic effects of the dinoflagellate *Gyrodinium corsicum* (Paulmier) on the ingestion and survival rates of the copepods *Acartia grani* and *Euterpina acutifrons*. *J. Exp. Mar. Biol. Ecol.* **2005**, *322*, 177–183. [CrossRef]

28. de Salas, M.F.; Bolch, C.J.S.; Hallegraeff, G.M. *Karlodinium australe* sp. nov. (Gymnodiniales, Dinophyceae), a new potentially ichthyotoxic unarmoured dinoflagellate from lagoonal habitats of south-eastern Australia. *Phycologia* **2005**, *44*, 640–650. [CrossRef]

29. Garcés, E.; Fernandez, M.; Penna, A.; Lenning, K.V.; Gutierrez, A.; Camp, J.; Zapata, M. Characterization of NW Mediterranean *Karlodinium* spp. (Dinophyceae) strains using morphological, molecular, chemical, and physiological methodologies. *J. Phycol.* **2006**, *42*, 1096–1112. [CrossRef]

30. Adolf, J.E.; Bachvaroff, T.R.; Deeds, J.R.; Place, A.R. Ichthyotoxic *Karlodinium veneficum* (Ballantine) J Larsen in the upper Swan River estuary (Western Australia): Ecological conditions leading to a fish kill. *Harmful Algae* **2015**, *48*, 83–93. [CrossRef]

31. Van Wagoner, R.M.; Deeds, J.R.; Tatters, A.O.; Place, A.R.; Tomas, C.R.; Wright, J.L.C. Structure and relative potency of several karlotoxins from *Karlodinium veneficum*. *J. Nat. Prod.* **2010**, *73*, 1360–1365. [CrossRef]

32. Cai, P.; He, S.; Zhou, C.; Place, A.R.; Haq, S.; Ding, L.; Chen, H.; Jiang, Y.; Guo, C.; Xu, Y.; et al. Two new karlotoxins found in*Karlodinium veneficum* (strain GM2) from the East China Sea. *Harmful Algae* **2016**, *58*, 66–73. [CrossRef]

33. Krock, B.; Busch, J.A.; Tillmann, U.; Garcia-Camacho, F.; Sanchez-Miron, A.; Gallardo-Rodriguez, J.J.; Lopez-Rosales, L.; Andree, K.B.; Fernandez-Tejedor, M.; Witt, M.; et al. LC-MS/MS detection of karlotoxins reveals new variants in strains of the marine dinoflagellate *Karlodinium veneficum* from the Ebro Delta (NW Mediterranean). *Mar. Drugs* **2017**, *15*. [CrossRef]

34. Mooney, B.D.; de Salas, M.; Hallegraeff, G.M.; Place, A.R. Survey for karlotoxin production in 15 species of gymnodinioid dinoflagellates (Kareniaceae, Dinophyta). *J. Phycol.* **2009**, *45*, 164–175. [CrossRef]

35. Mooney, B.D.; Hallegraeff, G.M.; Place, A.R. Ichthyotoxicity of four species of gymnodinioid dinoflagellates (Kareniaceae, Dinophyta) and purified karlotoxins to larval sheepshead minnow. *Harmful Algae* **2010**, *9*, 557–562. [CrossRef]

36. Andersen, A.; De, L.M.; Binzer, S.B.; Rasmussen, S.A.; Hansen, P.J.; Nielsen, K.F.; Jørgensen, K.; Larsen, T.O. HPLC-HRMS quantification of the ichthyotoxin karmitoxin from *Karlodinium armiger*. *Mar Drugs* **2017**, *15*, 278. [CrossRef] [PubMed]

37. Rasmussen, S.A.; Binzer, S.B.; Hoeck, C.; Meier, S.; de Medeiros, L.S.; Andersen, N.G.; Place, A.; Nielsen, K.F.; Hansen, P.J.; Larsen,T.O. Karmitoxin: An amine-containing polyhydroxy-polyene toxin from the marine dinoflagellate *Karlodinium armiger*. *J. Nat. Prod.* **2017**, *80*, 1287–1293. [CrossRef] [PubMed]

38. Bergholtz, T.; Daugbjerg, N.; Moestrup, Ø.; Fernández-Tejedor, M. On the identity of *Karlodinium veneficum* and description of *Karlodinium armiger* sp. nov. (Dinophyceae), based on light and electron microscopy, nuclear-encoded LSU rDNA, and pigment composition. *J. Phycol.* **2005**, *42*, 170–193. [CrossRef]

39. Toldrà, A.; Jauset-Rubio, M.; Andree, K.B.; Fernández-Tejedor, M.; Diogène, J.; Katakis, I.; O'Sullivan, C.K.; Campàs, M. Detection and quantification of the toxic marine microalgae *Karlodinium veneficum* and *Karlodinium armiger* using recombinase polymerase amplification and enzyme-linked oligonucleotide assay. *Anal. Chim. Acta* **2018**, *1039*, 140–148. [CrossRef] [PubMed]

40. Berge, T.; Hansen, P.J.; Moestrup, O. Prey size spectrum and bioenergetics of the mixotrophic dinoflagellate *Karlodinium armiger*. *Aquat. Microb. Ecol.* **2008**, *50*, 289–299. [CrossRef]

41. Berge, T.; Poulsen, L.K.; Moldrup, M.; Daugbjerg, N.; Hansen, P.J. Marine microalgae attack and feed on metazoans. *Isme J.* **2012**,*6*, 1926–1936. [CrossRef]

42. Leong, S.C.Y.; Lim, L.P.; Chew, S.M.; Kok, J.W.K.; Teo, S.L.M. Three new records of dinoflagellates in Singapore's coastal waters, with observations on environmental conditions associated with micro-algal growth in the Johor Straits. *Raffles Bull. Zool.* **2015**, 24–36.

43. De Salas, M.F.; Laza-Martínez, A.; Hallegraeff, G.M. Novel unarmored dinoflagellates from the toxigenic family Kareniaceae (Gymnodiniales): Five new species of *Karlodinium* and one new *Takayama* from the Australian sector of the southern ocean. *J. Phycol.* **2008**, *44*, 241–257. [CrossRef]

44. Benico, G.; Takahashi, K.; Lum, W.M.; Yniguez, A.T.; Iwataki, M. The harmful unarmored dinoflagellate *Karlodiniumin* Japan and Philippines, with reference to ultrastructure and micropredation of *Karlodinium azanzae* sp. nov. (Kareniaceae, Dinophyceae). *J. Phycol.* **2020**, *56*, 1264–1282. [CrossRef]

45. Escobar-Morales, S.; Hernandez-Becerril, D.U. Free-living marine planktonic unarmoured dinoflagellates from the Gulf of Mexico and the Mexican Pacific. *Botanica Marina* **2015**, *58*, 9–22. [CrossRef]

46. Zingone, A.; Siano, R.; D'Alelio, D.; Sarno, D. Potentially toxic and harmful microalgae from coastal waters of the Campania region (Tyrrhenian Sea, Mediterranean Sea). *Harmful Algae* **2006**, *5*, 321–337. [CrossRef]

47. Paulmier, G.; Berland, B.; Billard, C.; Nezan, E. *Gyrodinium corsicum* nov. sp. (Gymnodiniales, Dinophycean), responsible organism of the green water in salt-water lake of Diana (Corsica), in April 1994. *Cryptogamie Algol.* **1995**, *16*, 77–94.

48. Cen, J.; Wang, J.; Huang, L.; Ding, G.; Qi, Y.; Cao, R.; Cui, L.; Lv, S. Who is the "murderer" of the bloom in coastal waters of Fujian, China, in 2019? *J. Oceanol. Limnol.* **2020**, *38*, 722–732. [CrossRef]

49. Yang, Z.B.; Takayama, H.; Matsuoka, K.; Hodgkiss, I.J. *Karenia digitata* sp. nov. (Gymnodiniales, Dinophyceae), a new harmful algal bloom species from the coastal waters of west Japan and Hong Kong. *Phycologia* **2000**, *39*, 463–470. [CrossRef]

50. Cen, J.; Wang, J.; Huang, L.; Lin, Y.; Ding, G.; Qi, Y.; Lv, S. *Karlodinium elegans* sp. nov. (Gymnodiniales, Dinophyceae), a novel species isolated from the East China Sea in a dinoflagellate bloom. *J. Oceanol. Limnol.* **2020**, 1–17. [CrossRef]

51. Nezan, E.; Siano, R.; Boulben, S.; Six, C.; Bilien, G.; Cheze, K.; Duval, A.; Le Panse, S.; Quere, J.; Chomerat, N. Genetic diversity of the harmful family Kareniaceae (Gymnodiniales, Dinophyceae) in France, with the description of *Karlodinium gentienii* sp. nov.: A new potentially toxic dinoflagellate. *Harmful Algae* **2014**, *40*, 75–91. [CrossRef]

52. Ballantine, D. Two new marine species of *Gymnodinium* isolated from the Plymouth area. *J. Mar. Biol. Assoc. UK* **1956**, *35*, 467–474. [CrossRef]

53. Hallett, C.S.; Valesini, F.J.; Clarke, K.R.; Hoeksema, S.D. Effects of a harmful algal bloom on the community ecology, movements and spatial distributions of fishes in a microtidal estuary. *Hydrobiologia* **2016**, *763*, 267–284. [CrossRef]

54. Granéli, E.; Turner, J.T. *Ecology of Harmful Algae*; Springer: Berlin/Heidelberg, Germany, 2006; pp. 1–413.
55. Zhu, X.; Zhou, C.; Meng, R.; Li, S.; Fang, K.; Luo, Z.; Xu, J.; He, S.; Luo, Q.; Yan, X. Biochemical characteristics support the recently described species *Karlodinium zhouanum* (Gymnodiniales, Dinophyceae). *Phycol. Res.* **2019**, *68*, 14–22. [CrossRef]
56. Luo, Z.; Wang, L.; Chan, L.; Lu, S.; Gu, H. *Karlodinium zhouanum*, a new dinoflagellate species from China, and molecular phylogeny of *Karenia digitata* and *Karenia longicanalis* (Gymnodiniales, Dinophyceae). *Phycologia* **2018**, *57*, 401–412. [CrossRef]
57. Li, Z.; Shin, H.H. Morphology and phylogeny of an unarmored dinoflagellate, *Karlodinium jejuense* sp. nov. (Gymnodiniales), isolated from the northern East China Sea. *Phycol. Res.* **2018**, *66*, 318–328. [CrossRef]
58. Sheng, J.; Malkiel, E.; Katz, J.; Adolf, J.E.; Place, A.R. A dinoflagellate exploits toxins to immobilize prey prior to ingestion. *Proc. Natl Acad. Sci. USA* **2010**, *107*, 2082–2087. [CrossRef] [PubMed]
59. Berge, T.; Hansen, P.J.; Moestrup, O. Feeding mechanism, prey specificity and growth in light and dark of the plastidic dinoflagellate *Karlodinium armiger*. *Aquat. Microb. Ecol.* **2008**, *50*, 279–288. [CrossRef]
60. Stoecker, D.K. Mixotrophy among dinoflagellates. *J. Eukaryot. Microbiol.* **1999**, *46*, 397–401. [CrossRef]
61. Schnepf, E.; Elbrächter, M. Nutritional strategies in dinoflagellates: A review with emphasis on cell biological aspects. *Eur. J. Protistol.* **1992**, *28*, 3–24. [CrossRef]
62. Biecheler, B. Recherches sur les Péridiniens. *Bull. Biol. Fr. Belg.* **1952**, *36*, 1–149.
63. Calado, A.J.; Moestrup, Ø. Feeding in *Peridiniopsis berolinensis* (Dinophyceae): New observations on tube feeding by an omnivorous, heterotrophic dinoflagellate. *Phycologia* **1997**, *36*, 47–59. [CrossRef]
64. Hansen, P.J.; Calado, A.J. Phagotrophic mechanisms and prey selection in free-living dinoflagellates. *J. Eukaryot. Microbiol.* **1999**,*46*, 382–389. [CrossRef]
65. Jacobson, D.M.; Anderson, D.M. Thecate heterophic dinoflagellates: Feeding behavior and mechanisms. *J. Phycol.* **1986**, *22*, 249–258. [CrossRef]
66. Jeong, H.J.; Latz, M.I. Growth and grazing rates of the heterotrophic dinoflagellates *Protoperidinium* spp. on red tide dinoflagellates. *Mar. Ecol. Prog. Ser.* **1994**, *106*, 173–185. [CrossRef]
67. Elbrächter, M. Food uptake mechanisms in phagotrophic dinoflagellates and classification. In *The Biology of Free-Living Heterotrophic Flagellates*; Patterson, D.J., Larsen,J., Eds.; Clarendon Press: Oxford, UK, 1991; Volume 45, pp. 303–312.

68. Schnepf, E.; Deichgräber, G. "Myzocytosis", a kind of endocytosis with implications to compartmentation in endosymbiosis. *Naturwissenschaften* **1984**, *71*, 218–219. [CrossRef]

69. Spero, H.J. Phagotrophy in *Gymnodinium fungiforme* (Pyrrhophyta): The peduncle as an organelle of ingestion. *J. Phycol.* **1982**, *18*, 356–360. [CrossRef]

70. Wilcox, L.W.; Wedemayer, G.J. Phagotrophy in the freshwater, photosynthetic dinoflagellate *Amphidinium cryophilum*. *J. Phycol.* **1991**, *27*, 600–609. [CrossRef]

71. Schnepf, E.; Deichgräber, G.; Drebes, G. Food uptake and the fine structure of the dinophyte *Paulsenella* sp., an ectoparasite of marine diatoms. *Protoplasma* **1985**, *124*, 188–204. [CrossRef]

72. Place, A.R.; Bowers, H.A.; Bachvaroff, T.R.; Adolf, J.E.; Deeds, J.R.; Sheng, J. *Karlodinium veneficum* —The little dinoflagellate with a big bite. *Harmful Algae* **2012**, *14*, 179–195. [CrossRef]

73. Song, X.; Hu, Z.; Shang, L.; Leaw, C.P.; Lim, P.T.; Tang, Y.Z. Contact micropredation may play a more important role than exotoxicity does in the lethal effects of *Karlodinium australe* blooms: Evidence from laboratory bioassays. *Harmful Algae* **2020**, *99*, 101926. [CrossRef]

74. Vogelbein, W.K.; Lovko, V.J.; Shields, J.D.; Reece, K.S.; Mason, P.L.; Haas, L.W.; Walker, C.C. *Pfiesteria shumwayae* kills fish by micropredation not exotoxin secretion. *Nature* **2002**, *418*, 967–970. [CrossRef]

75. Lafferty, K.D.; Kuris, A.M. Trophic strategies, animal diversity and body size. *Trends Ecol. Evol.* **2002**, *17*, 507–513. [CrossRef]

76. Spero, H.J. Chemosensory capabilities in the phagotrophic dinoflagellate *Gymnodinium fungiforme*. *J. Phycol.* **1985**, *21*, 181–184. [CrossRef]

77. Sheng, J.; Malkiel, E.; Katz, J.; Adolf, J.; Belas, R.; Place, A.R. Digital holographic microscopy reveals prey-induced changes in swimming behavior of predatory dinoflagellates. *Proc. Natl Acad. Sci. USA* **2007**, *104*, 17512–17517. [CrossRef] [PubMed]

78. Hansen, P.J. Prey size selection, feeding rates and growth dynamics of heterotrophic dinoflagellates with special emphasis on *Gyrodinium-spirale*. *Mar. Biol.* **1992**, *114*, 327–334. [CrossRef]

79. Schnepf, E.; Winter, S.; Mollenhauer, D. *Gymnodinium aeruginosum* (Dinophyta): A blue-green dinoflagellate with a vestigial, anucleate, cryptophycean endosymbiont. *Plant Syst. Evol.* **1989**, *164*, 75–91. [CrossRef]

80. Skovgaard, A. Role of chloroplast retention in a marine dinoflagellate. *Aquat. Microb. Ecol.* **1998**, *15*, 293–301. [CrossRef]

81. Wilcox, L.W.; Wedemayer, G.J. *Gymnodinium acidotum* Nygaard (Pyrrophyta), a dinoflagellate with an endosymbiotic cryptomonad. *J. Phycol.* **2010**, *20*, 236–242. [CrossRef]
82. Hansen, P.J. Phagotrophic mechanisms and prey selection in mixotrophic phytoflagellates. *Physiol. Ecol. Harmful Algal Blooms***1998**, 525–537.
83. Jakobsen, H.H.; Hansen, P.J.; Larsen, J. Growth and grazing responses of two chloroplast-retaining dinoflagellates: Effect of irradiance and prey species. *Mar. Ecol. Prog. Ser.* **2000**, *201*, 121–128. [CrossRef]
84. Hansen, P.J. The role of photosynthesis and food uptake for the growth of marine mixotrophic dinoflagellates. *J. Eukaryot. Microbiol.* **2011**, *58*, 203–214. [CrossRef]
85. Adolf, J.E.; Stoecker, D.K.; Harding, L.W., Jr. The balance of autotrophy and heterotrophy during mixotrophic growth of *Karlodinium micrum* (Dinophyceae). *J. Plankton Res.* **2006**, *28*, 737–751. [CrossRef]
86. Adolf, J.E.; Bachvaroff, T.; Place, A.R. Can cryptophyte abundance trigger toxic *Karlodinium veneficum* blooms in eutrophic estuaries? *Harmful Algae* **2008**, *8*, 119–128. [CrossRef]
87. Bachvaroff, T.R.; Adolf, J.E.; Place, A.R. Strain variation in *Karlodinium veneficum* (Dinophyceae): Toxin profiles, pigments, and growth characteristics. *J. Phycol.* **2009**, *45*, 137–153. [CrossRef] [PubMed]
88. Yang, H.; Hu, Z.; Xu, N.; Tang, Y.Z. A comparative study on the allelopathy and toxicity of four strains of *Karlodinium veneficum* with different culturing histories. *J. Plankton Res.* **2019**, *41*, 17–29. [CrossRef]
89. Stoecker, D.K.; Gustafson, D.E. Cell-surface proteolytic activity of photosynthetic dinoflagellates. *Aquat. Microb. Ecol.* **2003**, *30*, 175–183. [CrossRef]
90. Solomon, C.M.; Glibert, P.M. Urease activity in five phytoplankton species. *Aquat. Microb. Ecol.* **2008**, *52*, 149–157. [CrossRef]
91. Croft, M.T.; Warren, M.J.; Smith, A.G. Algae need their vitamins. *Eukaryot. Cell* **2006**, *5*, 1175–1183. [CrossRef]
92. Hellebust, J.A. Excretion of some organic compounds by marine phytoplankton. *Limnol. Oceanogr.* **1965**, *10*, 192–206. [CrossRef]
93. Kamjunke, N.; Tittel, J. Utilisation of leucine by several phytoplankton species. *Limnologica* **2008**, *38*, 360–366. [CrossRef]
94. Tittel, J.; Wiehle, I.; Wannicke, N.; Kampe, H.; Poerschmann, J.; Meier, J.; Kamjunke, N. Utilisation of terrestrial carbon by osmotrophic algae. *Aquat. Sci.* **2009**, *71*, 46–54. [CrossRef]

95. Beamud, S.G.; Karrasch, B.; Pedrozo, F.L.; Diaz, M.M. Utilisation of organic compounds by osmotrophic algae in an acidic lake of Patagonia (Argentina). *Limnology* **2014**, *15*, 163–172. [CrossRef]

96. Dąbrowska, A.; Nawrocki, J.; Szeląg-Wasielewska, E. Appearance of aldehydes in the surface layer of lake waters. *Environ. Monit. Assess.* **2014**, *186*, 4569–4580. [CrossRef]

97. Nygaard, K.; Tobiesen, A. Bacterivory in algae—A survival strategy during nutrient limitation. *Limnol. Oceanogr.* **1993**, *38*, 273–279. [CrossRef]

98. Li, A.S.; Stoecker, D.K.; Coats, D.W.; Adam, E.J. Ingestion of fluorescently labeled and phycoerythrin-containing prey by mixotrophic dinoflagellates. *Aquat. Microb. Ecol.* **1996**, *10*, 139–147. [CrossRef]

99. Adolf, J.E.; Stoecker, D.K.; Harding, L.W. Autotrophic growth and photo acclimation in *Karlodinium micrum* (Dinophyceae) and *Storeatula major* (Cryptophyceae). *J. Phycol.* **2003**, *39*, 1101–1108. [CrossRef]

100. Li, J.; Glibert, P.M.; Alexander, J.A.; Molina, M.E. Growth and competition of several harmful dinoflagellates under different nutrient and light conditions. *Harmful Algae* **2012**, *13*, 112–125. [CrossRef]

101. Claessen, D.; de Roos, A.M.; Persson, L. Population dynamic theory of size-dependent cannibalism. *Proc. R. Soc. Lond. Ser. B Biol. Sci.* **2004**, *271*, 333–340. [CrossRef] [PubMed]

102. Martel, C.M.; Flynn, K.J. Morphological controls on cannibalism in a planktonic marine phagotroph. *Protist* **2008**, *159*, 41–51. [CrossRef]

103. Spero, H.J.; Moree, M.D. Phagotrophic feeding and its importance to the life cycle of the holozoic dinoflagellate, *Gymnodinium fungiforme*. *J. Phycol.* **1981**, *17*, 43–51. [CrossRef]

104. Schnepf, E.; Drebes, G. Chemotaxis and appetence of *Paulsenella* sp. (Dinophyta), an ectoparasite of the marine diatom *Streptotheca thamesis* Shrubsole. *Planta* **1986**, *167*, 337–343. [CrossRef]

105. Feinstein, T.N.; Traslavina, R.; Sun, M.Y.; Lin, S.J. Effects of light on photosynthesis, grazing, and population dynamics of the heterotrophic dinoflagellate *Pfiesteria piscicida* (Dinophyceae). *J. Phycol.* **2002**, *38*, 659–669. [CrossRef]

106. Jeong, H.J.; Lee, C.W.; Yih, W.H.; Kim, J.S. *Fragilidium* cf *mexicanum*, a thecate mixotrophic dinoflagellate which is prey for and a predator on co-occurring thecate heterotrophic dinoflagellate *Protoperidinium* cf. *divergens*. *Mar. Ecol. Prog. Ser.* **1997**, *151*, 299–305.

107. Naustvoll, L.J. Growth and grazing by the thecate heterotrophic dinoflagellate *Diplopsalis lemicula* (Diplopsalidaceae, Dino- phyceae). *Phycologia* **1998**, *37*, 1–9. [CrossRef]
108. Margulis, L. *Origins of Sex*; Yale University Press: New Haven, CT, USA, 1986.
109. Raven, J.A.; Beardall, J.; Flynn, K.J.; Maberly, S.C. Phagotrophy in the origins of photosynthesis in eukaryotes and as a complementary mode of nutrition in phototrophs: Relation to Darwin's insectivorous plants. *J. Exp. Bot.* **2009**, *60*, 3975–3987. [CrossRef] [PubMed]
110. Leles, S.G.; Mitra, A.; Flynn, K.J.; Stoecker, D.K.; Hansen, P.J.; Calbet, A.; McManus, G.B.; Sanders, R.W.; Caron, D.A.; Not, F.; et al. Oceanic protists with different forms of acquired phototrophy display contrasting biogeographies and abundance. *Proc. R. Soc. Lond. Ser. B Biol. Sci.* **2017**, *284*, 20170664. [CrossRef]
111. Margulis, L.; Schwartz, K.V. Five Kingdoms; Freeman, W.H. and Company: San Francisco, CA, USA, 1982; pp. 1–338.
112. Porter, K.G. Phagotrophic phytoflagellates in microbial food webs. *Hydrobiologia* **1988**, *159*, 89–97. [CrossRef]
113. Stoecker, D.K.; Tillmann, U.; Granéli, E. Phagotrophy in harmful algae. In *Ecology of Harmful Algae*; Granéli, E., Turner, J.T., Eds.; Springer: Berlin, Germany, 2006; pp. 177–188.
114. Mitra, A.; Flynn, K.J.; Tillmann, U.; Raven, J.A.; Caron, D.; Stoecker, D.K.; Not, F.; Hansen, P.J.; Hallegraeff, G.; Sanders, R.; et al. Defining planktonic protist functional groups on mechanisms for energy and nutrient acquisition: Incorporation of diverse mixotrophic strategies. *Protist* **2016**, *167*, 106–120. [CrossRef] [PubMed]
115. Stoecker, D.K.; Hansen, P.J.; Caron, D.A.; Mitra, A.; Annual, R. Mixotrophy in the marine plankton. *Ann. Rev. Mar. Sci.* **2017**, *9*, 311–335. [CrossRef] [PubMed]
116. Mitra, A.; Flynn, K.J.; Burkholder, J.M.; Berge, T.; Calbet, A.; Raven, J.A.; Graneli, E.; Glibert, P.M.; Hansen, P.J.; Stoecker, D.K.; et al. The role of mixotrophic protists in the biological carbon pump. *Biogeosciences* **2014**, *11*, 995–1005. [CrossRef]
117. Flynn, K.J.; Mitra, A.; Anestis, K.; Anschütz, A.A.; Calbet, A.; Ferreira, G.D.; Gypens, N.; Hansen, P.J.; John, U.; Martin, J.L.; et al. Mixotrophic protists and a new paradigm for marine ecology: Where does plankton research go now? *J. Plankton Res.* **2019**, *4*, 375–391. [CrossRef]
118. Legrand, C.; Granéli, E.; Carlsson, P. Induced phagotrophy in the photosynthetic dinoflagellate *Heterocapsa triquetra*. *Aquat. Microb. Ecol.* **1998**, *15*, 65–75. [CrossRef]

119. Berge, T.; Hansen, P.J. Role of the chloroplast in the predatory dinoflagellate *Karlodinium armiger*. *Mar. Ecol. Prog.* **2016**, *519*, 41–54. [CrossRef

120. Naik, R.K.; Chitari, R.R.; Anil, A.C. *Karlodinium veneficum* in India: Effect of fixatives on morphology and allelopathy in relation to *Skeletonema costatum*. *Curr. Sci. India.* **2010**, *99*, 1112–1116.

121. Tittel, J.; Bissinger, V.; Zippel, B.; Gaedke, U.; Bell, E.; Lorke, A.; Kamjunke, N. Mixotrophs combine resource use to outcompete specialists: Implications for aquatic food webs. *Proc. Natl Acad. Sci. USA* **2003**, *100*, 12776–12781. [CrossRef] [PubMed]

122. Granéli, E.; Edvardsen, B.; Roelke, D.L.; Hagström, J.A. The ecophysiology and bloom dynamics of *Prymnesium* spp. *Harmful Algae* **2012**, *14*, 260–270. [CrossRef]

123. Cabrerizo, M.J.; Manuel Gonzalez-Olalla, J.; Hinojosa-Lopez, V.J.; Peralta-Cornejo, F.J.; Carrillo, P. A shifting balance: Responses of mixotrophic marine algae to cooling and warming under UVR. *N. Phytol.* **2019**, *221*, 1317–1327. [CrossRef]

124. Okada, M.; Huston, C.D.; Mann, B.J.; Petri, W.A.; Kita, K.; Nozaki, T. Proteomic analysis of phagocytosis in the enteric protozoan parasite *Entamoeba histolytica*. *Eukaryot.Cell* **2005**, *4*, 827–831. [CrossRef]

125. Gotthardt, D.; Blancheteau, V.; Bosserhoff, A.; Ruppert, T.; Delorenzi, M.; Soldati, T. Proteomics finger printing of phagosome maturation and evidence for the role of a G alpha during uptake. *Mol. Cell. Proteomics* **2006**, *5*, 2228–2243. [CrossRef]

126. Jacobs, M.E.; DeSouza, L.V.; Samaranayake, H.; Pearlman, R.E.; Siu, K.W.M.; Klobutcher, L.A. The *Tetrahymena thermophile* phagosome proteome. *Eukaryot. Cell* **2006**, *5*, 1990–2000. [CrossRef]

127. Shevchuk, O.; Batzilla, C.; Haegele, S.; Kusch, H.; Engelmann, S.; Hecker, M.; Haas, A.; Heuner, K.; Gloeckner, G.; Steinert, M. Proteomic analysis of Legionella-containing phagosomes isolated from Dicyostelium. *Int. J. Med. Microbiol.* **2009**, *299*, 489–508. [CrossRef]

128. Boulais, J.; Trost, M.; Landry, C.R.; Dieckmann, R.; Levy, E.D.; Soldati, T.; Michnick, S.W.; Thibault, P.; Desjardins, M. Molecular characterization of the evolution of phagosomes. *Mol. Syst. Biol.* **2010**, *6*, 423. [CrossRef]

129. Burns, J.A.; Paasch, A.; Narechania, A.; Kim, E. Comparative genomics of a bacterivorous green alga reveals evolutionary causalities and consequences of phago-mixotrophic mode of nutrition. *Genome Biol. Evol.* **2015**, *7*, 3047–3061. [CrossRef]

130. Lie, A.A.Y.; Liu, Z.; Terrado, R.; Tatters, A.O.; Heidelberg, K.B.; Caron, D.A. Effect of light and prey availability on gene expression of the mixotrophic chrysophyte, *Ochromonas* sp. *BMC Genomics* **2017**, *18*, 163. [CrossRef] [PubMed]

CHAPTER-3

ALLELOPATHIC INHIBITION BY THE BACTERIA *BACILLUS CEREUS* BE23 ON GROWTH AND PHOTOSYNTHESIS OF THE MACROALGA ULVA PROLIFERA

1. INTRODUCTION

Allelopathic interactions are considered to be important factors that affect the growth or survival of organisms within the same ecological habit. Allelo-chemicals are secondary metabolites from plants, algae, or bacteria [1]. They may have positive benefits (positive allelopathy) or may be detrimental (negative allelopathy) [2]. Allelopathy has been considered to be one potential control mechanism for harmful algae blooms (HABs) [3]. The inhibition effects of allelopathic compounds on algae include destroying the cell structure [4,5], altering production of the reactive oxygen species (ROS) [6], impacting intracellular enzymatic activities [7], or altering the photosynthesis system [8] and related gene expression [9]. External stress can induce the production of ROS, i.e., hydrogen peroxide (H2O2) and superoxide radical ($O2^{\bullet-}$), and can induce the regulation of the anti-oxidative defense or the photoprotection system [10,11].

A number of bacteria-derived algicidal compounds have drawn wide attention as a control for HABs [12–14] and the algicidal compounds belonging to the *Cytophaga-Flavobacterium-Bacteroides* (CFB) phylum have been identified [15]. Among this phylogenetic profile, the genus of *Bacillus* shows promise in controlling HABs, as negative effects have been demonstrated on the diatom *Skeletonema costatum*, the raphidophyte *Heterosigma akashiwo*, the dinoflagellate *Prorocentrum donghaiense* [16], the prymnesiophyte *Phaeocystis globosa* [16,17], and the cyanobacterium *Microcystis aeruginosa* [18]. The potential allelochemicals that have been isolated and identified from *Bacillus* sp. include terpene, steroids, and alkaloids [19,20]. The active compounds and mechanisms remain to be identified due to the species-specific response to algicidal bacteria [21].

The green tides caused by blooms of *Ulva prolifera* have occurred in the Yellow Sea of China since 2007 [22–26]. These massive blooms negatively impact the local communities, aquaculture operations, and tourism, causing great damage to the local ecosystem service and enormous economic loss [27]. The rapid growth of *U. prolifera*, on the other hand, makes it the strongest competitor for nutrients and light [28,29] in the bloom area, thereby driving the great impact on the marine biodiversity and structure of the community [30–32]. There are currently no effective measures to control these blooms.

The *Bacillus* sp.-derived control of HABs is promising, but limited exploration has been undertaken in mitigating the green tides. As a complicating factor, the life stage of thalli has been reported to be an important factor in green tide development [27]. Therefore, a series of experiments were performed to understand the extent to which bacterial allelopathy may be effective in controlling the thalli of *U. prolifera*. Specifically, the following questions were addressed: (1) does the cell-free filtrate of *Bacillus* sp. inhibit the growth of *U. prolifera* and if so, what is the effective dose? (2) What is the mechanism by which negative allelopathy occurs, particularly with respect to the anti-oxidative defense system and the photosynthetic system II (PSII) response? (3) What are the potential allelochemicals in the filtrate of *Bacillus* sp. that cause negative effects on *U. prolifera*?

2. MATERIALS AND METHODS

2.1 Algal Culture and Identification

Asexual isolates of *Ulva prolifera* were provided by Zhejiang Xiangshan Xuwen Algal Exploitation Company, China, in October 2018. Specimens were subsequently transferred to the laboratory on ice, sterilized with 0.7% potassium iodide (KI) for 5 min, and then rinsed with autoclaved seawater.

The pre-sterilized thalli were maintained in sterilized f/2 medium [33], with salinity of 30, temperature of20°C, and light of 60 μmol·m^2·s^{-1} (12/12 h of light/ dark cycle). The media were replaced every 5 days.

To minimize the interference of carry-over epiphytic bacteria in *U. prolifera*, cultures were pretreated before each exposure experiment by antibiotic mixtures of penicillin (100 mg/L), polymixin (0.75 mg/L), and neomycin (0.9 mg/L) for 48 h [34].

The macroalga was identified using the method described in Li et al. [35]. Total DNA was extracted with a commercial Plant DNA Mini Kit (TaKaRa, China). ITS and

5S sequences were amplified by the corresponding PCR primers (Table 1) and the conducted BLAST analyses in the NCBI database.

Table 1. Sequences of primer pairs for *Ulva prolifera* analysis.

Primer	Sequence (5[1]–3[1])
5S	F: 5[1]-GGTTGGGCAGGATTAGTA-3[1]
	R: 5[1]-GCTGCGTTCTTCATCGWTG-3[1]
ITS	R: 5[1]-AGGCTTAAGTTGCGAGTT-3[1]
	F: 5[1]-TCGTAACAAGGTTTCCGTAGG-3[1]

2.2 Experiment 1: Bacteria-Derived Allelopathic Inhibition on *U. prolifera*

2.2.1 Preparation of Cell-Free Filtrate from *Bacillus cereus*

The bacterium strain *Bacillus cereus* BE23 was previously isolated from the mangrove area in Hainan province, China, and maintained in Luria Bertani (LB) broth (peptone 10.0 g/L, yeast extract5.0 g/L, sea salt 32 g/L, dissolved in dH2O) at 28 °C with shaking at 180 rpm/min. The strain was identified by the 16S rDNA gene and 1439 bp sequence that was acquired by PCR amplification. The bacteria were transferred from stock culture, with the initial concentration of 10^{10}/mL, in 500 mL of LB medium. In 5 days, cell density of *Bacillus cereus* BE23 reached approximately 1×10^{12}/mL,then cell-free filtrates were prepared by centrifuging 450 mL of the culture and filtering the supernatant through a Millipore™ (Burlington, MA, USA) Membrane Filter, 0.22 µm pore size.

2.2.2 Preparation of the Exposure Treatment

Triplicate intact macroalga thalli (approximately 1.25 g/L) were cultured in bacterial-free conditions with different ratios of *Bacillus cereus* BE23 filtrate to total media (filtrate + seawater, in volumes of 0:1, 1:100, 1:80, 1:60, 1:40, 1:20, and 1:10, hereafter identified as Control, T1:100, T1:80, T1:60, T1:40, T1:20,and T1:10, respectively) to a total of 400 mL each in 500 mL flasks. Then, stock f/2 medium was added to each flask. All final media were at f/2 levels, assuming that no or low nutrients were carried over by the filtrate. The concentration of bacteria cells in each treatment was 2.5×10^9, 1.25×10^{10}, 1.65×10^{10},2.5×10^{10}, 5×10^{10}, and 1×10^{11}, respectively. The control treatment of *U. prolifera* was cultured in f/2 medium only, without a bacterial filtrate. All experiments were conducted in the same culture environment under a light intensity of 60 µmol·m^{2}·s^{-1}, and with a light/dark cycle of 12/12 h, salinity of30, and temperature of 20 °C. The experiments were conducted in 500 mL flasks containing 400 mL of culture medium. Nutrients (equivalent to the nitrogen and phosphate level in f/2 media) were added every 48 h to exclude any effects of nutrient limitation, and pH values were monitored simultaneously.

The culture flasks were randomly changed in terms of incubator position every day to balance the effect of illumination. Sterile conditions were used throughout.

Specimens of macroalga were harvested after 192 h (8 days) of exposure for biomass, photosynthesis, and antioxidant analysis.

2.2.3 Growth

The wet weight biomass of the macroalga was determined (±0.0001 g) at 0 and 192 h, respectively. Samples were treated by blotting with 3 layers of filter paper and conditioning for 10 min at room temperature. The relative growth rates (G) were calculated as

$$Gx = (Wx - Wc)/Wc$$

where Wc is the initial wet weight (g) of thalli and Wx is the fresh thalli wet weight (g) after treatment X. The inhibition rate (IR) by the bacterium filtrates was calculated as

$$IR = (Gc - Gx)/Gc$$

where Gx is the relative growth rate (%) of *U. prolifera* after treatment X, and Gc is the relative growth rate (%) after 192 h in control.

2.2.4 The Antioxidant Defense System

Macroalgal samples (0.2~0.3 g wet weight) were homogenized in a bath of liquid nitrogen and extracted with commercial potassium phosphate buffer (pH = 7.2~7.4, Solarbio, China). Then, the extract was centrifuged at 10,000 rpm/min for 10 min yielding material for further analysis of total soluble protein (TSP), H2O2, and the enzymes superoxide dismutase (SOD) and catalase (CAT). Genes associated antioxidant activity, manganese superoxide dismutase (*up*MnSOD) and catalase (*up*CAT), were also quantified.

The TSP content was measured using the Coomassie blue dye binding assay [36]. Fifty microliters of extracts was homogenized with the Coomassie blue dye for 10 min and absorbance was measured at 595 nm. The results of TSP were expressed as g protein per liter (prot·g/L). One hundred microliters was mixed with the reaction reagents and detected at 405 nm. The concentration of ROS was measured as hydrogen peroxide (H2O2) and measured with a commercial assay kit (Jiancheng, Nanjing, China) following the manufacturer's protocols. Concentrations of H2O2 were determined based on the decomposition of H2O2 by peroxidase and the results were expressed as mmol H2O2 per g of TSP (mmol/gprot). The activity of SOD was measured according to the method of Sun et al. [37]. Samples (20 μL) and reaction reagents were mixed in the microliter 96-well flat-bottom plates and put into the plate reader (Tecan, Switzerland) for incubation at 37 °C. After 20 min incubation, the mixtures were detected at 450 nm. One unit of SOD was defined as the amount of enzyme required to generate 50% inhibition of reduction of WST-1 [2-(4-Iodophenyl)-3-(4-nitrophenyl)-5-(2,4-disulfophenyl)-2H-tetrazolium, monosodium salt]. The activity of CAT was assayed with the method described by Dhindsa et al. [38]. Briefly, a reaction mixture was composed of 50 μL extracts, 15 mM hydrogen peroxide, and 50 mM phosphate buffer. After addition of the enzyme extract, absorbance at 240 nm was recorded for 1 min. One unit of CAT activity is the amount of enzyme necessary to degrade 1 μmol H_2O_2 per mg of protein per sec.

The antioxidant enzyme coding genes (*up*MnSOD and *up*CAT) were amplified with gene-specific Primerpairs (Table 2). RNA extraction and real-time PCR were performed the same as the photosynthetic genes.

Table 2. Sequences of primer pairs in *Ulva prolifera* for real-time PCR.

Primer	Sequence (5-3)	Product Length
Tubulin	F: 5-CAAGGATGTCAATGCTGCTGT-3 R: 5-GACCGTAGGTGGCTGGTAGTT-3	112
PsbS	F: 5-AACAGGTTCATCCATCACGG-3 R: 5-TTGCCTCAAACTCATCCTCTG-3	121
LhcSR	F: 5-CTATGCGAAGACTCTCAACG-3 R: 5-CCTCGCGGTAGCGCTTAACT-3	83
PsbA	F: 5- CTTTATGGGCTCGCTTTTGT-3 R: 5- TGGAACTACAGCACCAGAAA-3	103
PsbD	F: 5- CAGGAAGTGTTCAACCAGTA-3 R: 5- AGCAGCGATGTGATGAGACG-3	167
upMnSOD	F: 5-ATCACCAGGCGTATGTCACC-3 R: 5-TTCAAGTGCCCTCCACCGTT-3	94
upCAT	F: 5-CTCTCAAGCCCAATCCTCGT-3 R: 5 -AGTTCAGTGGGATGCCAACA-3	95

2.2.5 Photosynthesis System

Concentrations of chlorophyll *a* (Chl *a*) and *b* (Chl *b*) were determined according to Zhao et al. [39]. Macro algae (0.2 g) were grounded in liquid nitrogen and extracted in 90% *v*/*v*) acetone buffer (5 mL) for 12 h. Then, the mixture was centrifuged at 4 °C, 10,000 rpm/min for 10 min. The supernatant was collected for chlorophyll analyses, and optical densities were measured with an ultraviolet–visible spectrophotometer (HITACHI, U2900, Japan) at 663 and 645 nm wavelength. Concentrations of Chl *a*= and *b* were then calculated as follows, and reported as units of mg/g fresh weight (mg/g FW):

Chl a = 12.7 OD663 – 2.69 OD645 Chl b = 22.9 OD645 – 4.68 OD663

Parameters associated the photosynthesis system II (PSII) were measured using an Imaging-PAM (Walz, Germany). These parameters included the effective quantum yield (Y(II)), non-photochemical quenching (NPQ), relative electron transport rate (rETR), and photochemical quenching (qP). The actinic light was set to be similar to the cultivation light (56 μmol·m^{-2}·s^{-1}). Subsamples of *U. prolifera* were dark-acclimated for 20 min prior to all measurements. All parameters were calculated according to the relationships in Table 3.

Table 3. Fluorescence parameters calculated from PAM in *Ulva prolifera* after exposure.

Parameter	Definition	Equation
Fv/Fm	maximum quantum yield of PSII	(Fm – F0)/Fm
Y(II)	effective quantum yield of PSII	(F'm – Ft)/F'm
NPQ	non-photochemical quenching	(Fm – F'm)/F'm
rETR	relative electron transport rate	0.5 × Y(II) × PAR × IA
qP	photochemical quenching	(F'm – Ft)/(F'm – F'0)

Four genes were selected for characterization: *Psb*S, *Lhc*SR, *Psb*A, and *Psb*D. *Psb*S and *Lhc*SR are associated with photo protection and non-photochemical quenching (NPQ). *Psb*A and *Psb*D are indicators of the D1 and D2 protein of the PSII apparatus, respectively. The tubulin gene was deployed as a housekeeping gene to standardize the expression variations of target genes [39].

These genes were amplified with gene-specific primer pairs (Table 2). Samples of *U. prolifera* were quickly frozen in liquid nitrogen and stored at –80 °C until RNA extraction. Total RNA was extracted by a commercial Mini BEST Plant Total RNA Extraction Kit (TaKaRa, Dalian, China) and the reverse transcripts cDNA were analyzed using a Prime Script™ II 1st stand cDNA Synthesis kit (TaKaRa, Dalian, China). Real-time PCR was performed using the "TB GreenTM Fast qPCR Mix" kit (TaKaRa, Dalian, China). The amplification program of real-time PCR was set at 94 °C for 30 s, following 40 cycles of 94 °C for 5 s and 60 °C for 10 s in Light Cycler® 480 System (Roche, Germany). Dissociation curve analysis of the amplification products was carried out to verify the single PCR production at the end of each thermal program.

2.3 Experiment 2: Isolation and Identification the Potential Allelopathic Compounds from Cell-Free Filtrate of Bacillus cereus BE23

2.3.1 Step 1: Solid Phase and Liquid Phase Extraction of Potential Allelopathic Compounds

Cell-free filtrate (10 L; approximately 1×10^{16} bacteria cells) of the *Bacillus cereus* BE23 cs culture was collected after 5 days of growth by centrifuging at 10,000 rpm/min for 10 min and filtering with a 0.22 μm membrane. The filtrate was eluted by solid phase extraction (SPE) with the resin Diaion® HP20 (particle size of 20–60 mesh) and the remaining residuals were rinsed off by methanol. After resuspending the residuals in Milli-Q water, they were used for liquid phase extraction (LPE). Three extracting agents, cyclohexane, ethyl acetate, and 1-butanol, were considered as selection agents for different polarity fragments. Sub-residuals of LPE were extracted from each agent 3 times and concentrated in a rotary evaporator (IKA, RV8V, Germany) in a 30~40 °C water bath (Figure 1). The sub-residuals were identified as cyclohexane (Ech), ethyl acetate (Eea), and 1-butanol seriatim (Ebs). These sub-residuals, Ech, Eea, and Ebs, were weighted with an electron balance (±0.0001 g), dissolved in 20 mL dimethyl sulfoxide (DMSO), and stored at 4 °C for further bioassay experimentation.

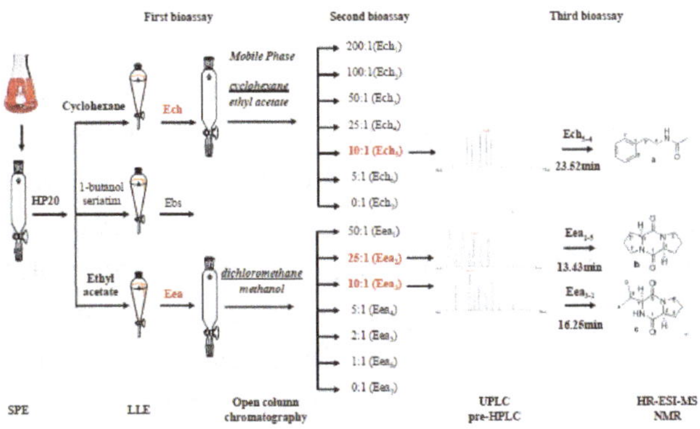

Figure 1. Isolation and bioassay program for potential allelopathic compounds from crude extraction of cell-free filtrate of *Bacillus cereus* BE23.

The first *U. prolifera* bioassay experiment was performed in 6-well plates by filling them with macroalgae (approximately 0.05 g) and crude extraction (5 mg/L) or DMSO (control) in 10 mL f/2medium. Each treatment was conducted in triplicate for 192 h under the same environmental conditions as the primary *U. prolifera* culture. Growth and inhibition rates were used to determine the potential allelopathic activities in each treatment (Supplementary Figure S2). Of the three extracting agents, extractions in cyclohexane (Ech) and in ethyl acetate (Eea) had an inhibition effect (Supplementary Figure S2), therefore, these extractions were used for further investigation.

Step 2: Open Column Chromatography to Select the Potential Allelopathic Compounds:

To further purify the potential allelopathic compounds, Ech and Eea were eluted through an open silica gel column chromatography (170 × 30 mm in dimension and with a silica particle size of 200–300 mesh), respectively, and the effluents from each mobile phase were collected. As for extractions in cyclohexane (Ech), the mobile phase was cyclohexane and ethyl acetate with ratios of 200:1, 100:1, 50:1, 25:1, 10:1, 5:1, and 0:1 (hereafter named as Ech1, Ech2, etc.). For extraction in ethyl acetate (Eea), the mobile phase was dichloromethane and methanol with ratios of 50:1(Eea1), 25:1(Eea2), 10:1(Eea3), 5:1(Eea4), 2:1(Eea5), 1:1(Eea6), and 0:1(Eea7), respectively.

Then, a second bioassay was performed in 6-well plates by adding 0.05 g of *U. prolifera* (wet weight) and the corresponding extracted compounds (5 mg/L) in 10 mL of f/2 medium. Each treatment was conducted in triplicate for 192 h under the same environmental conditions as the primary *U. prolifera* culture. The extractions with significant inhibition, Ech5, Eea2, and Eea3 (Supplementary Figure S3), were collected for further detection.

Step 3: Ultra- and High-Performance Liquid Chromatography to Select the Potential Allelopathic Compounds

The bioactive fractions were collected separately and analyzed by analytical ultra-performance liquid chromatography (UPLC, ultimate 3000, Thermo Fisher Scientific, USA) with a C18 column(250 × 4.6 mm, 5 μm, Agilent, China) at a flow rate of 1 mL/min and the UV detection at 210 nm. The mobile phase was methanol or acetonitrile/water (10/90, v/v) -100% methanol with an elution time of 35 min. The dominant components (highest peaks), including 5 components from Ech5, 7 components from Eea2, and 8 components from Eea3, were chosen and the optimal UPLC conditions were retrieved for a further preparative step.

The fractions were then purified and collected by preparative high-performance liquid chromatography (HPLC, Shimadzu, AP20, Japan) with a C18 column (250 × 21.2 mm, 5 μm, NanoMicro, China) at a flow rate of 10 mL/min for different times up to 35 min for Ech5, Eea2, and Eea3, separately, using the recorded optimized mobile phase (Figure 1).

The third bioassay was conducted with the 20 components. Three compounds, Ech5-4, Eea2-5, and Eea3-2, were collected at 23.52, 13.43, and 16.25 min in each extraction run (Supplementary Figure S4).

2.3 Structure Identification

The three potential allelochemicals, Ech5-4, Eea2-5, and Eea3-2, were preliminarily analyzed by an Agilent 6230 time-of-flight liquid chromatography–mass spectrometer (TOF LC-MS) (Agilent, CA, USA) to determine the molecular weight. Then, structures were identified by a pulse Fourier transform nuclear magnetic resonance spectroscope (NMR, 600 MHz, JNM-ECZR, JEOL, Japan). Deutero methanol or deutero dimethyl sulfoxide solutions containing trimethylsilyl were used as reference substances and acted as solvents to record ^{1}H and ^{13}C NMR spectra. All chemical shifts were exhibited as relative values.

2.4. Statistical Analysis

All data were presented as mean ± standard error and were analyzed by one-way ANOVA with a significant level of 0.05 (Sigma plot 12.5, Systat Software Inc., London, UK). A phylogenetic tree was constructed using the neighbor-joining algorithm with the MEGA 7.0 program. Relative gene expression levels were analyzed following the $2^{-\Delta\Delta Ct}$ method.

3. RESULTS

3.1 Identification of Macroalga and Bacteria

The 5S sequence of the macroalga, 418 bp, was 100% identical to *Ulva prolifera* (GenBankID:HM584772.1) and the ITS sequence, 614 bp, was 99% identical to *U. prolifera* (GenBankID:KF130870.1). Thus, the macroalga deployed in the present study was identified as *U. prolifera*.

The 16S rDNA sequence of the bacterial strain BE23 (GenBank accession number: MN814015) was 100% identical, with few genetic distance differences, to that of *Bacillus cereus* strain ATCC14597(Supplementary Figure S1). Thus, bacterial strain BE23 was identified as *Bacillus cereus*.

3.2 Inhibition on the Growth of U. prolifera

To simplify the treatment and response analysis of *U. prolifera*, two major treatment groups of *B. cereus* filtrates were classified. They are herein separated as high-concentration (HC), i.e., the T1:10 and T1:20 treatments, and low-concentration (LC), i.e., the T1:40, T1:60, T1:80, and T1:100 treatments.

Cell-free filtrates of *Bacillus cereus* BE23 were used as the source of the allelopathic compounds tested on *U. prolifera*. These cell-free filtrates induced growth of *U. prolifera* at LC, i.e., T1:100~T1:40 (ANOVA, $p < 0.05$), with growth rates of 105% ± 11% on average ($n = 12$), but inhibited growth at HC treatments (T1:20 and T1:10), with inhibition rates of 67% and 75%, respectively (Figure 2). Values of pH were monitored during the exposure in all treatments (Supplementary Table S1) and variation of the pH value was within the optimal range for *U. prolifera* growth [40].

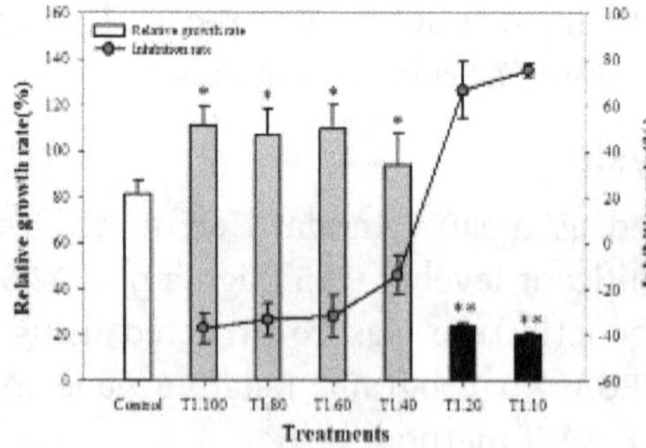

Figure 2. Relative growth rates and inhibition rates of *Ulva prolifera* under the exposure of different amounts of cell-free filtrate of *Bacillus cereus* BE23. T1:100, and T1:80~T1:10 indicate the treatments of volume ratio of cell-free filtrate of *Bacillus cereus* BE23 to f/2 medium. Values are means ± SD (n = 3). * indicates a significant difference ($p < 0.05$) and ** indicates a significant difference ($p < 0.001$) compared to control.

3.3 Response of Antioxidant System of U. prolifera

A significant amount of H2O2 (ANOVA, $p < 0.001$) was produced in the HC treatments, ranging from 38.21 to 50.33 mmol/gprot (Figure 3) after 192 h of exposure. The production of ROS was associated with changes in activities of SOD (ANOVA, $p < 0.05$) and CAT (ANOVA, $p < 0.001$),with concentrations of T1:40 eliciting a response in SOD activity (Figure 4a) but only the highest dosage, T1:10, elicited a response in CAT (Figure 4b). The antioxidant enzyme genes, *up*CAT and *up*MnSOD, were upregulated gradually in response to the increased dosage of cell-free extracts (Figure 4a,b), indicating the initiation of the antioxidant defense system under the stress of the filtrate of *Bacillus cereus* BE23.

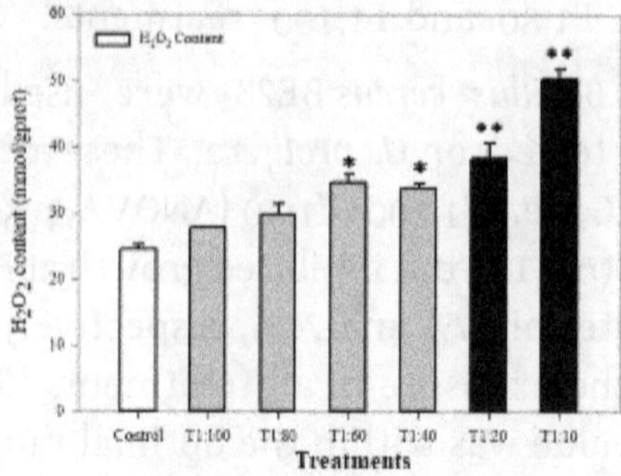

Figure 3. H2O2 content of *Ulva prolifera* under the exposure of different amounts of cell-free filtrate of *Bacillus cereus* BE23. T1:100, and T1:80~T1:10 indicate the treatments of volume ratio of cell-free filtrate of *Bacillus cereus* BE23 relative to f/2 medium. Values are means ± SD (n = 3). * indicates a significant difference ($p < 0.05$) and ** indicates a significant difference ($p < 0.001$) compared to control.

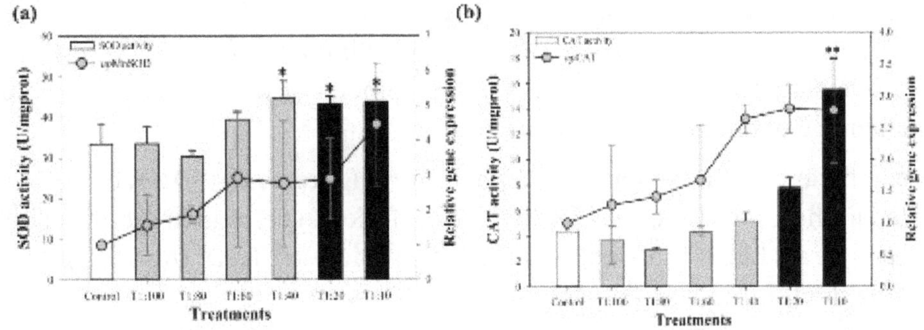

Figure 4. (a) Superoxide dismutase (SOD) activity and relative gene expression of manganese superoxide dismutase (*up*MnSOD), and (b) catalase (CAT) activity and catalase gene expression (*up*CAT) of *Ulva prolifera* under the exposure of different amounts of cell-free filtrate of *Bacillus cereus* BE23. T1:100, and T1:80~T1:10 indicate the treatments of volume ratio of cell-free filtrate of *Bacillus cereus* BE23 relative to f/2 medium. Values are means ± SD (n = 3). * indicates a significant difference ($p < 0.05$) and ** indicates a significant difference ($p < 0.001$) compared to control.

Response of PSII System of U. prolifera

To investigate the effects of the *Bacillus cereus* BE23 filtrate on the photosynthetic pigments of the macroalga, Chl *a* and *b* contents were quantified (Figure 5a). No significant changes of either Chl *a* or *b*were observed in the LC treatments, but significant decreases were observed (ANOVA, $p < 0.001$) in the HC exposures, from 0.41 to ~0.13 mg/g FW for Chl *a*, and from 0.57 to ~0.24 mg/g FW for Chl *b* (Figure 5a).

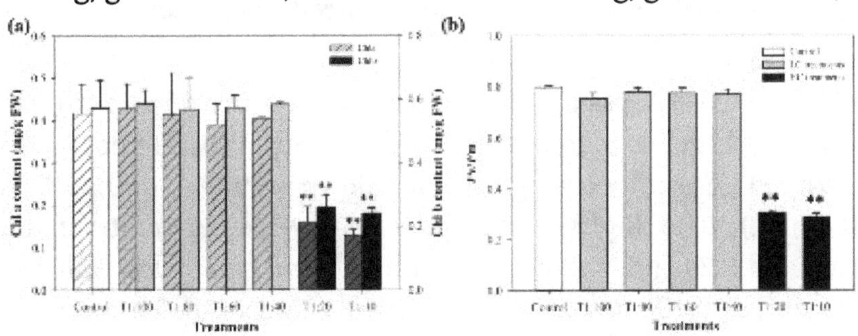

Figure 5. (a) The chlorophyll *a* and *b* content, and (b) the maximum quantum yields of PSII (Fv/Fm) of *Ulva prolifera* under the exposure of different amounts of cell-free filtrate of *Bacillus cereus* BE23. Values are means ± SD (n = 3). ** indicates a significant difference ($p < 0.001$) compared to control.

The photosynthetic response of *U. prolifera* under the stress of cell-free filtrate of *B. cereus* BE23 was significant (Figure 5b, Figure 6, Figure 7). The maximum photochemical quantum yields of PSII(Fv/Fm) were reduced in the HC treatments, from 0.80 to ~0.29 (n = 6, Figure 5b). Accordingly, values ofY(II), the effective quantum yield of PSII, were significantly downregulated (ANOVA, $p < 0.001$),from 0.22 to 0.15 in the HC treatments (Figure 6a). Similar responses were found in the relative electron transport rates (rETR), coincident with a sharp reduction in photochemical quenching (qP) (Figure 6b).

A significant enhancement of NPQ activity (Figure 6b) (ANOVA, $p < 0.001$) was recorded in the LC treatments, from 0.18 to 0.44. However, high doses of the filtrate of *Bacillus cereus* BE23 induced a down regulation of NPQ (ANOVA, $p < 0.001$), indicating photoinhibition damage.

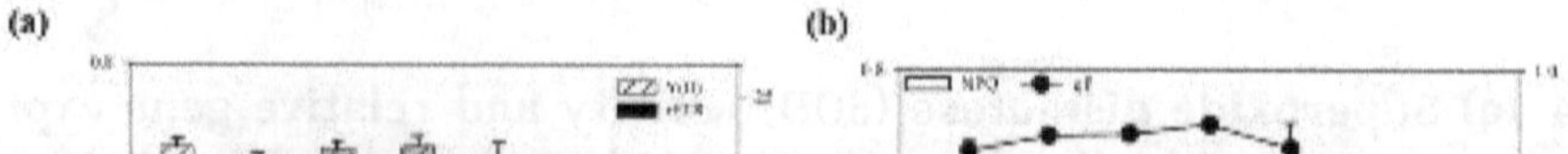

Figure 6. Photosynthetic system II parameters of *Ulva prolifera* under the exposure of different amounts of cell-free filtrate of *Bacillus cereus* BE23: (a) quantum yield (Y(II)) and relative electron transport rate (rETR), and (b) non-photochemical quenching (NPQ) and photochemical (qP). T1:100, and T1:80~T1:10 indicate the volume ratio of cell-free filtrate of *Bacillus cereus* BE23 relative to f/2 medium in the different treatments. Values are means ± SD (n = 3). ** indicates a significant difference ($p < 0.001$) compared to control.

The expression of the two assayed photo-protection-related genes, *Psb*S and *Lhc*SR, varied in response to cell-free filtrate exposure (Figure 7a). The relative expressions of both genes increased with the bacterial filtrate dosage from 1:100 (T1:100) to 1:40 (T1:40) but were significantly down regulated in the HC treatments (T1:20 and T1:10). The highest *Psb*S and *Lhc*SR were in treatments of T1:40, reaching2.66 and 5.29 times that of the control, and the lowest value was in the T1:10 treatment, at 0.75 and 0.72 of the control (Figure 7a). The response of *Psb*A and *Psb*D was not as clear, but a substantial degradation of *Psb*A was observed in the HC treatment, with a value of 0.59 of the control in T1:10 (Figure 7b).

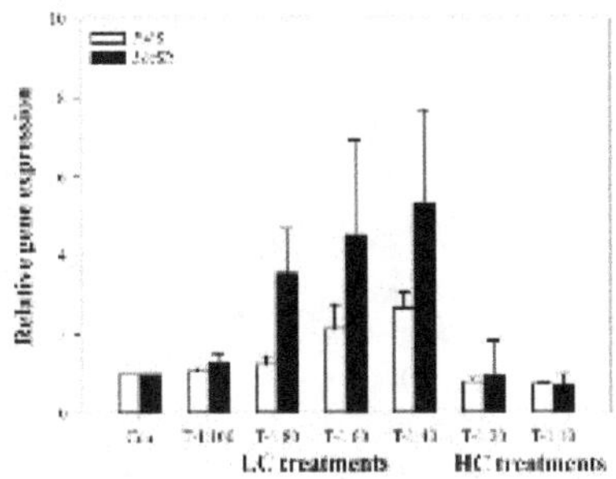

(a)

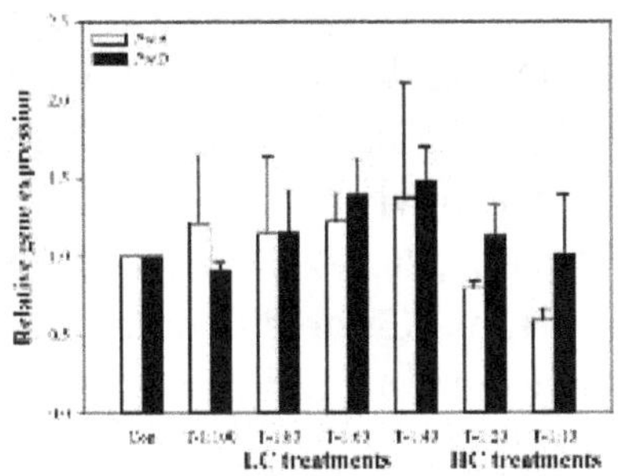

(b)

Figure 7. Relative expression of the genes (a) *Psb*S and *Lhc*SR, and (b) *Psb*A and *Psb*D of *Ulva prolifera* under the exposure of different amounts of cell-free filtrate of *Bacillus cereus* BE23. T1:100, and T1:80~T1:10 indicate the treatments of volume ratio of cell-free filtrate of *Bacillus cereus* BE23 relative to f/2 medium. Values are means ± SD (n = 3).

3.5 Identification of Allelochemicals from *Bacillus cereus* BE23 Filtrate

To isolate the bioactive compounds, five steps of extraction and insolation (solid phase–liquid phase–open column–UPLC–preHPLC) were conducted. After each isolation, the separated groups were tested for bioactivity (Figures S2–S4). Three bioactive compounds in the cell-free filtrates of *Bacillus cereus* BE23 were identified by high-resolution mass spectrometric data and NMR spectroscopic analysis. The molecular formula $C_{10}H_{13}NO$ of compound Ech5-4 was deduced from its ion at m/z 164.1072 $[M+H]^+$ (Supplementary Figure S5a, calculated for $C_{10}H_{14}NO$, 164.1075) and its ^{13}C data. The ^{13}C-NMR spectrum (600 MHz, DMSO-d6) of Ech5-4 displayed signals at δC 169.5 (C=O), 140.0 (C,C-1), 129.1 (CH, C-3, C-5), 128.8 (CH, C-2, C-6), 126.5 (CH, C-4), 40.7 (CH_2, C-7), 35.7 (CH_2, C-8),and 23.09 (CH_3) (Supplementary Figure S5b,c). The ^{1}H-NMR signals were observed at δH 7.92 (1H, brs, NH), 7.27–7.30 (2H, t, J = 8.0 Hz, Ar-H), 7.18–7.20 (3H, m, Ar-H), 3.22–3.26 (2H, m, H-7), 2.69 (2H, t,J = 7.5 Hz, H-8), and 1.78 (3H, s, -CH_3). Based on these data and the comparison with the reported data [41], compound Ech5-4 was identified as N-phenethylacetamide (Figure 8a).

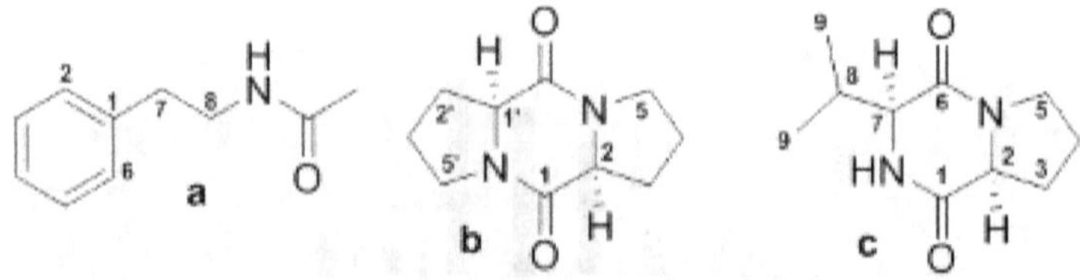

Figure 8. Structures of the compounds Ech5-4 (a), Eea2-5 (b), and Eea3-2 (c) isolated from the crude extract of *Bacillus cereus* BE23 filtrate.

The molecular formula of C10H14N2O2 for compound Eea2-5 was determined based on its m/z 217.0953 $[M+Na]^+$ (Supplementary Figure S6a, calculated for C10H14N2NaO2, 217.0953). The ^{13}C and^1H NMR spectra of Eea2-5 showed signals for the functional groups of carbonyl (δC 168.1), methine (δC 61.2; δH 4.34, 1H, t, J = 9.0 Hz), and methelene (δC 45.7, 28.2, 23.7; δH 3.45–3.53, 2H, m, 2.25–2.30,1H, m, 1.99–2.09, 2H, m, 1.91–1.97, 1H, m,) (Supplementary Figure S6b,c). These data and comparison with the reference data [42] indicated that compound Eea2-5 was cyclo (L-Pro-L-Pro) (Figure 8b).

The compound Eea3-2 has the molecular formula of C10H10N2O2 deduced from its m/z 219.1103 [M+Na] (Supplementary Figure S7a, calculated for C10H10N2NaO2, 219.1109). The ^{13}C-NMR spectrum(600 MHz, Methanol-d4) of Eea3-2 exhibited 10 carbon signals, resonating at δC172.8 (C, C-1), 167.8 (C, C-6), 61.8 (CH, C-7), 60.3 (CH, C-2), 46.4 (CH2, C-5), 30.1 (CH, C-8), 29.8 (CH2, C-3), 23.5 (CH2, C-4),19.1a (CH3, C-10), and 16.9 (CH3, C-9). The 1H NMR spectrum displayed signals at δH 4.20 (1H, t, J = 8.6 Hz, H-2), 4.05 (1H, br t, H-7), 3.56 (1H, m, H-5a), 3.48 (1H, m, H-5b), 2.48 (1H, m, H-3a), 2.31 (1H,m, H-8), 2.02 (1H, m, H-3b), 1.91–1.96 (2H, m, H-4), 1.08b (3H, d, J = 7.3 Hz, H-9), and 0.95b (3H, d,J = 7.3 Hz, H-10). Thus, the compound Eea3-2 was identified as cyclo (L-Pro-L-Val) (Figure 8c) [43].

4. DISCUSSION

Bacteria-derived interactions play important roles in species distribution and abundance [44], succession of algal blooms [45], and biomass control of microorganisms [46] and macroalgae [47]. Such allelopathic interactions consist of two pathways, direct (bacterial and algal cell contact) and indirect (release of natural products) [12,32]. The present study demonstrated the potential mechanisms of allelopathic stress on *U. prolifera* by products of *B. cereus* BE23 in indirect ways.

The low dosage (i.e., T1:100~T1:40) of *B. cereus* BE23 filtrate promoted the growth of *U. prolifera*, whereas the high dosage (T1:20 and T1:10) inhibited biomass production (Figure 2). The response of the macroalgae in the LC treatments may have resulted from a hormesis effect [48] and adaption to the low concentrations of allelochemicals [49]. The upregulation of physiological activity of *U. prolifera* (Figures 4–6) in the LC treatments contributed to the growth-promotive effect. Meanwhile, the nutrients, including the inorganic nutrient from f/2 + artificial seawater and the nutrient carrying over by the *B. cereus* BE23 filtrate (4~40 mL), contributed to the growth of macroalga. Inorganic nitrogen, i.e., nitrate or ammonium, has been reported to be rapidly taken up by *Ulva* [28], and within 192 h, the addition of inorganic nutrient of f/2 medium was calculated to be suffcient to the thalli of *U. prolifera* [50,51]. The carried-over inorganic nutrient was low (less than 10%), therefore, the effects of nutrients in *B. cereus* BE23 filtrate were minimal to the growth of *Ulva* in the present study.

A general stress response in algae is the production of ROS [52,53] and it can be produced in response to abiotic and allelopathic stresses [54–56]. Here, ROS was produced in response to BE23 cell-free filtrates (Figure 3). The source of ROS may include two main pathways: the intrinsic oxidization by allelochemicals, and inactivation of the electron transport in the PSII systems. The production of ROS is also a signal of the pressure from the excitation energy collected by the PSII light-harvesting complex [57,58]. To regulate the extra ROS, algae have a series of antioxidant defense mechanisms, including the ability to vary antioxidant enzymes or genes. Variations in activities of the enzymes SOD and CAT are important in alleviating oxidative damage [59,60]. In general, SOD scavenges the cellular ROS first, catalyzing $O2^{\bullet-}$ to H2O2. Then, the CAT enzyme decomposes H2O2 to O2 and H2O [61]. MnSOD, one of the total SODs, was selected as the representative enzyme; it is mostly detected in the cytosol and thylakoid membrane [62].

Here, a small amount of ROS (H2O2) was produced in the LC treatments, i.e., T1:60 and T1:40, but no significant variation was observed in the quantum effciency of photosynthesis (Fv/Fm), indicating *U. prolifera* may activate photoprotection to defend against such allelopathic stress. However, a significant increase in ROS concentration (ANOVA, $p < 0.001$) was recorded in the HC treatments, accompanied by the decline in rETR, indicating normal electron transport in PSII was disturbed and excess energy likely contributed to the ROS generation in HC treatments. High production of ROS induced oxidative stress in the algae and final inhibited the photosynthesis systems. To moderate the oxidative damage, *U. prolifera* upregulated the activity of SOD and CAT, supported herein by the gene expression level of *up*MnSOD and *up*CAT in the LC treatments (Figure 5). Similar responses have been noted in *Cylindrospermopsis raciborskii* under hyper-salinity or light-stress conditions [63,64], and linoleic acid stress [65]. The upregulation of the transcript levels of FeSOD and CAT genes in *U. Prolifera* have also been reported in response to salicylic acid and hyper-temperature [66]. In the present study, however, the enhanced CAT activities were not suffcient to scavenge the sudden increased H2O2 and this likely caused extensive oxidative stress in this macroalga.

External stresses, including allelopathic stressors, can alter the algal energy flux of PSII by reducing the photosynthetic effciency [67–69], and by enhancing non-photochemical quenching (NPQ) [65]. The maximum quantum yield (Fv/Fm) is an effective indicator of the effciency of photochemical stress. In *Ulva* sp., changes in Fv/Fm have been observed when the algae are exposed to internal or external stresses [70] such as light [71], desiccation [72], salinity [73], and allelopathy [50].

Significant declines in Fv/Fm (Figure 5b), growth rate (Figure 2), and Chl *a* and *b* (Figure 5a) were shown after 192 h exposure to high concentrations of *B. cereus* BE23 filtrate, suggesting disruption of the PSII reaction centers' (RCs) complexes [67] including the electron transport chain [74]. Reduced rETR and Y(II) indicate a reduction in the electron transport rate and CO2 assimilative capacity [75]. Therefore, one mechanism by which *U. prolifera* responds to allelopathic stress is a lowering of the photosynthetic performance, which directly impacts carbon fixation and therefore the growth rate [76]. The significant decreases in the Chl *a* and *b* concentrations in the HC treatments may also be considered as an adaptive strategy which decreases the absorption of photons, thereby leading to less ROS production [67].

The NPQ pathways are photoprotective mechanisms for phototrophs [77]. In the present study, no significant variation in Fv/Fm (Figure 5b) or rETR (Figure 5a) was observed in the LC treatments; however, a significant increase in NPQ was recorded as the concentrations of the LC treatments increased, namely T1:40 and T1:60. Under the HC treatments, a substantial decrease in NPQ was observed, indicating that allelopathic stress may hinder the operation of photoprotective mechanisms, and thus the macroalgae dissipated excess energy through non-regulated pathways [78]. At high levels of bacterial filtrate, *U. prolifera* was unable to self-protect against photodamage [39]. The significant decrease in qP in the treatments with high concentrations of filtrate indicated a high level of energy dissipation and potential damage to the PSII reaction centers. Thus, the decrease in the effciency of PSII was associated with a simultaneous decrease in the photochemical and non-photochemical pathways in the HC treatments, reflecting a complete disruption of normal energy pathways.

Previous studies have suggested that *Ulva* sp. can modulate NPQ levels by adjusting the copy number of *Lhc*SR or *Psb*S and regulation of the xanthophyll cycle [79,80]. It thus appears that low levels of exposure to *B. cereus* BE23 filtrate induced an upregulation of *Lhc*SR and *Psb*S in *U. prolifera* and activated the photoprotection mechanism that enables the self-regulation of external allelopathic stress without loss of electron transfer effciency of photosynthesis and growth. An upregulated transcriptlevel of both selected genes and a triggering of *Lhc*SR-dependent NPQ was also previously reported in *Ulva* sp. [80]. High amounts of filtrate, in contrast, inhibited the photosynthetic effciency and the capability of self-regulation of *U. prolifera*, as evidenced by the down regulation of Fv/Fm, qP, and NPQ activity, and finally the inhibition of growth. Therefore, the low value of NPQ was a result of the loss of the photoprotection of *U. prolifera* and a failure of self-regulation under allelopathic stress [81].

Allelopathic damage to the PSII systems is also suggested by the responses of the genes located in the D1-D2 protein [54,82]. *Psb*A and *Psb*D, encoding the D1 and D2 subunits of the PSII complex, constitute the heterodimeric photochemical reaction center [80]. Here, no clear variation in *Psb*A and *Psb*D gene expression was observed after 192 h exposure in the LC treatments (Figure 7b), suggesting the excess absorbed electrons (Figure 4a) were dissipated by the upregulated NPQ, together with the upregulation of *Lhc*SR and *Psb*S transcript levels (Figure 7a). In contrast, clear down regulation of *Psb*A expression levels was recorded in the HC treatments, suggesting that the *B. cereus* BE23 filtrate suppressed *Psb*A expression and may have blocked the elector transport on the PSII receptor side from QA to QB [81].

In summary, the inhibition effect on the PSII of *Ulva* due to bacteria-derived stress may go through two main steps: (1) the inhibition of the electron transport chain, and (2) the deleterious effects on PSIIRCs' complexes [83,84]. In the present study, the upregulated expression of *Psb*S and *Lhc*SR under LC levels of cell-free filtrate might indicate the successful regulation of stress via regulated NPQ [85,86], but failure in the HC treatments. The depletion of the transcript pools of *Lhc*SR and *Psb*S contributed directly to the decrease in NPQ activity and likely inactivated the PSII RCs' complexes. Down regulation of Chl *a* and *b* corresponded to the down regulation of *Psb*A expression levels, suggesting the BE23 filtrate degraded the absorption of light energy and blocked the electron transport on the PSII receptor side [65,80]. Surplus electrons exceeded the electron transport chain capacity of *U. prolifera* and induced additional ROS production (Figure 3) that, in turn, damaged the PSII systems [16]. Together, these data clearly document the photooxidative stress in *U. Prolifera* upon allelopahtic stress in HC treatments.

Using ESI and NMR, three potential allelopathic chemicals were isolated and identified from the cell-free filtrate of *B. cereus* BE23. The chemical cyclo (L-Pro-L-Pro) (Figure 8b), extracted from Eea2, displayed the largest inhibitory effect on *U. prolifera* , and has previously been shown to yield a strong algicidal effect on *Microcystis aeruginosa* [55] and *Phaeocystis globosa* [54] by inhibiting the operation of the photosynthesis and antioxidant systems of target algae. In the present study, the diketopiperazine derivatives decreased the gene expression of *Psb*A [54,87], directly impacting the PSII electron acceptor sides, resulting in the failure of the photosynthetic process. Given that cyclo (L-Pro-L-Pro) is easily biodegradable [88], it may be a good candidate as an environmentally friendly algicide for green algae bloom control.

5. CONCLUSIONS

The high concentration of the cell-free filtrate of *B. cereus* BE23 (approximately 1×10^{11}/mL) yielded significant inhibition of growth of *U. prolifera* via degradation of the photosynthetic system as shown by changes in biomass accumulation, photosynthetic responses, gene regulation, and enzyme activities. The potential allelopathic compounds inhibited growth by means of reduction of Fv/Fm,rETR, and NPQ, resulting in *U. prolifera*'s failure to dissipate the excess energy through regulated NPQ pathways. This alteration of energy dissipation caused excess cellular ROS accumulation and the antioxidative defense system was generated. This ROS production also inhibited the PSII reaction center apparatus. The potential allelochemicals were further isolated and identified as N-phenethylacetamide, cyclo (L-Pro-L-Val), and cyclo (L-Pro-L-Pro). The diketopiperazines derivative, cyclo (L-Pro-L-Pro), exhibited the highest inhibition effect on *U. prolifera* and further study on its potential as an algicidal product for green algae bloom control is warranted.

Supplementary Materials: The following are available online at http://www.mdpi.com/2077-1312/8/9/718/s1, Figure S1. Phylogenetic tree of *Bacillus cereus* BE23. Figure S2. Relative growth rates and inhibition rates of*Ulva prolifera* of the first bioassay test. Figure S3. Relative growth rates and inhibition rates of *Ulva prolifera* in the second bioassay test. Figure S4. Relative growth rates and inhibition rates of *Ulva prolifera* in the third bioassay test. Figure S5. High-resolution electrospray ionization mass spectrometry (HRESIMS) spectrum (**a**), ^{13}C NMR spectrum (**b**), and ^{1}H NMR spectrum (**c**) of compound Ech5-4. Figure S6. High-resolution electrospray ionization mass spectrometry (HRESIMS) spectrum (**a**), ^{13}C NMR spectrum (**b**), and ^{1}H NMR spectrum (**c**) of compound Eea2-5. Figure S7. High-resolution electrospray ionization mass spectrometry (HRESIMS) spectrum (**a**), ^{13}C NMR spectrum (**b**), and ^{1}H NMR spectrum (**c**) of compound Eea3-2. Table S1. Changes of pH values with culture time in exposed experiments.

REFERENCES

1. Wang, R.; Wang, J.T.; Xue, Q.N.; Tan, L.J.; Cai, J.; Wang, H.Y. Preliminary analysis of allelochemicals produced by the diatom *Phaeodactylum tricornutum*. *Chemosphere* **2016**, *165*, 298–303. [CrossRef] [PubMed]

2. Gross, E.M.; Hilt, S.; Lombardo, P.; Mulderij, G. Searching for allelopathic effects of submerged macrophyteson phytoplankton—State of the art and open questions. *Hydrobiologia* **2007**, *584*, 77–88. [CrossRef]

3. Zhang, Y.W.; Wang, J.T.; Tan, L.J. Characterization of allelochemicals of the diatom *Chaetoceros curvisetus* and the effects on the growth of *Skeletonema costatum*. *Sci. Total Environ.* **2019**, *660*, 269–276. [CrossRef] [PubMed]

4. Zhang, H.; Peng, Y.; Zhang, S.; Cai, G.; Li, Y.; Yang, X.; Yang, K.; Chen, Z.; Zhang, J.; Wang, H.; et al. Algicidal effects of prodigiosin on the harmful algae *Phaeocystis globosa*. *Front. Microbiol.* **2016**, *7*, 602. [CrossRef] [PubMed]

5. Zhou, S.; Yin, H.; Tang, S.Y.; Peng, H.; Yin, D.G.; Yang, Y.X.; Liu, Z.H.; Ding, Z. Physiological responses of *Microcystis aeruginosa* against the algicidal bacterium *Pseudomonas aeruginosa*. *Ecotoxicol. Environ. Saf.* **2016**, *127*, 214–221. [CrossRef]

6. Zhang, F.X.; Ye, Q.; Chen, Q.L.; Yang, K.; Zhang, D.Y.; Chen, Z.R.; Lu, S.S.; Shao, X.P.; Fan, X.Y.; Yao, L.M.; et al. Algicidal Activity of novel marine bacterium *Paracoccus* sp. Strain Y42 against a harmful algal-bloom-causing dinoflagellate, *Prorocentrum donghaiense*. *Appl. Environ Microbiol.* **2018**, *84*. [CrossRef]

7. Qian, H.F.; Xu, J.H.; Lu, T.; Zhang, Q.; Qu, Q.; Yang, Z.P.; Pan, X.L. Responses of unicellular alga *Chlorella pyrenoidosa* to allelochemical linoleic acid. *Sci. Total Environ.* **2018**, *625*, 1415–1422. [CrossRef]

8. Zhao, W.; Zheng, Z.; Zhang, J.L.; Roger, S.F.; Luo, X.Z. Allelopathically inhibitory effects of eucalyptusextracts on the growth of *Microcystis aeruginosa*. *Chemosphere* **2019**, *225*, 424–433. [CrossRef]

9. Yu, Y.; Zeng, Y.D.; Li, J.; Yang, C.Y.; Zhang, X.H.; Luo, F.; Dai, X.Z. An algicidal *Streptomyces amritsarensis* strain against *Microcystis aeruginosa* strongly inhibits microcystin synthesis simultaneously. *Sci. Total Environ.* **2019**, *650*, 34–43. [CrossRef]

10. Arora, A.; Sairam, R.K.; Srivastava, G.C. Oxidative stress and antioxidative system in plants. *Curr. Sci.* **2002**,*82*, 1227–1239.

11. Apel, K.; Hirt, H. Reactive oxygen species: Metabolism, oxidative stress, and signal transduction. *Annu. Rev. Plant Biol.* **2004**, *55*, 373–399. [CrossRef] [PubMed]

12. Mayali, X.; Azam, F. Algicidal bacteria in the sea and their impact on algal blooms. *J. Eukaryot. Microbiol.***2004**, *51*, 139–144. [CrossRef] [PubMed]

13. Zheng, N.N.; Ding, N.; Gao, P.K.; Han, M.X.; Liu, X.X.; Wang, J.G.; Li, S.; Fu, B.Y.; Wang, R.J.; Zhou, J. Diverse algicidal bacteria associated with harmful bloom-forming *Karenia mikimotoi* in estuarine soil and seawater. *Sci. Total Environ.* **2018**, *631*, 1415–1420. [CrossRef]

14. Sun, R.; Sun, P.; Zhang, J.; Esquivel-Elizondo, S.; Wu, Y. Microorganisms-based methods for harmful algal blooms control: A review. *Bioresour. Technol.* **2018**, *248*, 12–20. [CrossRef] [PubMed]

15. Lu, X.H.; Zhou, B.; Xu, L.; Liu, L.L.; Wang, G.Y.; Liu, X.D.; Tang, X.X. A marine algicidal *Thalassospira* and its active substance against the harmful algal bloom species *Karenia mikimotoi*. *Appl. Microbiol. Biotechnol.* **2016**, *100*, 5131–5139. [CrossRef]

16. Hou, S.L.; Shu, W.J.; Tan, S.; Zhao, L.; Yin, P.H. Exploration of the antioxidant system and photosynthetic system of a marine algicidal *Bacillus* and its effect on four harmful algal bloom species. *Can. J. Microbiol.***2016**, *62*, 49–59. [CrossRef]

17. Hu, X.L.; Yin, P.H.; Zhao, L.; Yu, Q.M. Characterization of cell viability in *Phaeocystis globosa* cultures exposed to marine algicidal bacteria. *Biotechnol. Bioprocess Eng.* **2015**, *20*, 58–66. [CrossRef]

18. Shao, J.H.; He, Y.X.; Chen, A.W.; Peng, L.; Luo, S.; Wu, G.Y.; Zou, H.L.; Li, R.H. Interactive effects of algicidal effciency of *Bacillus* sp. B50 and bacterial community on susceptibility of *Microcystis aeruginosa* with different growth rates. *Int. Biodeterior. Biodegrad.* **2015**, *97*, 1–6. [CrossRef]

19. Jeong, S.Y.; Ishida, K.; Ito, Y.; Okada, S.; Murakami, M. Bacillamide, a novel algicide from the marine bacterium, *Bacillus* sp. SY-1, against the harmful dinoflagellate, *Cochlodinium polykrikoides*. *Tetrahedron Lett.* **2003**, *44*, 8005–8007. [CrossRef]

20. Wu, L.M.; Wu, H.J.; Chen, L.N.; Xie, S.S.; Zang, H.Y.; Borriss, R.; Gao, X.W. Bacilysin from *Bacillus amyloliquefaciens* FZB42 has specific bactericidal activity against harmful algal bloom species. *Appl. Environ. Microbiol.* **2014**, *80*, 7512–7520. [CrossRef]

21. Skerratt, J.H.; Bowman, J.P.; Hallegraeff, G.; James, S.; Nichols, P.D. Algicidal bacteria associated with bloomsof a toxic dinoflagellate in a temperate Australian estuary. *Mar. Ecol. Prog. Ser.* **2002**, *244*, 1–15. [CrossRef]

22. Liu, D.Y.; Keesing, J.K.; Xing, Q.G.; Shi, P. World's largest macroalgal bloom caused by expansion of seaweed aquaculture in China. *Mar. Pollut. Bull.* **2009**, *58*, 888–895. [CrossRef] [PubMed]

23. Wang, Z.L.; Xiao, J.; Fan, S.L.; Li, Y.; Liu, X.Q.; Liu, D.Y. Who made the world's largest green tide in China?—An integrated study on the initiation and early development of the green tide in Yellow Sea. *Limnol. Oceanogr.* **2015**, *60*, 1105–1117. [CrossRef]

24. Ye, N.H.; Zhuang, Z.Z.; Jin, X.; Wang, Q.; Zhang, X.; Li, D.M.; Wang, H.X.; Mao, Y.Z.; Jiang, Z.J.; Li, B.; et al. China is on the track tackling *Enteromorpha* spp forming green tide. *Nat. Preced.* **2008**. [CrossRef]

25. Ye, N.H.; Zhang, X.W.; Mao, Y.Z.; Liang, C.W.; Xu, D.; Zou, J.; Zhuang, Z.Z.; Wang, Q.Y. 'Green tides' are overwhelming the coastline of our blue planet: Taking the world's largest example. *Ecol. Res.* **2011**, *26*, 477–485. [CrossRef]

26. Huo, Y.Z.; Han, H.B.; Shi, H.H.; Wu, H.L.; Zhang, J.H.; Yu, K.F.; Xu, R.; Liu, C.C.; Zhang, Z.L.; Liu, K.F.; et al. Changes to the biomass and species composition of *Ulva* sp. on *Porphyra* aquaculture rafts, along the coastal radial sandbank of the Southern Yellow Sea. *Mar. Pollut. Bull.* **2015**, *93*, 210–216. [CrossRef]

27. Zhang, J.H.; Huo, Y.Z.; Wu, H.; Yu, K.; Kim, J.K.; Yarish, C.; Qin, Y.T.; Liu, C.C.; Xu, R.; He, P.M. The origin of the *Ulva* macroalgal blooms in the Yellow Sea in 2013. *Mar. Pollut. Bull.* **2014**, *89*, 276–283. [CrossRef]

28. Li, H.M.; Zhang, Y.Y.; Chen, J.; Zheng, X.; Liu, F.; Jiao, N.Z. Nitrogen uptake and assimilation preferences of the main green tide alga *Ulva prolifera* in the Yellow Sea, China. *J. Appl. Phycol.* **2018**, *31*, 625–635. [CrossRef]

29. Xiao, J.; Zhang, X.H.; Gao, C.L.; Jiang, M.J.; Li, R.X.; Wang, Z.L.; Li, Y.; Fan, S.L.; Zhang, X.L. Effect of temperature, salinity and irradiance on growth and photosynthesis of *Ulva prolifera*. *Acta Oceanol. Sin.* **2016**,*35*, 114–121. [CrossRef]

30. Liu, Q.; Yan, T.; Yu, R.C.; Zhang, Q.C.; Zhou, M.J. Interactions between selected microalgae and microscopic propagules of *Ulva prolifera*. *J. Mar. Biol. Assoc. UK* **2017**, *98*, 1571–1580. [CrossRef]

31. Fan, X.; Xu, D.; Wang, Y.T.; Zhang, X.W.; Cao, S.N.; Mou, S.L.; Ye, N.H. The effect of nutrient concentrations, nutrient ratios and temperature on photosynthesis and nutrient uptake by *Ulva prolifera*: Implications for the explosion in green tides. *J. Appl. Phycol.* **2014**, *26*, 537–544. [CrossRef]

32. Sun, X.; Wu, M.Q.; Xing, Q.G.; Song, X.D.; Zhao, D.H.; Han, Q.Q.; Zhang, G.Z. Spatio-temporal patterns of *Ulva prolifera* blooms and the corresponding influence on chlorophyll-a concentration in the Southern Yellow Sea, China. *Sci. Total Environ.* **2018**, *640*, 807–820. [CrossRef] [PubMed]

33. Guillard, R.R.L. Culture of Phytoplankton for Feeding Marine Invertebrates. In *Culture of Marine Invertebrate Animals*; Springer: Boston, MA, USA, 1975.

34. Jin, Q.; Dong, S.L.; Wang, C.Y. Allelopathic growth inhibition of *Prorocentrum micans* (Dinophyta) by *Ulva pertusa* and *Ulva linza* (Chlorophyta) in laboratory cultures. *Eur. J. Phycol.* **2005**, *40*, 31–37. [CrossRef]

35. Li, H.; Huang, H.J.; Li, H.Y.; Liu, J.S.; Yang, W.D. Genetic diversity of *Ulva prolifera* population in Qingdao coastal water during the green algal blooms revealed by: Microsatellite. *Mar. Pollut. Bull.* **2016**, *111*, 237–246. [CrossRef] [PubMed]

36. Bradford, M.M. A rapid method for the quantitation of microgram quantities of protein utilizing the principle of protein-dye binding. *Anal. Biochem.* **1976**, *72*, 248–254. [CrossRef]

37. Sun, X.; Lu, Z.; Liu, B.; Zhou, Q.; Zhang, Y.; Wu, Z. Allelopathic effects of pyrogallic acid secreted by submerged macrophytes on *Microcystis aeruginosa*: Role of ROS generation. *Allelopath. J.* **2014**, *33*, 121–130.

38. Dhindsa, R.S.; Plumb-Dhindsa, P.; Thorpe, T.A. Leaf senescence: Correlated with increased levels of membrane permeability and lipid peroxidation, and decreased levels of superoxide dismutase and catalase. *J. Exp. Bot.* **1981**, *32*, 93–101. [CrossRef]

39. Zhao, X.Y.; Tang, X.X.; Zhang, H.; Qu, T.F.; Wang, Y. Photosynthetic adaptation strategy of *Ulva prolifera* floating on the sea surface to environmental changes. *Plant Physiol. Biochem.* **2016**, *107*, 116–125. [CrossRef]

40. Wang, J.W.; Yan, B.L.; Lin, A.P.; Hu, J.P.; Shen, S.D. Ecological factor research on the growth and induction of spores release in *Enteromorpha Prolifera* (Chlorophyta). *Mar. Sci. Bull.* **2007**, *26*, 60–66.

41. Zhao, P.J.; Wang, H.X.; Li, G.H.; Li, H.D.; Liu, J.; Shen, Y.M. Secondary metabolites from endophytic *Streptomyces* sp. Lz531. *Chem. Biodivers.* **2007**, *4*, 899–904. [CrossRef]

42. Li, T.; Wang, G.C.; Huang, X.J.; Ye, W.C. ChemInform Abstract: Whitmanoside A (I), a New α-Pyrone Glycoside from the *Leech Whitmania pigra*. *J. Cheminform.* **2013**, *44*. [CrossRef]

43. Furtado, N.A.J.C.; Pupo, M.T.; Carvalho, I.; Campo, V.L.; Duarte, M.C.T.; Bastos, J.K. Diketopiperazines produced by an *Aspergillus fumigatus* Brazilian strain. *J. Braz. Chem. Soc.* **2005**, *16*, 1448–1453. [CrossRef]

44. Tilney, C.L.; Pokrzywinski, K.L.; Coyne, K.J.; Warner, M.E. Effects of a bacterial algicide, IRI-160AA, ondinoflagellates and the microbial community in microcosm experiments. *Harmful Algae* **2014**, *39*, 210–222. [CrossRef]

45. Meyer, N.; Bigalke, A.; Kaulfuss, A.; Pohnert, G. Strategies and ecological roles of algicidal bacteria. *FEMS Microbiol. Rev.* **2017**, *41*, 880–899. [CrossRef]

46. Hare, C.E.; Demir, E.; Coyne, K.J.; Craig Cary, S.; Kirchman, D.L.; Hutchins, D.A. A bacterium that inhibits the growth of *Pfiesteria piscicida* and other dinoflagellates. *Harmful Algae* **2005**, *4*, 221–234. [CrossRef]

47. Zozaya-Valdes, E.; Egan, S.; Thomas, T. A comprehensive analysis of the microbial communities of healthy and diseased marine macroalgae and the detection of known and potential bacterial pathogens. *Front. Microbiol.* **2015**, *6*, 9–18. [CrossRef]

48. Perveen, S.; Mushtaq, M.N.; Yousaf, M.; Sarwar, N. Allelopathic hormesis and potent allelochemicals from multipurpose tree Moringa oleifera leaf extract. *Plant Biosyst.* **2020**, *18*, 1–6. [CrossRef]

49. Wang, C.X.; Zhu, M.X.; Chen, X.H.; Qu, B. Review on allelopathy of exotic invasive plants. *Procedia Eng.***2011**, *18*, 240–246.

50. Li, N.C.; Tong, M.M.; Glibert, P.M. Effect of allelochemicals on photosynthetic and antioxidant defense system of *Ulva prolifera*. *Aquat. Toxicol.* **2020**, *224*, 105513. [CrossRef]

51. Xu, D.; Gao, Z.Q.; Zhang, X.W.; Fan, X.; Wang, Y.T.; Li, D.M.; Wang, W.; Zhuang, Z.; Ye, N. Allelopathic interactions between the opportunistic species *Ulva prolifera* and the native macroalga *Gracilaria lichvoides*. *PLoS ONE* **2012**, *7*, e33648. [CrossRef]

52. Zhou, Q.X.; Hu, X.G. Systemic stress and recovery patterns of rice roots in response to graphene oxide nanosheets. *Environ. Sci. Technol.* **2017**, *51*, 2022–2030. [CrossRef] [PubMed]

53. Wang, Y.; Zhao, X.Y.; Tang, X.X. Antioxidant system responses in two co-occurring green-tide algae under stress conditions. *J. Ocean Univ.* **2016**, *34*, 102–108. [CrossRef]

54. Tan, S.; Hu, X.L.; Yin, P.H.; Zhao, L. Photosynthetic inhibition and oxidative stress to the toxic *Phaeocystis globosa* caused by a diketopiperazine isolated from products of algicidal bacterium metabolism. *J. Microbiol.* **2016**, *54*, 364–375. [CrossRef] [PubMed]

55. Guo, X.L.; Liu, X.L.; Pan, J.L.; Yang, H. Synergistic algicidal effect and mechanism of two diketopiperazines produced by *Chryseobacterium* sp. strain GLY-1106 on the harmful bloom-forming *Microcystis aeruginosa*. *Sci. Rep.* **2015**, *5*, 14720. [CrossRef] [PubMed]

56. Zhou, Q.X.; Xu, J.R.; Cheng, Y. Quantitative analyses of relationships between ecotoxicological effects and combined pollution. *Plant Soil* **2004**, *261*, 155–162. [CrossRef]

57. Hess, F.D. Light-dependent herbicides: An overview. *Weed Sci.* **2000**, *48*, 160–170. [CrossRef]

58. Ni, L.T.; Rong, S.Y.; Gu, G.X.; Hu, L.L.; Wang, P.F.; Li, D.Y.; Yue, F.F.; Wang, N.; Wu, H.Q.; Li, S.Y. Inhibitory effect and mechanism of linoleic acid sustained-release microspheres on *Microcystis aeruginosa* at different growth phases. *Chemosphere* **2018**, *212*, 654–661. [CrossRef]

59. Wang, G.X.; Zhang, Q.; Li, J.L.; Chen, X.Y.; Lang, Q.L.; Kuang, S.P. Combined effects of erythromycin and enrofloxacin on antioxidant enzymes and photosynthesis-related gene transcription in *Chlorella vulgaris*. *Aquat. Toxicol.* **2019**, *212*, 138–145. [CrossRef]

60. Zhou, Q.X.; Yue, Z.K.; Li, Q.Z.; Zhou, R.R.; Liu, L. Exposure to PbSe nanoparticles and male reproductive damage in a rat model. *Environ. Sci. Technol.* **2019**, *53*, 13408–13416. [CrossRef]

61. Kurama, E.E.; Fenille, R.C.; Rosa, V.E., Jr.; Rosa, D.D.; Ulian, E.C. Mining the enzymes involved in the detoxification of reactive oxygen species (ROS) in sugarcane. *Mol. Plant Pathol.* **2010**, *3*, 251–259. [CrossRef]

62. Fan, M.H.; Sun, X.; Xu, N.J.; Liao, Z.; Wang, R.X. cDNA cloning, characterization and expression analysis of manganese superoxide dismutase in *Ulva prolifera*. *J. Appl. Phycol.* **2015**, *28*, 1391–1401. [CrossRef]

63. Cruces, E.; Rautenberger, R.; Cubillos, V.M.; Ramirez-Kushel, E.; Rojas-Lillo, Y.; Lara, C.; Montory, J.A.; Gomez, I. Interaction of photoprotective and acclimation mechanisms in *Ulva rigida* (Chlorophyta) in response to diurnal changes in solar radiation in Southern Chile. *J. Phycol.* **2019**, *55*, 1011–1027. [CrossRef]

64. Sung, M.S.; Hsu, Y.T.; Wu, T.M.; Lee, T.M. Hypersalinity and hydrogen peroxide upregulation of gene expression of antioxidant enzymes in *Ulva fasciata* against oxidative stress. *Mar. Biotechnol.* **2009**, *11*, 199–209. [CrossRef]

65. Xu, S.; Yang, S.Q.; Yang, Y.J.; Xu, J.Z.; Shi, J.Q.; Wu, Z.X. Influence of linoleic acid on growth, oxidative stress and photosynthesis of the cyanobacterium *Cylindrospermopsis raciborskii*. *N. Z. J. Mar. Freshw. Res.* **2017**, *51*, 223–236. [CrossRef]

66. Fan, M.H.; Sun, X.; Liao, Z.; Wang, J.X.; Cui, D.L.; Xu, N.J. Full-length cDNA cloning, characterization of catalase from *Ulva prolifera* and antioxidant response to diphenyliodonium. *J. Appl. Phycol.* **2018**, *30*, 3361–3372. [CrossRef]

67. Long, M.; Tallec, K.; Soudant, P.; Le Grand, F.; Donval, A.; Lambert, C.; Sarthou, G.; Jolley, D.F.; Hégaret, H. Allelochemicals from *Alexandrium minutum* induce rapid inhibition of metabolism and modify the membranes from *Chaetoceros muelleri*. *Algal Res.* **2018**, *35*, 508–518. [CrossRef]

68. Wang, X.; Szeto, Y.T.; Jiang, C.; Wang, X.; Tao, Y.; Tu, J.; Chen, J. Effects of *Dracontomelon duperreanum* leaf litter on the growth and photosynthesis of *Microcystis aeruginosa*. *Bull. Environ. Contam. Toxicol.* **2018**, *100*, 690–694. [CrossRef]

69. Yu, S.M.; Li, C.; Xu, C.C.; Effong, K.; Xiao, X. Understanding the inhibitory mechanism of antialgalallelo chemical flavonoids from genetic variations: Photosynthesis, toxin synthesis and nutrient utility. *Ecotox. Environ. Saf.* **2019**, *177*, 18–24. [CrossRef]

70. Maxwell, K.; Johnson, G.N. Chlorophyll fluorescence—A practical guide. *J. Exp. Bot.* **2000**, *51*, 659–668. [CrossRef]

71. Zheng, Z.Z.; Gao, S.; Wang, G.C. Far red light induces the expression of LHCSR to trigger nonphotochemical quenching in the intertidal green macroalgae *Ulva prolifera*. *Algal Res.* **2019**, *40*, 101512. [CrossRef]

72. Gao, S.; Shen, S.D.; Wang, G.C.; Niu, J.F.; Lin, A.P.; Pan, G.H. PSI-driven cyclic electron flow allows intertidal macro-algae *Ulva* sp. (Chlorophyta) to survive in desiccated conditions. *Plant Cell Physiol.* **2011**, *52*, 885–893. [CrossRef] [PubMed]\

73. Gao, S.; Chi, Z.; Chen, H.L.; Zheng, Z.B.; Weng, Y.X.; Wang, G.C. A Super complex, of approximately 720 kDa and composed of both photosystem reaction centers, dissipates excess energy by PSI in green macroalgae under salt stress. *Plant Cell Physiol.* **2019**, *60*, 166–175. [CrossRef] [PubMed]

74. Lelong, A.; Haberkorn, H.; Le Goïc, N.; Hégaret, H.; Soudant, P. A new insight into allelopathic effects of *Alexandrium minutum* on photosynthesis and respiration of the diatom *Chaetoceros neogracile* revealed by photosynthetic-performance analysis and flow cytometry. *Microb. Ecol.* **2011**, *62*, 919–930. [CrossRef] [PubMed]

75. Genty, B.; Briantais, J.M.; Baker, N.R. The relationship between the quantum yield of photosynthetic electron transport and quenching of chlorophyll fluorescence. *Biochim. Biophys. Acta Gen. Subj.* **1989**, *990*, 87–92. [CrossRef]

76. Mhatre, A.; Patil, S.; Agarwal, A.; Pandit, R.; Lali, A.M. Influence of nitrogen source on photochemistry and antenna size of the photosystems in marine green macroalgae, *Ulva lactuca*. *Photosynth. Res.* **2019**, *139*, 539–551. [CrossRef]

77. Peers, G.; Truong, T.B.; Ostendorf, E.; Busch, A.; Elrad, D.; Grossman, A.R.; Hippler, M.; Niyogi, K.K. An ancient light-harvesting protein is critical for the regulation of algal photosynthesis. *Nature* **2009**, *462*, 518–521. [CrossRef]

78. Figueroa, F.L.; Celis-Plá, P.S.M.; Martínez, B.; Korbee, N.; Trilla, A.; Arenas, F. Yield losses and electron transport rate as indicators of thermal stress in *Fucus serratus* (Ochrophyta). *Algal Res.* **2019**, *41*, 101560. [CrossRef]

79. Dong, M.T.; Zhang, X.W.; Zhuang, Z.Z.; Zou, J.; Ye, N.H.; Xu, D.; Mou, S.L.; Liang, C.W.; Wang, W.Q. Characterization of the LhcSR gene under light and temperature stress in the green alga *Ulva linza*. *Plant Mol. Biol. Rep.* **2011**, *30*, 10–16. [CrossRef]

80. Mou, S.L.; Zhang, X.W.; Dong, M.; Fan, X.; Xu, J.; Cao, S.; Xu, D.; Wang, W.; Ye, N.H. Photoprotection in the green tidal alga *Ulva prolifera*: Role of LhcSR and PsbS proteins in response to high light stress. *Plant Biol.* **2013**, *15*, 1033–1039. [CrossRef]

81. Kommalapati, M.; Hwang, H.J.; Wang, H.L.; Burnap, R.L. Engineered ectopic expression of the psbA gene encoding the photosystem II D1 protein in *Synechocystis* sp. PCC6803. *Photosynth. Res.* **2007**, *92*, 315–325. [CrossRef]

82. Barati, B.; Lim, P.E.; Gan, S.Y.; Poong, S.W.; Phang, S.M. Gene expression profile of marine *Chlorella* strains from different latitudes: Stress and recovery under elevated temperatures. *J. Appl. Phycol.* **2018**, *30*, 3121–3130.[CrossRef]

83. Ohnishi, N.; Allakhverdiev, S.I.; Takahashi, S.; Higashi, S.; Watanabe, M.; Nishiyama, Y.; Norio, M. Two-step mechanism of photodamage to photosystem II: Step 1 occurs at the oxygen-evolving complex and step 2 occurs at the photochemical reaction center. *Biochemistry* **2005**, *44*, 8494–8499. [CrossRef] [PubMed]

84. Hakala, M.; Tuominen, I.; Keränen, M.; Tyystjärvi, T.; Tyystjärvi, E. Evidence for the role of the oxygen-evolving manganese complex in photoinhibition of Photosystem II. *Biochim. Biophys. Acta Bioenergy* **2005**, *1706*, 68–80. [CrossRef] [PubMed]

85. Correa-Galvis, V.; Redekop, P.; Guan, K.; Griess, A.; Truong, T.B.; Wakao, S.; Niyogi, K.K.; Jahns, P. Photosystem II Subunit PsbS is involved in the induction of LHCSR protein-dependent energy dissipation in *Chlamydomonas reinhardtii*. *J. Biol. Chem.* **2016**, *291*, 17478–17487. [CrossRef]

86. Pinnola, A.; Cazzaniga, S.; Alboresi, A.; Nevo, R.; Levin-Zaidman, S.; Reich, Z.; Bassi, R. Light-Harvesting Complex stress-eelated proteins catalyze excess energy dissipation in both photosystems of *physcomitrella patens*. *Plant Cell* **2015**, *27*, 3213–3227. [CrossRef] [PubMed]

87. Li, Y.; Zhu, H.; Lei, X.; Zhang, H.; Cai, G.; Chen, Z.; Fu, L.; Xu, H.; Zheng, T.L. The death mechanism of the harmful algal bloom species *Alexandrium tamarense* induced by algicidal bacterium *deinococcus* sp. Y35. *Front. Microbiol.* **2015**, *6*, 992–997. [CrossRef]

88. Perzborn, M.; Syldatk, C.; Rudat, J. Enzymatical and microbial degradation of cyclic dipeptides (diketopiperazines). *AMB Express* **2013**, *3*, 51. [CrossRef] [PubMed]

CHAPTER-4

ISOLATION, IDENTIFICATION, AND BIOCHEMICAL CHARACTERISTICS OF A COLD-TOLERANT *CHLORELLA VULGARIS* KNUA007 ISOLATED FROM KING GEORGE ISLAND, ANTARCTICA

1. INTRODUCTION

Green microalgae (Chlorophyta) can be found in almost every conceivable environment from polar regions to deserts; they play a pivotal role in global carbon, nitrogen, and phosphorus cycles [1–3]. Due to their remarkable tolerance to harsh conditions combined with rapid growth rates, phototrophs are often dominant in the cryosphere, representing a large portion of the total ecosystem biomass [4–6].

Microalgae have unique characteristics that enable them to succeed in frozen environments, including a high tolerance to adverse weather conditions and a lack of unfrozen water.

Accordingly, microalgae that inhabit the cryosphere, including cyanobacteria, have received considerable interest. Recent studies have assessed algal diversity in Antarctica using polymerase chain reaction-based methods [7–9]. Molecular phylogenetic techniques have been successfully applied to investigate the hidden microbial communities within Antarctic ecosystems; however, the isolation and characterization of individual algal strains remain important for many areas of research and various applications [10–12].

Chlorella (Chlorophyta) was first isolated by Dutch microbiologist Beijerinck in 1890. The genus is widely distributed and can be found thriving in freshwater, marine, and soil environments. There are currently 37 taxonomically acknowledged species in the genus *Chlorella*; these single-celled organisms consist of a spherical or oval cell, containing a single cup-shaped chloroplast with pyrenoids. *Chlorella vulgaris* range from 2 to 10 μm in diameter, lack flagella, and reproduce asexually, forming daughter cells within the parental cells through mitosis [13–15]. *Chlorella* species are regarded as promising biological resources due to their high photosynthetic abilities and rapid growth rates. In particular, *C. vulgaris* is commercially produced for applications in the food industry due to its high protein content, as well as to create specialty oils used in the cosmetic and nutraceutical industries [16]. Due to the increasing demand for the bioengineering of microalgae in recent years, whole-genome and transcriptome studies have been extensively conducted using next generation technologies in order to reveals potential genes for better utilization of high-value compounds from commercially important microalgae [15,17–20].

However, members of the genus *C. vulgaris* can be diffcult to differentiate from the genus *Micractinium* because they share similar morphological traits. The characteristic bristles that are important species-specific characteristics of the genus *Micractinium* can be easily missed since some *Micractinium* species do not produce bristles when they are not exposed to grazing zooplankton pressure or harsh conditions [21]. The similarities of these taxa can cause confusion for those attempting to identify the two genera using morphology alone. Despite the ecological and economic importance of *Chlorella*, the relevant research has been shown little attention. In addition to the commonly used region for molecular identification, recent studies have discovered that a secondary structure of internal transcribed spacer 2 (ITS-2) can be used to define precisely between the two morphologically ambiguous genera [22]. Precise identification based on characteristics studied in various aspects provides concrete evidence for further research and potential applicability. As such, polyphasic studies to delineate culturable cold-tolerant Antarctic microalgae will provide a better understanding of algal diversity and physiology as well as biotechnological potential.

In this paper, we describe the isolation of a unicellular microalga from temporal meltwater on King George Island and the determination of its phylogenetic position using several molecular markers and morphological traits. The phylogenetic position was investigated using small subunit (SSU) rRNA sequence analysis and ITS-2 secondary structure prediction. Physicochemical and chemotaxonomic characteristics were also analyzed to determine the isolate's potential for biotechnological applications.

2. MATERIALS AND METHODS

2.1 Sample Collection and Isolation

Antarctic freshwater bloom samples were collected in January 2010 from temporal water runoff near King Sejong Station (62° 13'S, 58° 47'W) located on the Barton Peninsula, King George Island, South Shetland Islands, West Antarctica. Samples were then transported to the laboratory and 1 mL of each sample was inoculated into 100 mL BG-11 medium [23]. The broad-spectrum antibiotic imipenem (JW Pharmaceutical, Seoul, Korea) was added to the medium at a concentration of 100 μg mL^{-1} to prevent bacterial growth. The inoculated flasks were incubated on an orbital shaker (Vision Scientific, Bucheon, Korea) at 160 rpm and 15 °C under cool fluorescent light (approximately 70 μmol photon m^{-2} s^{-1}) under a 16:8 light/dark cycle until algal growth was apparent. Well-developed algal cultures (1.5 mL) were centrifuged at 3000× *g* for 15 min to harvest the algal biomass. The resulting pellets were streaked onto BG-11 agar supplemented with imipenem (20 μg mL^{-1}) and incubated under the aforementioned conditions. A single colony was then aseptically re-streaked onto a fresh BG-11 plate to obtain an axenic culture.

2.2 Morphological and Molecular Identification

The isolate was grown in BG-11 medium for 20 days. Live cells were harvested by centrifugation at 3000× *g* for 5 min, washed twice with sterile distilled water, and examined at 1000 × magnification using a Zeiss Axioskop 2 light microscope (Carl Zeiss, Standort Göttingen, Vertrieb, Germany) equipped with differential interference contrast optics.

For scanning electron microscope (SEM) analysis, 10 mL of cells were fixed in osmium tetroxide (OsO_4; Electron Microscopy Sciences, EMS hereafter, Hatfield, PA, USA) for 10 min at a final concentration of 2% (*v*/*v*) in distilled water. Fixed cells were collected on a 3 μm pore polycarbonate membrane filter without additional pressure, then rinsed with distilled water to remove residual salts. Cells were dehydrated in an ethanol series (Merck, Darmstadt, Germany) and dried using a critical point dryer (CPD 300, Bal-Tec, Balzers, Liechtenstein). The dried filters were mounted on a stub and coated with gold–palladium using a sputter coater (SCD 005; Bal-Tec). Cell images were analyzed with a field emission (FE)-SEM (S-4800; Hitachi, Hitachinaka, Japan).

For molecular analysis, genomic DNA was extracted using a DNeasy Plant Mini kit (Qiagen, Hilden, Germany). The universal primers NS1 and NS8 described by White et al. [24] were used for 18S rRNA sequence analysis. The D1–D2 region of the large subunit (28S) rRNA gene was amplified using NL1 and NL4 primers [25], and the internal transcribed spacer (ITS) region was amplified using ITS1 and ITS4 primers [24]. The DNA sequences obtained in this study were deposited in the

National Center for Biotechnology Information database under accession numbers KJ148623 to KJ148625 (Table 1). Phylogenetic analysis was performed with the 18S rDNA string sequences using the software package MEGA version X [26]. The combined datasets of *C. vulgaris* strains and *Neochloris* sp. (as an outgroup) were aligned with those of the 20 chlorellacean microalgae strains using ClustalW in MEGA X. The 18S rDNA sequence lengths ranged from 1697 bp to 1732 bp. The best-fit nucleotide substitution model K2 was selected using MEGA X based on the Bayesian information criterion. This model was used to build a maximum likelihood (ML) phylogenetic tree with 1000 bootstrap replicates [27]. The ITS-2 secondary structures were constructed using Mfold [28] according to Germond et al. [29].

Table 1. Results from BLAST searches using sequences of small subunit rRNA (18S rRNA), internal transcribed spacer (ITS), and large subunit rRNA (LSU) genes from strain KNUA007.

Marker Gene	Accession no.	Product Size (bp)	Closet Match (GenBank Accession no.)	Query Cover (%)	Identification (%)
18S rRNA	KJ148623	1771	*C. vulgaris* CCAP 211/19 (MK541792)	100	99.89
ITS	KJ148624	783	*C. vulgaris* ATFG2 (MT137382)	100	99.87
LSU	KJ148625	612	*C. vulgaris* NIES:227 (AB237642)	100	99.84

2.3 Temperature Testing

Late exponential-phase cultures of *C. vulgaris* KNUA007 (1 mL each) were inoculated into BG-11 medium in triplicate and incubated for 20 days. The survival and growth of KNUA007 cells maintained at temperatures ranging from 5 to 35 °C (at intervals of 5 °C) were examined to determine the optimum culture temperature. Algal cell density was determined by measuring the optical density (OD) of the cultures at 680 nm with an Optimizer 2120UV spectrophotometer (Mecasys, Daejeon, Korea). We also incubated samples in BG-11 medium in a PhotoBiobox (Shinhwa Science, Daejeon, Korea) at temperatures ranging from 6 to 28 °C and light intensity ranging from 50–400 μmol photon m^{-2} s^{-1} with cycles of 24 h light for 3 days to monitor the optimal conditions. OD at 680 nm was measured using a Spectra Max i3x microplate reader (Molecular Devices, San Jose, CA, USA) and OD values were visualized on a heat map by SigmaPlot 13 software (Systat Software, San Jose, CA, USA) [30].

2.4 Biomass Characterization

The isolates were autotrophically grown in BG-11 medium for 20 days and cells were harvested by centrifugation at 2063 *g* (1580R, Labogene, Daejeon, Korea). The freeze-dried biomass samples were pulverized with a mortar and pestle and sieved through ASTM No. 230 mesh (63 μm holes). Ultimate analysis was conducted to determine the carbon (C), hydrogen (H), nitrogen (N), and sulfur (S) contents using a Flash 2000 elemental analyzer (Thermo Fisher Scientific, Milan, Italy). Gross calorific value (GCV) was estimated using the following equation developed by Friedl et al. [31], and protein content was calculated from the N content using a conversion factor of 6.25 ×.

$$GCV = 3.55C^2 - 232C - 2230H + 51.2C \times H + 131N + 20{,}600\ (MJ\ kg^{-1})$$

2.6 Gas Chromatography/Mass Spectrometry (GC/MS) Analysis

Extracted fatty acid methyl ester (FAME) composition was analyzed using a gas chromatograph–mass spectrometer (GC/MS 7890A; Agilent, Santa Clara, CA, USA) equipped with a mass selective detector (5975C; Agilent) and DB-FFAP column (30 m, 250 μm ID, 0.25 μm film thickness; Agilent). The initial temperature of 50 °C was maintained for 1 min. The temperature was then increased to 200 °C at a rate of 10 °C min^{-1} for 30 min, then increased to 240 °C at a rate of 10 °C min^{-1} and held for 20 min. The injection volume was 1 μL, with a 20:1 split injection ratio. Helium was supplied as the carrier gas at a constant flow rate of 1 mL min^{-1}. For the mass spectrometer parameters, the injector and source temperatures were 250 and 230 °C, respectively, and the electron impact mode was used for sample ionization at an acceleration voltage of 70 eV, with an acquisition range of 50–550 mass to charge ratio [32]. All compounds were identified by analyzing their mass spectra with the Wiley/NBS libraries.

3. RESULTS

3.1 Identification of Strain KNUA007

The cells were solitary, non-motile, and spherical, with a diameter of approximately 4–6 μm (Figure 1) and a prominent cup-shaped chloroplast. The cell walls were composed of a single smooth layer. The strain KNUA007 exhibited morphology typical of the genus *Chlorella*.

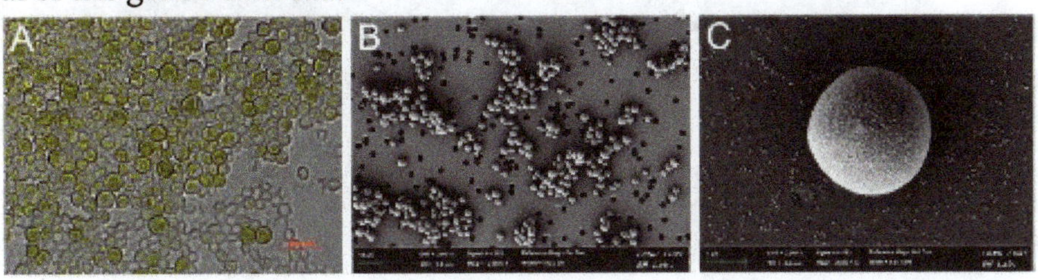

Figure 1. Light microscopy and FE-SEM images of *C. vulgaris* KNUA007.

Molecular characterization inferred from sequence analyses of the genes for 18S rRNA, 28S rRNA, and the ITS region confirmed that the isolate belonged to the *C. vulgaris* group (Table 1). As illustrated in Figure 2, strain KNUA007 was clustered with *C. vulgaris* in a different clade from *Micractinium* and *Parachlorella*. Therefore, the isolate was tentatively identified as *C. Vulgaris* KNUA007.

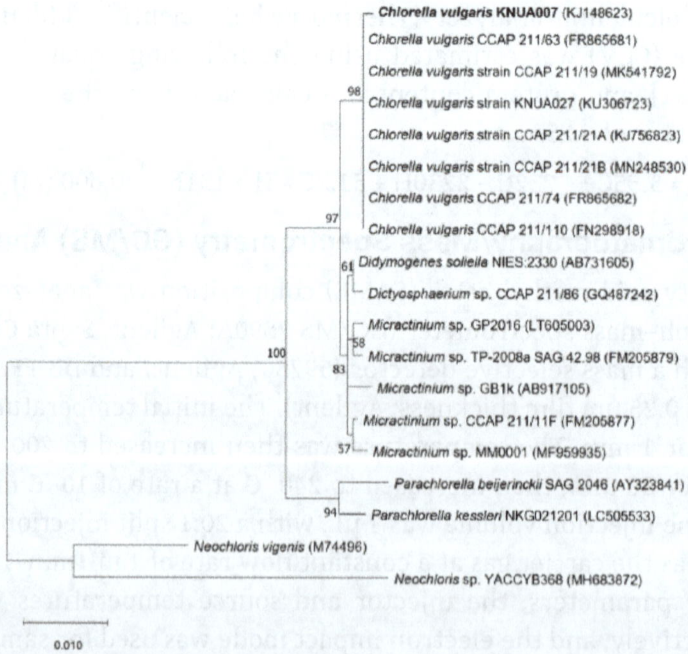

Figure 2. Maximum likelihood (ML) tree of small subunit rRNA genes. The numbers at the nodes indicate bootstrap probabilities (>50%) of ML analyses (1000 replicates). The scale bar represents a 0.1% difference.

To accurately differentiate between the morphologically ambiguous genera *Micractinium* and *Chlorella*, the secondary structure of ITS-2 was examined. Helix 3 of ITS-2 has a distinct molecular signature that separates *Chlorella* and *Micractinium*. Specifically, in Figure 3, the paired nucleotides indicated in the orange box are a synapomorphy of *Micractinium* [22]. Therefore, strain KNUA007 was tentatively identified as *C. vulgaris* KNUA007.

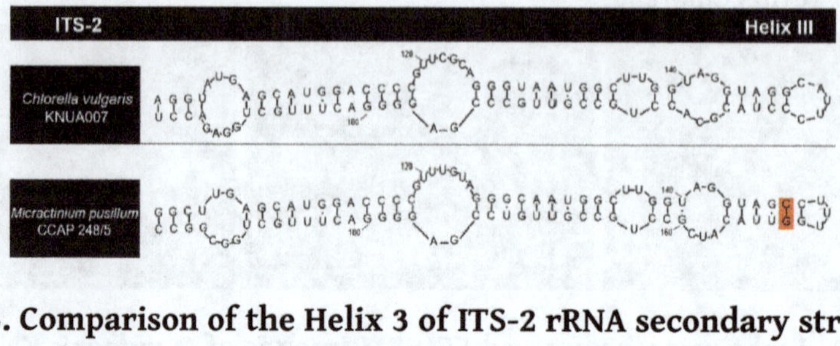

Figure 3. Comparison of the Helix 3 of ITS-2 rRNA secondary structures.

3.2 Cold Tolerance of Strain KNUA007

C. vulgaris KNUA007 could grow at temperatures ranging from 5 to 30 °C; maximum growth was observed at 20 °C. The cells grew and survived at 5, 10, 15, and 25 °C, but relatively slow growth was observed (Figure 4). However, the isolate did not survive at 35 °C. This strain also grew at a wide range of light intensity, from 50 to 400 μmol photon m¯2 s¯1; maximum growth was observed at 150 μmol photon m¯2 s¯1, and the strain preferred relatively low light intensity.

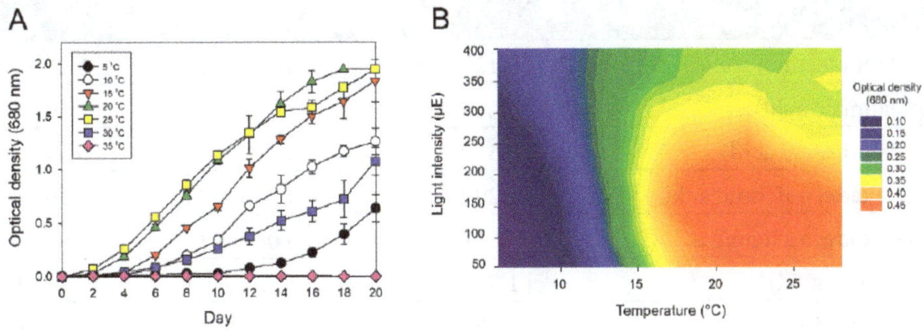

Figure 4. (A) Growth curves for KNUA007 cells maintained at various temperatures. (B) Optimal light intensity and temperature of KNUA007.

3.3 Biomass Properties

The C, H, N, and S contents were determined by elemental analysis (Table 2). C and N decreased in inverse proportion to the temperature increase, S increased with temperature, and H content was independent of the temperature gradient. This result affected the GCV values determined based on elemental analysis, and consequently, the GCV decreased with increasing temperature.

Table 2. Ultimate analysis and gross calorific value of *C. vulgaris* KNUA007.

	Contents (wt%)		
	10 °C	20 °C	30 °C
C H N S	45.36 ± 0.06	42.94 ± 0.15	42.72 ± 0.26
Gross calorific value (MJ kg¯1)	6.64 ± 0.03	6.42 ± 0.05	6.54 ± 0.07
	6.17 ± 0.00	5.84 ± 0.03	5.25 ± 0.02
	0.76 ± 0.07	1.10 ± 0.03	1.14 ± 0.02
	18.7 ± 0.02	17.7 ± 0.06	17.5 ± 0.11

3.4 GC/MS Analysis of Strain KNUA007

The FAME profile of *C. vulgaris* KNUA007 is summarized in Table 3; values are presented as the average ± standard deviation of three determinations. The major cellular fatty acids of the microalgae were C16:3 ω3, C16:0, C18:2 ω6, and C18:3 ω3. The C18:3 composition was approximately 43% at 10 °C, decreasing to 36 and 20% with increasing temperatures. C16:3 also decreased more than five-fold with the temperature increase, similar to C18:3. Meanwhile, C16:2 and C18:2, which have two unsaturated chains, increased two and three times, respectively, and the other fatty acid contents did not change. The protein contents calculated using ultimate analysis were 38.56, 36.5, and 32.81% at 10, 20, and 30 °C, respectively.

Table 3. FAME profiles of *C. vulgaris* KNUA007.

	Contents (wt%)		
	10 °C	20 °C	30 °C
C15:0 (Pentadecanoic acid)	0.12 ± 0.01	0.15 ± 0.03	N.D
C16:0 (Palmitic acid)	18.55 ± 0.02	22.35 ± 0.48	25.02 ± 0.03
C16:1 (Palmitoleic acid)	0.51 ± 0.02	0.53 ± 0.08	0.55 ± 0.04
C16:2 (Hexadecadienoic acid)	2.29 ± 0.07	4.25 ± 0.33	5.16 ± 0.04
C16:3 (Hexadecatrienoic acid)	17.31 ± 0.19	8.08 ± 0.31	2.93 ± 0.11
C18:0 (Stearic acid)	1.30 ± 0.10	1.52 ± 0.05	2.36 ± 0.28
C18:1 (Oleic acid)	8.05 ± 0.16	9.00 ± 0.12	13.78 ± 0.46
C18:2 (Linoleic acid)	8.52 ± 0.08	17.97 ± 0.84	29.54 ± 0.17
C18:3 (α-Linolenic acid)	43.35 ± 0.05	36.14 ± 2.01	20.67 ± 0.42
	N.D: Not detected.		

4. DISCUSSION

In this study, a cold-tolerant *C. vulgaris* strain, KNUA007, was axenically isolated using a combination of imipenem treatment and physical separation techniques. We analyzed the morphological, molecular, physiological, and biochemical characteristics of the isolates to identify the strain. The small and rounded cells lacked flagella and possessed the morphological features typical of *C. vulgaris*, including smooth cell wall layers and the presence of chloroplast pyrenoids (Figure 1). In addition, the phylogenetic position of the isolate was confirmed through sequence analysis of the SSU, ITS, and LSU regions (Table 1). As shown in Figure 2, the isolated strain was clearly separated from the genus *Micractinium* and was closely related to *C. vulgaris*. However, *Micractinium* has similar morphological features to the isolated strain [22], so the Helix 3 structure of the ITS-2 region was compared for further classification (Figure 3). Our analysis of the Helix 3 structure revealed that the strain used in this experiment did not possess the characteristic paired-nucleotide synapomorphy of *Micractinium* [22,29,33]. The isolate had a 100% sequence homology with *C. vulgaris* CCAP 211/63 (accession no. FR865681) (Figure 2) and the same ITS-2 overall sequence (data not shown); therefore, the strain used in this experiment was named *C. vulgaris* KNUA007.

The GCV was calculated to estimate the potential of this microalga as a biofuel feedstock. The results indicated that the GCV was similar to terrestrial energy crops and cellulosic biomass (17.0–20.0 MJ kg^{-1}) (Table 2). A number of previous studies have reported the GCVs of *Chlorella* strains under various autotrophic growth conditions, and the biomass samples were characterized in the range of 20.0–30.0 MJ kg^{-1} [34,35]. However, given the high photosynthetic effciency and growth rate of this *C. vulgaris* strain, it could be a superior bioenergy source for biofuel production. We also measured protein using ultimate analysis; the protein content was over 30% at all temperature ranges and was highest at 10 °C. These results demonstrate the potential of *C. vulgaris* KNUA007 microalgal biomass for producing animal feedstock or bio-fertilizer.

Antarctica is one of the most extreme environments on the planet. Microalgae in Antarctic ecosystems are continuously exposed to direct sunlight during the summer months and to freezing temperatures during the winter months. Thus, Antarctic microalgae have developed a number of protection strategies to survive under these harsh conditions. In general, microalgae protect themselves from excessive light and oxidative damage by dissipating excess energy as heat, which has been termed non-photochemical quenching of chlorophyll fluorescence [36,37]. Moreover, some *Chlorella* species have been reported to express anti-freezing genes in response to low-temperature stress, enhancing their tolerance to freezing temperatures [38–40]. One of the most common self-defense mechanisms observed in microalgae to acquire cold tolerance is to increase the fluidity of lipid membrane, carried out by fatty acid desaturation, which increases the production of unsaturated fatty acids with lower chain lengths that remain in a liquid state at low temperatures [41,42]. The majority of fatty acids in *C. vulgaris* are C16 and C18, and the sequential desaturation of these fatty acids usually involves the following steps: C16:0 → C18:0 → C18:1 → C18:2 → C18:3 [43,44]. Analysis of the cellular fatty acid composition of the strain KNUA007 revealed that it was rich in C16:3, C18:2, and C18:3 unsaturated fatty acids (Table 3). In particular, C16:3 and C18:3, which have three unsaturated chains, increased in inverse proportion to the temperature; the C16:3 and C18:3 contents were more than twice as high at 10 °C than at 30 °C. As mentioned above, the content of polyunsaturated fatty acids is thought to increase as a defense mechanism to cope with low-temperature stress. Meanwhile, decreases in C18:1 and C18:2 were observed in the 10 °C culture of KNUA007 compared to those of the 20 °C culture of KNUA007. We presume that these slight decreases in C18:1 and C18:2 accounted for the degree of unsaturation, since C18:1 and C18:2 are intermediate forms of desaturation in the biosynthetic pathway of the major fatty acids by desaturases. On the other hand, the most unsaturated form of the major C18 fatty acids, C18:3, showed a significant increase in response to the low temperature. We assume C18:3 plays an important role in the fluidity modification capacity of *C. vulgaris* KNUA007. This kind of trend was also observed in *Scenedesmus acutus* [45]. Its fatty acid profile changed in response to temperature changes and the same pattern (increase in C18:3 composition in lower temperatures and increase in C18:1 and C18:2 compositions in higher temperatures, respectively) as our study with *C. Vulgaris* was also observed in their study. Numerous studies demonstrated that these essential PUFAs also have many beneficial health effects [46] and various commercial products are available worldwide.

Omega-3 PUFAs are typically derived from fish oils and omega-6 PUFAs are primarily obtained from plant sources such as sunflower, corn, and soybean oils. Therefore, this isolate may have the potential to be used as an alternative to fish- and/or plant-based oil sources. In addition, the biomass itself may serve as an excellent animal feed because of its adequate protein content.

Lastly, it should be stated that most polar microalgae are known to be psychrotolerant; however, truly psychrophilic microalgae are relatively rare [47,48], and temperature flexibility is more common than psychrophily [49,50]. Likewise, strain KNUA007 was shown to be a psychrotolerant and mesophilic microalga that could grow at temperatures as low as 5 °C; optimum growth was observed at 20 °C (Figure 4). Furthermore, the strain grew over a wide temperature range (5–30 °C). This wide temperature tolerance is similar to those previously reported for Antarctic *Chlorella* strains [38,51]. The eurythermal properties of *Chlorella* may be advantageous for outdoor mass cultivation. The *C. vulgaris* KNUA007 isolate may be used for the production of value-added products under unfavourable weather conditions, including the cold temperatures that occur in autumn and winter seasons since ambient temperatures (approximately 20–35 °C) were the optimal growth conditions for most of *C. vulgaris* cultures for PUFA production [52]. In general, cell growth is suppressed at lower temperatures and the metabolic rate is accelerated at higher temperatures, but it is reported that temperatures above 35 °C usually inhibit growth [13].

In conclusion, this Antarctic microalga could serve as a potential biological resource to produce compounds of biochemical interest. The real potential of the isolate described in this paper should be evaluated through further cultivation studies at molecular, laboratory, and field scales.

REFERENCES

1. Komárek, J.; Nedbalová, L. Green cryosestic algae. In *Algae and Cyanobacteria in Extreme Envionments*; Seckbach, J., Ed.; Springer: Dordrecht, The Netherlands, 2007; pp. 323–344.

2. Lewis, L.A.; Lewis, P.O. Unearthing the molecular phylodiversity of desert soil green algae (Chlorophyta). *Syst. Biol.* **2005**, *54*, 936–947. [CrossRef] [PubMed]

3. Vincent, W.F. Cyanobacterial dominance in the polar regions. In *The Ecology of Cyanobacteria: Their Diversity in Time and Space*; Whitton, B.A., Potts, M., Eds.; Kluwer Academic Publishers: Dordrecht, The Netherlands, 2000; pp. 321–340.

4. Broady, P.A. Diversity, distribution and dispersal of Antarctic terrestrial algae. *Biodivers. Conserv.* **1996**, *5*, 1307–1335. [CrossRef]

5. Vincent, W.F.; James, M.R. Biodiversity in extreme aquatic environments: Lakes, ponds and streams of the Ross Sea sector, Antarctica. *Biodivers. Conserv.* **1996**, *5*, 1451–1471. [CrossRef]

6. Wynn-Williams, D.D. Antarctic microbial diversity: The basis of polar ecosystem processes. *Biodivers. Conserv.* **1996**, *5*, 1271–1293. [CrossRef]

7. De Wever, A.; Leliaert, F.; Verleyen, E.; Vanormelingen, P.; Van Der Gucht, K.; Hodgson, D.A.; Sabbe, K.; Vyverman, W. Hidden levels of phylodiversity in Antarctic green algae: Further evidence for the existence of glacial refugia. *Proc. R. Soc. B Biol. Sci.* **2009**, *276*, 3591–3599. [CrossRef]

8. Jungblut, A.D.; Vincent, W.F.; Lovejoy, C. Eukaryotes in Arctic and Antarctic cyanobacterial mats. *FEMS Microbiol. Ecol.* **2012**, *82*, 416–428. [CrossRef]

9. Wood, S.A.; Rueckert, A.; Cowan, D.A.; Cary, S.C. Sources of edaphic cyanobacterial diversity in the Dry Valleys of Eastern Antarctica. *ISME J.* **2008**, *2*, 308–320. [CrossRef]

10. Ferrara, M.; Guerriero, G.; Cardi, M.; Esposito, S. Purification and biochemical characterisation of a glucose-6-phosphate dehydrogenase from the psychrophilic green alga *Koliella antarctica*. *Extremophiles* **2013**, *17*, 53–62. [CrossRef]

11. Pulz, O.; Gross, W. Valuable products from biotechnology of microalgae. *Appl. Microbiol. Biotechnol.* **2004**, *65*, 635–648. [CrossRef]

12. Rˇ ezanka, T.; Nedbalová, L.; Sigler, K. Unusual medium-chain polyunsaturated fatty acids from the snow alga Chloromonas brevispina. Microbiol. Res. **2008**, 163, 373–379. [CrossRef]

13. Ru, I.T.K.; Sung, Y.Y.; Jusoh, M.; Wahid, M.E.A.; Nagappan, T. *Chlorella vulgaris*: A perspective on its potential for combining high biomass with high value bioproducts. *Appl. Phycol.* **2020**, *1*, 2–11. [CrossRef]

14. Safi, C.; Zebib, B.; Merah, O.; Pontalier, P.Y.; Vaca-Garcia, C. Morphology, composition, production, processing and applications of *Chlorella vulgaris*: A review. *Renew. Sustain. Energy Rev.* **2014**, *35*, 265–278. [CrossRef]

15. Cecchin, M.; Marcolungo, L.; Rossato, M.; Girolomoni, L.; Cosentino, E.; Cuine, S.; Li-Beisson, Y.; Delledonne, M.; Ballottari, M. *Chlorella vulgaris* genome assembly and annotation reveals the molecular basis for metabolic acclimation to high light conditions. *Plant J.* **2019**, *100*, 1289–1305. [CrossRef] [PubMed]

16. Ahmad, M.T.; Shariff, M.; Md. Yusoff, F.; Goh, Y.M.; Banerjee, S. Applications of microalga *Chlorella vulgaris* in aquaculture. *Rev. Aquac.* **2020**, *12*, 328–346. [CrossRef]

17. Merchant, S.S.; Prochnik, S.E.; Vallon, O.; Harris, E.H.; Karpowicz, S.A.; Witman, G.B.; Terry, A.; Salamov, A.; Fritz-Laylin, L.K.; Maréchal-Drouard, L.; et al. The *Chlamydomonas* genome reveals the evolution of key animal and plant functions. *Natl. Inst. Heal.* **2007**, *318*, 245–250. [CrossRef]

18. Blanc, G.; Duncan, G.; Agarkova, I.; Borodovsky, M.; Gurnon, J.; Kuo, A.; Lindquist, E.; Lucas, S.; Pangilinan, J.; Polle, J.; et al. The *Chlorella variabilis* NC64A genome reveals adaptation to photosymbiosis, coevolution with viruses, and cryptic sex. *Plant Cell* **2010**, *22*, 2943–2955. [CrossRef]
19. Vieler, A.; Wu, G.; Tsai, C.H.; Bullard, B.; Cornish, A.J.; Harvey, C.; Reca, I.B.; Thornburg, C.; Achawanantakun, R.; Buehl, C.J.; et al. Correction: Genome, functional gene annotation, and nuclear transformation of the Heterokont oleaginous alga *Nannochloropsis oceanica* CCMP1779. *PLoS Genet.* **2017**, *13*, e1006802. [CrossRef]
20. Arriola, M.B.; Velmurugan, N.; Zhang, Y.; Plunkett, M.H.; Hondzo, H.; Barney, B.M. Genome sequences of *Chlorella sorokiniana* UTEX 1602 and *Micractinium conductrix* SAG 241.80: Implications to maltose excretion by a green alga. *Plant J.* **2018**, *93*, 566–586. [CrossRef]
21. Luo, W.; Pflugmacher, S.; Pröschold, T.; Walz, N.; Krienitz, L. Genotype versus phenotype variability in *Chlorella* and *Micractinium* (Chlorophyta, Trebouxiophyceae). *Protist* **2006**, *157*, 315–333. [CrossRef]
22. Luo, W.; Krienitz, L.; Pflugmacher, S.; Walz, N. Genus and species concept in *Chlorella* and *Micractinium* (Chlorophyta, Chlorellaceae): Genotype versus phenotypical variability under ecosystem conditions. *SIL Proc.* **2005**, *29*, 170–173. [CrossRef]
23. Rippka, R.; Deruelles, J.; Waterbury, J.B.; Herdman, M.; Stanier, R.Y. Generic assignments, strain histories and properties of pure cultures of cyanobacteria. *J. Gen. Microbiol.* **1979**, *111*, 1–61. [CrossRef]
24. White, T.J.; Bruns, T.; Lee, S.; Taylor, J. Amplification and direct sequencing of fungal ribosomal RNA genes for phylogenetics. In *PCR protocols: A Guide to Method and Applications*; Academic Press: Cambridge, MA, USA, 1990; pp. 315–322.
25. Marshall, M.N.; Cocolin, L.; Mills, D.A.; Vander Gheynst, J.S. Evaluation of PCR primers for denaturing gradient gel electrophoresis analysis of fungal communities in compost. *J. Appl. Microbiol.* **2003**, *95*, 934–948. [CrossRef] [PubMed]
26. Kumar, S.; Stecher, G.; Li, M.; Knyaz, C.; Tamura, K. MEGA X: Molecular evolutionary genetics analysis across computing platforms. *Mol. Biol. Evol.* **2018**, *35*, 1547–1549. [CrossRef] [PubMed]
27. Felsenstein, J. Confidence limits on phylogenies: An approach using the bootstrap. *Evolution* **1985**, *39*, 783–791. [CrossRef] [PubMed]
28. Zuker, M. Mfold web server for nucleic acid folding and hybridization prediction. *Nucleic Acids Res.* **2003**, *31*, 3406–3415. [CrossRef]

29. Germond, A.; Hata, H.; Fujikawa, Y.; Nakajima, T. The phylogenetic position and phenotypic changes of a *Chlorella*-like alga during 5-year microcosm culture. *Eur. J. Phycol.* **2013**, *48*, 485–496. [CrossRef]
30. Heo, J.; Cho, D.-H.; Ramanan, R.; Oh, H.-M.; Kim, H.-S. PhotoBiobox: A tablet sized, low-cost, high throughput photobioreactor for microalgal screening and culture optimization for growth, lipid content and CO2 sequestration. *Biochem. Eng. J.* **2015**, *103*, 193–197. [CrossRef]
31. Friedl, A.; Padouvas, E.; Rotter, H.; Varmuza, K. Prediction of heating values of biomass fuel from elemental composition. *Anal. Chim. Acta* **2005**, *544*, 191–198. [CrossRef]
32. Aussant, J.; Guihéneuf, F.; Stengel, D.B. Impact of temperature on fatty acid composition and nutritional value in eight species of microalgae. *Appl. Microbiol. Biotechnol.* **2018**, *102*, 5279–5297. [CrossRef]
33. Chae, H.; Lim, S.; Kim, H.S.; Choi, H.-G.; Kim, J.H. Morphology and phylogenetic relationships of *Micractinium* (Chlorellaceae, trebouxiophyceae) taxa, including three new species from antarctica. *Algae* **2019**, *34*, 267–275. [CrossRef]
34. Illman, A.M.; Scragg, A.H.; Shales, S.W. Increase in *Chlorella* strains calorific values when grown in low nitrogen medium. *Enzyme Microb. Technol.* **2000**, *27*, 631–635. [CrossRef]
35. Do, J.-M.; Jo, S.-W.; Kim, I.-S.; Na, H.; Lee, J.H.; Kim, H.S.; Yoon, H.-S. A feasibility study of wastewater treatment using domestic microalgae and analysis of biomass for potential applications. *Water* **2019**, *11*, 2294. [CrossRef]
36. Lunch, C.K.; Lafountain, A.M.; Thomas, S.; Frank, H.A.; Lewis, L.A.; Cardon, Z.G. The xanthophyll cycle and NPQ in diverse desert and aquatic green algae. *Photosynth. Res.* **2013**, *115*, 139–151. [CrossRef] [PubMed]
37. Müller, P.; Li, X.-P.; Niyogi, K.K. Non-photochemical quenching. A response to excess light energy. *Plant Physiol.* **2001**, *125*, 1558–1566. [CrossRef] [PubMed]
38. Chong, G.-L.; Chu, W.-L.; Othman, R.Y.; Phang, S.-M. Differential gene expression of an Antarctic *Chlorella* in response to temperature stress. *Polar Biol.* **2011**, *34*, 637–645. [CrossRef]
39. Li, H.; Liu, X.; Wang, Y.; Hu, H.; Xu, X. Enhanced expression of antifreeze protein genes drives the development of freeze tolerance in an Antarctica isolate of *Chlorella vulgaris*. *Prog. Nat. Sci.* **2009**, *19*, 1059–1062. [CrossRef]
40. Machida, T.; Murase, H.; Kato, E.; Honjoh, K.; Matsumoto, K.; Miyamoto, T.; Iio, M. Isolation of cDNAs for hardening-induced genes from *Chlorella vulgaris* by suppression subtractive hybridization. *Plant Sci.* **2008**, *175*, 238–246. [CrossRef]

41. Spijkerman, E.; Wacker, A.; Weithoff, G.; Leya, T. Elemental and fatty acid composition of snow algae in Arctic habitats. *Front. Microbiol.* **2012**, *3*, 1–15. [CrossRef] [PubMed]

42. Lyon, B.R.; Mock, T. Polar microalgae: New approaches towards understanding adaptations to an extreme and changing environment. *Biology* **2014**, *3*, 56–80. [CrossRef]

43. Harris, P.; James, A.T. The effect of low temperatures on fatty acid biosynthesis in plants. *Biochem. J.* **1969**, *112*, 325–330. [CrossRef]

44. Suga, K.; Honjoh, K.I.; Furuya, N.; Shimizu, H.; Nishi, K.; Shinohara, F.; Hirabaru, Y.; Maruyama, I.; Miyamoto, T.; Hatano, S.; et al. Two low-temperature-inducible *Chlorella* genes for Δ12 and ω-3 fatty acid desaturase (FAD): Isolation of *Δ12* and ω-*3* *fad* cDNA clones, expression of *Δ12* *fad* in *Saccharomyces cerevisiae*, and expression of ω-*3* *fad* in *Nicotiana tabacum*. *Biosci. Biotechnol. Biochem.* **2002**, *66*, 1314–1327. [CrossRef] [PubMed]

45. El-sheekh, M.; Abomohra, A.E.; El-azim, M.A.; Abou-shanab, R. Effect of temperature on growth and fatty acids profile of the biodiesel producing microalga *Scenedesmus acutus*. *Biotechno. Agron. Soc. Environ.* **2017**, *21*, 233–239.

46. Mehta, L.R.; Dworkin, R.H.; Schwid, S.R. Polyunsaturated fatty acids and their potential therapeutic role in multiple sclerosis. *Nat. Clin. Pract. Neurol.* **2009**, *5*, 82–92. [CrossRef]

47. Nadeau, T.-L.; Castenholz, R.W. Characterization of psychrophilic Oscillatorians (cyanobacteria) from antarctic meltwater ponds. *J. Phycol.* **2000**, *36*, 914–923. [CrossRef]

48. Tang, E.P.Y.; Vincent, W.F. Strategies of thermal adaptation by high-latitude cyanobacteria. *New Phytol.* **1999**, *142*, 315–323. [CrossRef]

49. Chevalier, P.; Proulx, D.; Lessard, P.; Vincent, W.F.; De La Noüe, J. Nitrogen and phosphorus removal by high latitude mat-forming cyanobacteria for potential use in tertiary wastewater treatment. *J. Appl. Phycol.* **2000**, *12*, 105–112. [CrossRef]

50. Tang, E.P.Y.; Tremblay, R.; Vincent, W.F. Cyanobacterial dominance of polar freshwater ecosystems: Are high-latitude mat-formers adapted to low temperature? *J. Phycol.* **1997**, *33*, 171–181. [CrossRef]

51. Hu, H.; Li, H.; Xu, X. Alternative cold response modes in *Chlorella* (Chlorophyta, Trebouxiophyceae) from Antarctica. *Phycologia* **2008**, *47*, 28–34. [CrossRef]

52. Hong, J.W.; Kim, O.H.; Jo, S.-W.; Kim, H.; Jeong, M.R.; Park, K.M.; Lee, K.I.; Yoon, H.-S. Biochemical composition of a Korean domestic microalga *Chlorella vulgaris* KNUA027. *Microbiol. Biotechnol. Lett.* **2016**, *44*, 400–407. [CrossRef]

CHAPTER-5

MULTIDISCIPLINARY ANALYSIS OF CYSTOSEIRA SENSU LATO (SE SPAIN) SUGGEST A COMPLEX COLONIZATION OF THE MEDITERRANEAN

1. INTRODUCTION

The family *Sargassaceae* Kützing (*Phaeophyceae*) inhabits all oceans, from polar waters to the warmest tropical seas [1]. The *Cystoseira sensu lato (sl)* species have their maximum diversity in the Mediterranean Sea, with two thirds of all species described found there [2–4]. Some of them have habitats as reduced as coastal platforms [5]. Outside the Mediterranean Sea, they are mainly found in the Northeast Atlantic. This may explain why they were reintroduced six million years ago into the Mediterranean Sea. During the Zanclean deluge, after the Messinian desiccation crisis, they may have been dragged by the currents [2,6,7], starting a new colonization process. Recent research corroborates the polyphyletic nature of these algae, distinguishing three genera [8–10]. They comprise *Cystoseira sensu stricto, Carpodesmia* and *Treptacantha.*

As an engineering species they form dense meadows on rocky substrates up to 100 m deep [2,5,11,12]. They are widely recognized as bioindicators for their restricted habitat in the sub-coastal zone and low tolerance to pollution [13,14]. Most Mediterranean species are protected by the Barcelona (Annex II, COM/2009/585) and Bern Conventions (Annex I) and the INDEMARES project [15], a contributor to the expansion of the Natura 2000 Network. They are also used to assess the ecological quality of the coast (CARLIT), a requirement under the Water Framework Directive (2000/60/EU) for the conservation of good water status [13].

Cystoseira spp. have a high morphological plasticity which, with recurrent hybridization processes [2,16], makes taxonomic assignment of some species rather diffcult [10,17,18]. *Cystoseira sl* communities are currently declining due to anthropogenic pressures [10,19–22]. Reforestation projects have been developed [20,22,23]. Even their economic value has been quantified to promote their preservation [24]. However, there is little public knowledge on the value of these ecosystems outside the phycological community.

SE Spain has a great diversity of *Cystoseira sl* populations due to the presence of rocky platforms from the Quaternary Period, which generate adequate niches for their development [25,26]. However, there is little research on the distribution patterns of *Cystoseira sl* in these areas, since most studies of Spanish Mediterranean populations have been carried out on the coasts of Catalonia and the Balearic Islands [27–33]. Without light limitations, the main factor modifying the abundance of coastal communities is nutrient availability, which depends largely on hydrodynamism [13,34]. Geomorphological characteristics of the substrate (lithology, slope, depth) may account for the environmental heterogeneity of the system, a key factor in algal distribution [35]. Inclination of the substrate can affect vertical zonation [36]. Previous studies indicate that the type of substrate and depth can also affect the distribution of *Cystoseira sl*. Decimetric blocks and pebbles displaced by storms can affect these communities and lead to their replacement by other species [37]. These authors also reported variation in patterns of distribution with the degree of exposure to waves. On the other hand, intraspecific variability and environmental conditions to which they are exposed may cause variations in the concentration of polyphenols, polysaccharides and pigments among other primary and secondary metabolites. These compounds can affect DNA extraction protocols [38]. Sequencing of various genes (e.g., mt23S, tRNA-Lys, psbA, COI) has clarified the phylogenetic relationships of these genera [2,9,10,39,40].

Our hypothesis stands that the width and the location of the platform of *Cystoseira sl* communities, and the degree of wave exposure, will affect their distribution. These factors are directly related to hydrodynamics, and therefore to nutrient availability. There will be a differentiation in horizons depending on the width of the platform and the degree of exposure to waves to which it is subjected. The aim of this study is to increase our knowledge of *Cystoseira sl* in SE Spain using ecological, morphological and molecular tools. We have chosen El Cabo de las Huertas, a well preserved natural coast amidst largely touristic developed shores. This area has been a Site of Community Importance since 2001 (ESZZ16008) because of the presence of a sandy seabed with well-preserved seagrasses (*Posidonia oceanica, Zostera marina* and *Cymodocea nodosa*), which despite the extensive touristic development indicates a good water status. This makes this site interesting to revaluate its degree of protection [41]. Unfortunately, a management plan for this area does not exist nor has it progressed its declaration as a Specially Protected Area [42]. In conclusion, our work could be a sound foundation to develop an ambitious *Cystoseira sl* protection plan in South-East (SE) Spain.

2. MATERIALS AND METHODS

2.1 Cartography of Cystoseira Sensu Lato

Our sampling method consisted of a walk along 14 km of the Alicante city coast (38º21'26.48" N, 0º24'31.37" W–38º19'29.64" N, 0º30'40.41" W). We divide the coastal strip into 3 horizons parallel to the coastline: the proximal is next to the midlittoral zone, the distal horizon is hit by waves and it is sometimes emerged; the medium horizon is between these two. *Cystoseira sl* species have been identified in situ and their semi-quantitative abundance scored visually, as in Ballesteros et al. (2007). The lowest value (1) corresponds to isolated individuals. Several individuals forming no patches score 2.

For isolated patches, the abundance value is 3. For patches forming a discontinuous horizon, the value is 4. A continuous horizon of the same species of *Cystoseira sl* scores 5. Sixty-one linear transects (30–80m) were performed along the coastline of study (2018, May–July), recording the abundance of *Cystoseira sl* communities per horizon.

Trails were georeferenced using GPSies⁺ (Klaus Bechtold, ©2017) and distribution maps drawn using QGIS v 2.18 with the WGS84 (EPSG:4326) coordinate system. An orthophoto of the region provided by the National Plan for Aerial Orthophotography (PNOA) of the National Geographic Institute [43] was used as model to draw a map according to the real geography.

2.2 Abiotic Factors and Spatial Variability of Cystoseira Spp.

Platform width, wave exposure and the sublittoral horizons were recorded as the main hydrodynamism variables (Table S1). Geomorphological characteristics of the substrate (lithology, slope, presence of pools or rifts) were also scored. Records have been analysed with Primer software (v.6.1) [44]. A three fixed factor PERMANOVA analysis has also been carried out with the Bray Curtis similarity matrix to study the influence of the main variables related to hydrodynamism on the abundance of *Cystoseira sl* species. A reduced model with 4999 permutations has been chosen. The distribution patterns of the samples have been analysed with a Principal Component Ordination Analysis (PCO), built with the same Bray–Curtis matrix.

Morphological Characterization

Specimens were collected in spring (2018) using chisel and hammer, avoiding damaging the basal structure and kept in 10% alcohol in seawater until analysis. A qualitative matrix using Primer 7 software [45] with 15 phenotypic characters of 37 *Cystoseira sl* was constructed (Table S2). Morphological data from samples described in Gómez-Garreta et al. (2000) [45], Cormaci et al. (2012) [46] and Orellana et al. (2019) [9] were recorded as reference algal groups. A cladogram was constructed using a dissimilarity Bray–Curtis matrix with the 'simple matching' method and a SIMPROF test carried out to distinguish statistical differences between individuals, adding a 'dummy variable' to improve the robustness of the cladogram.

2.4 DNA Extraction

Individuals collected were kept in cold seawater. They were cleaned of epiphytes and lyophilized in less than 24 h. Fine lyophilized thallus ground in liquid nitrogen was used for DNA extraction. DNA extractions were performed using a protocol based on the pre-treatment of Lane et al. (2006) [47] in combination with the cleaning steps of Rogers and Bendich (1989) [48]. All treatments were performed on ice to avoid DNA degradation. Lyophilized thallus is added to Buffer A (1.65 M sorbitol, 50 mM MES (sulphonic acid) pH 6.1, 10 mM EDTA, 2% (*w*/*v*) PVP-40, 0.1% (*w*/*v*) BSA (Bovine Serum Albumin) and 5 mM ß-mercaptoethanol) while stirring for 2–3 min. The mixture is filtered with Miracloth® and centrifuged at 3000× *g* for 2 min. Buffer A is removed, the pellet is resuspended in Buffer B (Buffer A without PVP) and centrifuged at 3000× *g* for 2 min. Buffer B is discarded, and the pellet is resuspended in DNA extraction buffer [48]. Samples were incubated on ice for 1 h then centrifugated at 15,000× *g* for 10 min. Aqueous phase is transferred into new tubes for phenol, chloroform and isoamyl alcohol extraction for 5 min and then centrifugation. This step was performed again without phenol for 2 min. DNA is treated with −20 °C isopropyl alcohol on ice for 20 min. After centrifugation for 10 min, the supernatant is discarded and the pellet washed with ethanol at −20 °C. Pellets containing DNA are resuspended in nuclease free water and stored at −80 °C.

2.5 Gene Amplification, Sequencing and Phylogenetic Analysis

Six sets of primers from previous work were used (Table S3). Some of them (psbA-FR2 and psbA-F2R) can be used in three different combinations [2]. PCR reactions included a preheating stage of 2 min at 92 °C, nine cycles of 30 s at 94 °C, 1 min at 45 °C with a final extension of 1 min at 72 °C plus 29 cycles of 30 s at 94 °C, 30 s at 50 °C with a final extension of 1 min at 72 °C. With the psbA-FR primer, final extensions were increased from 1 to 2 min. PCR reactions were repeated when required to supply enough DNA for Sanger sequencing. Amplicons were run in 2% agarose gels for quality testing, then treated with a DNA purification kit (Qiagen).

Sequencing was carried out by Macrogen Inc. (Korea). Consensus sequences were determined using Bioinformatics 'Reverse complement' [49] and Omega Cluster tools [50]. A sequence database has been set up based on previous studies [2,9]. Phylogeny analyses were performed using Mega X software [51], through a Muscle type alignment [52]. Both super tree and isolate trees for individual genes (mt23S, tRNA-Lys and psbA) have been built using the 'Neighbor-Joining' method [53,54] and phylogenetic test 'Bootstrap' with 1500 interactions [55]. All taxa used in the phylogenetic analyses are listed in Supplementary Table S4. Sequences were obtained from Draisma et al. (2010) [2], Orellana et al. (2019) [9] and Bruno de Sousa et al. (2019) [10], among others.

3. RESULTS

3.1 Cartography of Communities Associated with the Coastal Fringe

Cystoseira sl species were found over more than 4 km of the 14 km sampled (Figure 1a,b), in nearly uninterrupted communities at Cabo de las Huertas (Cape area) and separate individuals or patches on La Almadraba beach, Paseo Marítimo and La Playita bay (Figure 1c). *Cystoseira compressa*, *C. humilis*, C. foeniculacea, Carpodesmia amentacea var. stricta, C. brachycarpa var. balearica, Treptacantha algeriensis and *T. sauvageauana* were the species found in the area of study.

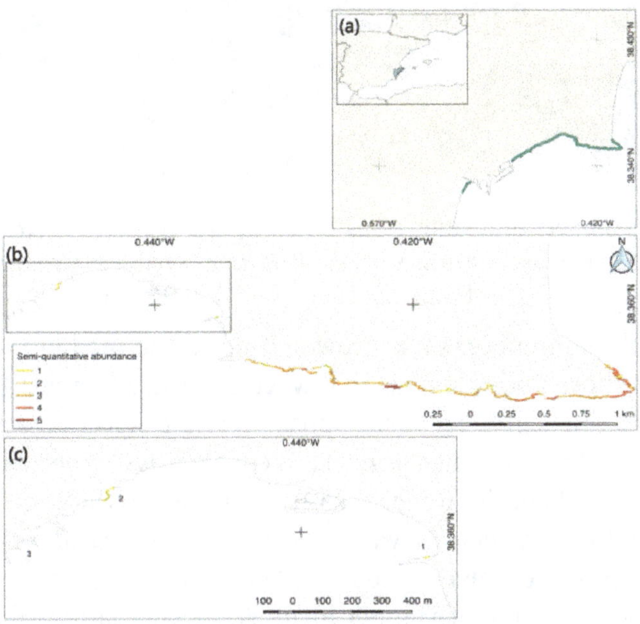

Figure 1. (a) Area of study: El Cabo de las Huertas (Alicante, Valencian Community, Spain). (b) Semi-quantitative abundance found of *Cystoseira sensu lato* in El Cabo de las Huertas. Legend: 1: One individual (isolated algae). 2: Several isolated individuals. 3: *Cystoseira sensu lato* patches. 4: Discontinuous belts of at least one

species. 5: Continuous belts of one or more species. (c) Individuals or isolated patches found near El Cabo de las Huertas. 1: La Almadraba beach. 2: Paseo Marítimo. 3: La Calita beach.

3.2 Distribution of Cystoseira Sensu Lato

Cystoseira sl communities are frequent in the area of study, in rocky shores with high hydro-dynamism (Figure 2a). They usually disappear abruptly when the rock substrate changes from platform to boulders (Figure 2b).

Figure 2. (a) Exposed platform of El Cabo de las Huertas with *C. amentacea* var. *stricta* and *T. Algeriensis* community. (b) Boulders without presence of *Cystoseira sl* in front of the exposed platform.

Cystoseira sl communities on narrow platforms and medium wave exposure are different from those under low or high wave exposure (Tables 1 and 2 and S5). In wide platforms, the degree of wave exposure conditions *Cystoseira sl* communities. For instance, *Treptacantha sauvageauana* likes low wave exposure because it is the third most abundant species under these conditions. On the contrary, other species such as *Carpodesmia amentacea* var. *stricta, Treptacantha algeriensis* or *Cystoseira compressa* prefer large wave exposure. The abundance of *Cystoseira sl* species depends on their location at the platform, since populations from proximal and distal horizons also differ significantly.

Table 1. PERMANOVA analysis of variables related with hydrodynamism (*p*-value < 0.05).

Source	df	SS	MS	Pseudo-F	P (perm)	Uniq. Perms
Width	3	15,532	5177.3	3679	0.0002	4986
Wave exposure	2	14,298	7148.9	5.08	0.0002	4983
Horizon	2	10,849	5424.7	3.8548	0.0016	4985
Width*WaveExp	5	20,474	4094.9	2.9098	0.0002	4972
Width*Horizon	6	5638.4	939.73	0.66778	0.8234	4974
Horizon*WaveExp	4	6531.8	1632.9	1.1604	0.3108	4982
Width*WaveExp*Horiz	10	6083.7	608.37	0.43231	0.9948	4982
Res	150	2.1109 × 105	1407.3			

Table 2. A posteriori PERMANOVA analysis of significant interactions of variables related with hydrodynamism (p-value < 0.05).

Width*Wave Exposure			
Wave Exposure	t	P(perm)	Perms
Width = 0 (No platform)			
Med–Low	1.9482	0.0438	2246
Width = Narrow			
Med–Low	1.8362	0.0408	4981
Med–Highq	2.5084	0.0008	4990
Low–High	1.0787	0.3138	4983
Width = Medium			
Med–Low	1.2407	0.2272	4994
Med–High	1.7067	0.0532	4989
Low–High	1.0416	0.3604	4992
Width =Wide			
Med–Low	2.1585	0.0034	4985
Med–High	2.1066	0.0052	4993
Low–High	2.7538	0.0002	4992
Horizon			
Horizon	*t*	*P(perm)*	*perms*
Prox–Med	1.4607	0.0978	4985
Prox–Distal	2783	0.0004	4986
Med–Distal	1314	0.1656	4987

A PCO of *Cystoseira sl* abundance with environmental factors was analysed using 183 samples from 61 transects recorded (Figure 3). Axes explain 79.9% of the total variability and vectors with high correlation are represented (cor > 0.2). Main hydrodynamism variables vectors are located to the lower right quadrant indicating the direction of higher values of hydrodynamism. Two geomorphologic variables have close correlation with hydrodynamism, 'Inclination' negatively and 'Rifts' positively. Canopy height is also directly linked. The lowest heights of *Cystoseira sl* have been located in La Calita, which is a cove with high anthropogenic impact (Figure S1). Thus, *Cystoseira sl* general abundance cannot be explained by a single environmental factor.

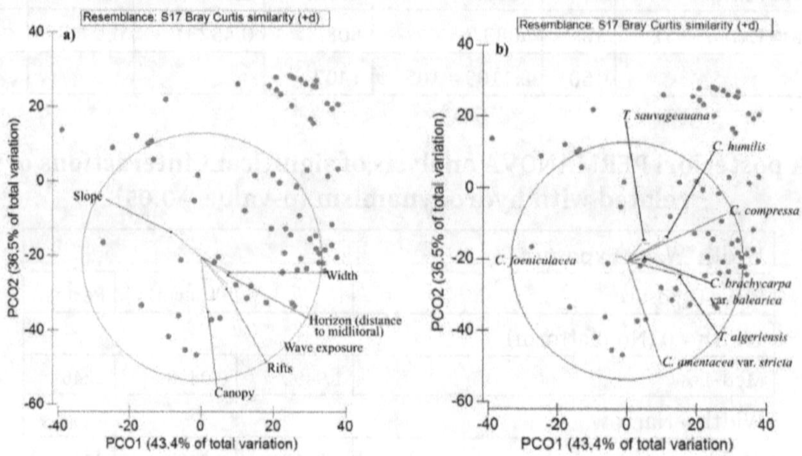

Figure 3. (a) Ordination Principal Component Ordination (PCO) plot of *Cystoseira sl* communities in SE Spain related to environmental variables. (b) Ordination PCO plot of *Cystoseira sl* communities in SE Spain with *Cystoseira sl* species (n = 183).

When *Cystoseira sl* individual species are considered, different patterns of distribution appear (Figure 4). For instance, *C. compressa* has a wide distribution, since this species is the most environmentally tolerant of all found. With the largest ecological range, it is the most abundant and is nearly homogeneously distributed in all horizons of El Cabo de las Huertas (Figure 4a). *C. amentacea* var. *stricta* and *T. algeriensis* are distributed positively with vectors, therefore to hydrodynamism. They have a slightly smaller ecological range, with irregular distribution (Figure 4b,c). Meanwhile, *C. humilis*, *T. sauvageauana* and *C. foeniculacea* have less correlation with hydro-dynamism, with reduced distribution usually located at proximal horizon (low wave exposure) (Figure 4d–f). The proximal horizon has a discontinuous presence of *C. compressa*, *C. amentacea* var. *stricta*, *T. algeriensis*, *T. sauvageauana* and *C. humilis* while more exposed horizons have a continuous community of these species (Figure S2).

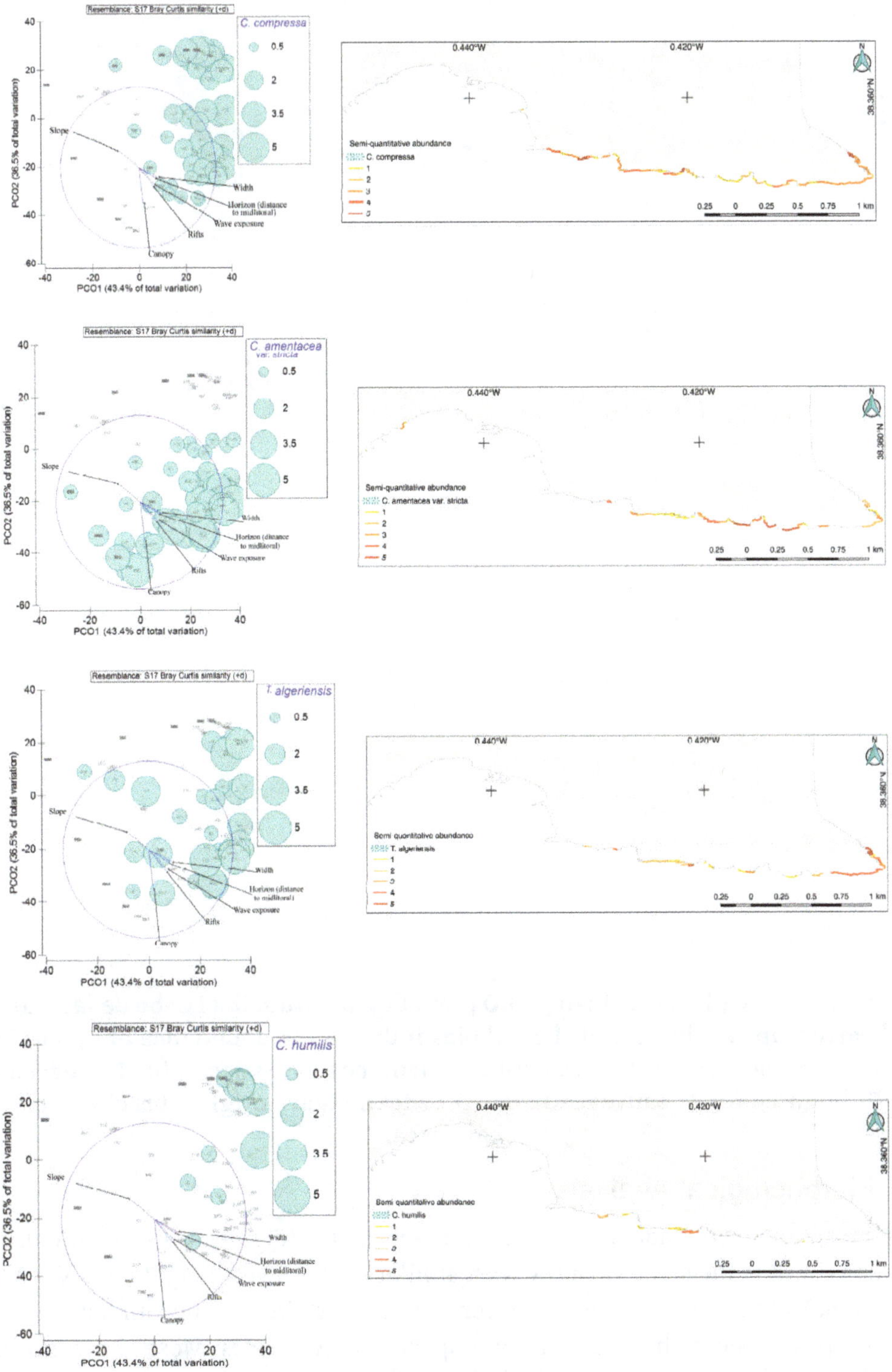

Figure 4. ***Cont.***

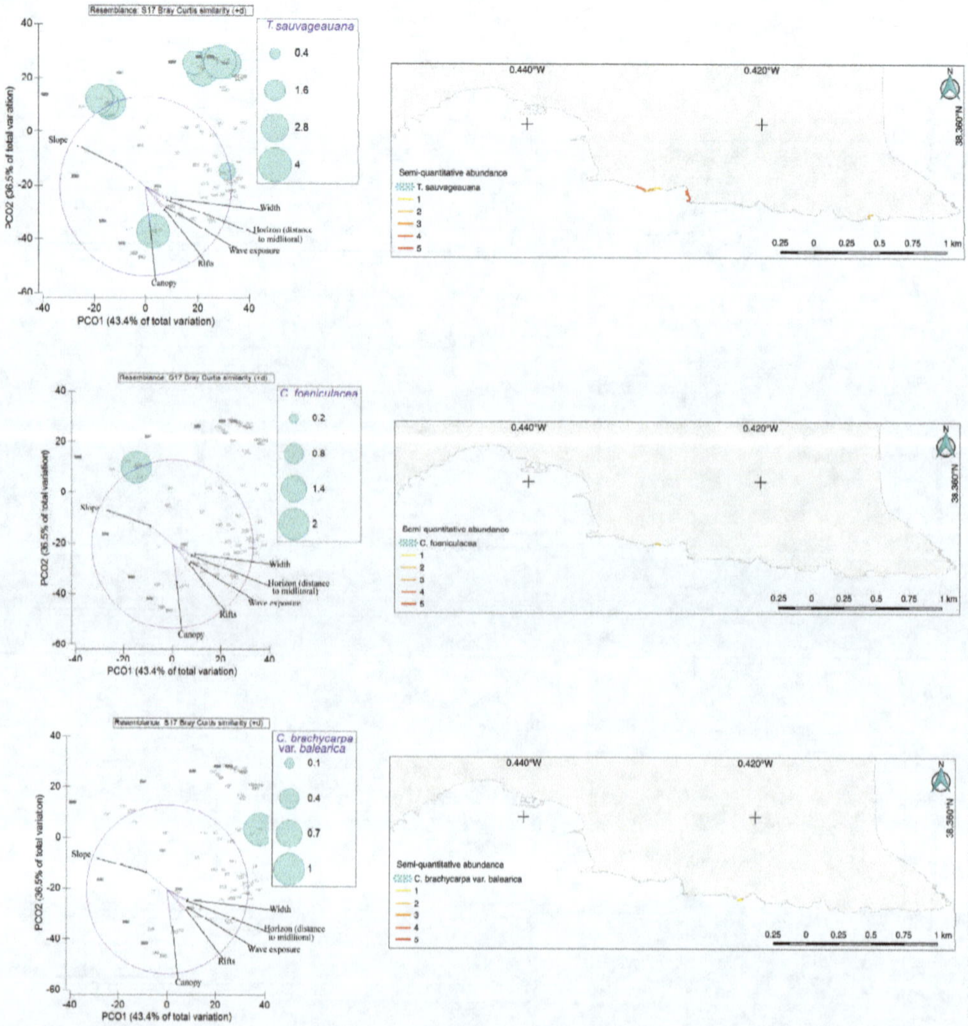

Figure 4. Bubble plots overlaying PCO plot of *Cystoseira sl* in El Cabo de las Huertas and cartography. The size of the bubbles indicate the abundance of *Cystoseira sl* in each sample. (a) *C. compressa.* (b) C. amentacea var. stricta. (c) T. algeriensis. (d) C. humilis. (e) T. sauvageauana. (f) C. foeniculacea. (g) C. brachycarpa var. balearica.

3.3 Morphological Analysis

Our morphological cladogram includes eight clades (*Cystoseira* I–VIII) (Figure 5). *Cystoseira* VIII is the clade with the least similarity, followed by *Cystoseira* VII. From *Cystoseira* VI to I, branches can be differentiated with increasing similarity values. Similarities of more than 85% prevent species assignments. Most of our samples lay close to their counterparts included in previous reference studies [9,45,46]. However, several species (*T. sauvageauana*, *C. foeniculacea*, *C. compressa*) display more variability within a given clade or are located in different clades.

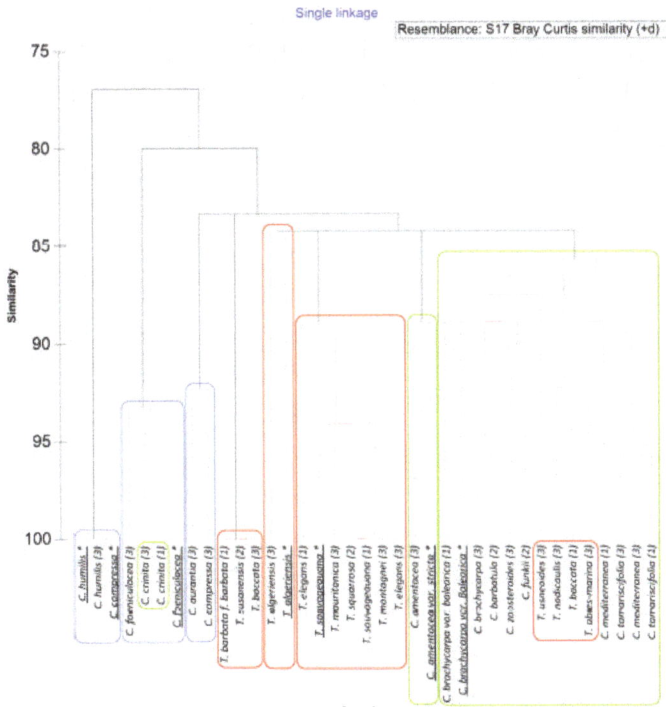

Figure 5. Single linkage cladogram based on data from *Cystoseira sl* built using the Bray–Curtis similarity matrix. Boxes correspond to current genus classification by Orellana et al., 2019 [9]: Blue = *Cystoseira sensu stricto* or Cystoseira-I; Red = *Treptacantha* or Cystoseira-II. Green = *Carpodesmia* or Cystoseira-III. Label legend: (1): Gómez-Garreta et al., 2010 [45]. (2): Cormaci et al., 2012 [46]. (3): Orellana et al., 2019 [9]. (*) Underlined: This study.

3.4 Phylogenetic Analysis

Our DNA protocol completely removed *Cystoseira sl* secondary metabolites and allowed template amplification with all primers used (Figures S3–S6). In this study, 30 *Cystoseira sl* DNA samples have been sequenced, corresponding to 3 genes. We have built a concatenated tree (supertree) with a total of 98 *Fucaceae* sequences from 41 species; 22 of these are *Cystoseira sl* species (Figure 6). The super tree is more robust (higher node stats values) than those trees built using single gene sequences (Figures S7–S9). Three clades (I–III) have been obtained. Cystoseira-I comprises *C. foeniculacea,* in a separate subclade, *C. compressa* and *C. humilis* together in a further subclade. Cystoseira-II includes four differentiated branches with *Treptacantha* spp. species (*T. abies-marina, T. baccata, T. usneoides, T. Susanensis* and *T. barbata*). The latter two species clustered in a separate subclade. Cystoseira-II includes sequences of *T. sauvageauana* and *T. algeriensis* for the first time. Both sequences appear separated from the rest. Cystoseira-III includes *C. amentacea, C.*

brachycarpa, C. crinita, C. funkii, C. mediterranea, C. tamariscifolia, and *C. zosteroides*, this last species as an outlier. Within clade III, *C. crinita* and *C. brachycarpa* can be clearly differentiated from the rest. Cystoseira-I and Cystoseira-II appear closer phylogenetically, unlike Cystoseira-III.

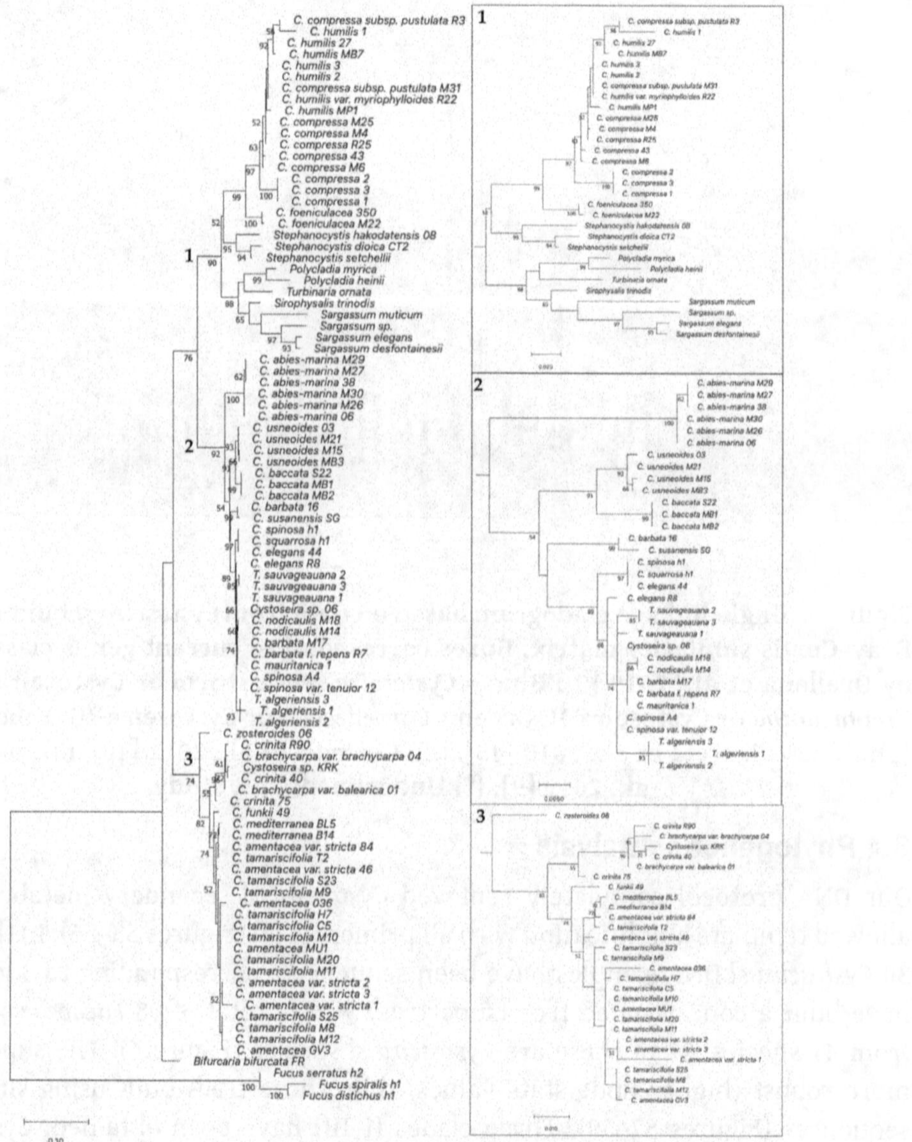

Figure 6. Maximum likelihood phylogenetic subtree obtained with Nearest Neighbor Interaction on concatenated mt23S-psbA-tRNA-Lys sequences of samples from the *Sargassaceae* family. Values of branches represent maximum likelihood bootstrap support values (>50%). Samples sequenced in this study labelled as 1, 2 or 3. Labels: 1. Cystoseira-I. 2. Cystoseira-II. 3. Cystoseira-III.

4. DISCUSSION

Hydrodynamism plays a key role in the *Cystoseira sl* population distribution from the SE Spanish coast. This has also been found in previous works [13,31,34,56]. In our study, *Cystoseira sl*, in general terms, is most abundant and widely distributed in zones with large exposure to waves. For instance, *C. amentacea* var. *stricta*, *T. algeriensis* and *C. compressa*, with large productive potential, require light saturation [57] and high hydrodynamism [13,27,31,37,58,59]. Consequently, they are frequently found in rifts and other exposed places. There are seasonal species which have their optimal development during spring [13]. The widespread distribution of *C. compressa* may reflect its high ecological tolerance, even to anthropogenic impact. This is perhaps the reason why it is the only *Cystoseira sl* species found in urban areas such as La Almadraba beach [13,19,60]. Other species (e.g., *T. sauvageauana*, *C. foeniculacea*) are distributed preferentially in more sheltered environments [61,62] where hydrodynamism is less relevant. They have a perennial behaviour, maintaining their thalluses all year [30,63]. Although *C. brachycarpa* inhabits sheltered environments [45,64], in our study it was found in an abrasion platform. The orientation on the coast, facing south, receives less impact from the dominant eastern waves [65] allowing *C. brachycarpa* to survive under these conditions. A coastline slope does not favour the presence of *Cystoseira sl* populations, as found in the Thyrrenian islands [66]. Slope degree has a close inverse relationship with those populations, since above 60° can highly reduce the probability of settlement of these communities. Human trampling is to be considered because it takes place on coastal platforms during the summer season. Those coves with a high tourist influx have lower values of abundance and canopy of *Cystoseira sl*, indicating that they are very sensitive to this anthropogenic disturbance. Even low intensities of trampling can highly affect the spatial distribution of algal communities [60,67], causing the simplification of the platform communities [12]. Therefore, future efforts could be applied studying this relation in order to apply management plans for *Cystoseira sl* protection.

Recent studies show that natural hybridization is operating in seaweed forests of *Cystoseira sl* [2,8,16]. The high variability of *T. sauvageauana, C. foeniculacea, C. compressa* in our morphological cladogram would agree with this. Each *Cystoseira sl* gender has morphological characters with various phenotypes [9], usually one of them more common than the rest. So, perhaps species that exhibit uncommon morphotypes could not be classified appropriately in the cladogram. Numerical taxonomy techniques are useful to tackle this intraspecific variability. For instance, the lack of spines in secondary and higher order branches of *C. crinita* would lead the transfer of this species from Cystoseira-III into Cystoseira-I, in which all species are characterized by the absence of this type of appendages. The latest studies indicate that there are many exceptions to this character, reinforcing the need to combine morphological with phylogenetical studies [9]. Like any other method of morphological identification, it needs to rely on molecular taxonomy to clarify the classification. Therefore, morphological data has been completed with phylogenetic studies [2,10].

Our molecular supertree of sequences of psbA, mt23S and tRNA-Lys spacer gene from *Cystoseira sl* from SE of Spain mostly agrees with previous studies [2,9,10]. In agreement with Orellana et al. [9], our clades Cystoseira-II (*Treptacantha*) and Cystoseira-I (*Cystoseira sensu stricto*) are closer phylogenetically and with other genera such as *Polycladia*, *Stephanocystys*, *Sargassum* and *Sirophysalis*, and far from Cystoseira-III (*Carpodesmia*). This supports the lack of a common ancestor of these genera [2,9,10]. Cystoseira-III (*Carpodesmia*) is the most chemically complex genus [68,69] and most members are Mediterranean [2,70], suggesting that the seasonality of this sea [71] may be driving a pressure on these algae enhancing the production of their metabolites to deal with stressors such as UV and temperature [56]. Neither Bruno de Sousa et al. [9] nor this study have succeeded in the resolution on *C. tamariscifolia*/*C. amentacea* clade. Therefore, future studies should be aimed to resolve these phylogenetic relationships. We have generated for the first-time sequence data from *T. algeriensis* and *T. sauvageauana* which support that both species are well differentiated and separated from the rest. This would also agree with their different morphology and ecological preferences [45,46]. The challenge of understanding their taxonomy may reside in the recent and continuous speciation of these genera in Mediterranean Sea [2,61,72], their adaptative convergence or possible genetic constraints [10]. There is an urgent need to keep investigating to understand their evolution and ecology in order to enhance their protection and maintain conservation efforts that help to preserve marine ecosystems for future generations.

This study provides the first bionomic cartography and evaluation of conservation of *Cystoseira sl* populations in SE Spain. Out of the 14 *Cystoseira sl* species cited in this area [73], half have been found in this study in a coastal fringe of just 4 km. In view of the diversity and abundance found, this region displays a relevant ecological quality [74]. The relative continuous algal canopy would also imply a good water status, as reported [75]. Therefore, because of its environmental characteristics and high algal diversity in the shores, this place should be a Special Conservation Area (SCA), integrated in the Natura 2000 Network (EU Directive 92/43/EEC). The inclusion of *Cystoseira*, *Treptacantha* and *Carpodesmia* genera in El Cabo de las Huertas Community Importance Site would also be a great achievement for the conservation of marine natural habitats of the Spanish Levant coast. The site may be useful for future extraction of specimens for transplants to former habitats which have undergone local extinctions [23]. There is also a pressing need to develop a management plan in view of the increasing disappearance of the prairies of eastern Mediterranean *Fucaceae* [19–22]. Long-term monitoring of these biocenosis would be desirable, which could be carried out using the CARLIT methodology, that employs *Cystoseira sl* populations as a key species establishing the health of the water body [13]. *Cystoseira* sl evolution data would generate time series to know the state of these communities, thus evaluating the health of our coast. This type of study is already being carried out in different areas of the Mediterranean Sea [20,31,59,76].

5. CONCLUSIONS

The bionomic cartography of *Cystoseira sl* assesses the good ecological quality and good water body status of El Cabo de las Huertas. Distribution analysis shows two levels of relation to hydrodynamism which predicts perennial or seasonal behaviours. Morphological cladogram detects inter-specifical variability between our species and previous reference studies of these genera. Our phylogenetic tree supports actual classification. Previously unpublished sequences of psbA, mt23S and tRNA-Lys of *T. algeriensis* and *T. sauvageauana* confirmed those are included in Bruno de Sousa's *Cystoseira* II group (2019), in agreement with morphological studies.

Supplementary Materials: The following are available online at: http://www.mdpi.com/2077-1312/8/12/961/s1. Table S1: Geomorphological characteristics of the substrate and its levels selected to PERMANOVA (*) and Principal Component Analysis (PCO) analyses. Table S2: List of qualitative phenotypical characters used for the building of Bray–Curtis dissimilarity matrix. Table S3: List of primers used for phylogenetic analysis. Table S4: GenBank accession numbers used in this study. Table S5: Abundance of *Cystoseira sl* dominant species attending on Wave exposure, Width of the platform and Horizons of the platform. Percentage of contribution of each species to the dissimilarity. Figure S1: Canopy height of *Cystoseira sl* recorded in the area of study. Figure S2: Semi-quantitative abundance of *Cystoseira sl* by horizons in the area of study. (a) Proximal horizon. (b) Medium horizon. (c) Distal horizon. Legend: 1: One isolated algae. 2: Several isolated individuals. 3: *Cystoseira sl* patches. 4: Discontinuous belts of at least one specie. 5: Continuous belts of one or more species. Figure S3: Quality electrophoresis DNA gel on 2% agarose. Legend: CA: *T. algeriensis.* CC: *C. compressa.* CH: *C. humilis.*

CS: *T. sauvageauana.* CST: *C. amentacea* var. *stricta.* C$^-$: Negative control. Figure S4: (a) Quality electrophoresis DNA gel of mt23S primer. (b) Quality electrophoresis DNA gel of mt23SB primer. Legend: CA: *T. algeriensis.* CC:

C. compressa. CH: C. humilis. CS: T. sauvageauana. CST: C. amentacea var. stricta. Figure S5: Quality electrophoresis DNA gel of tRNA-LysB primer.

Legend: CA: T. algeriensis. CC: C. compressa. CH: C. humilis. CS: T. sauvageauana. CST: C. amentacea var. stricta Figure S6: (a) Quality electrophoresis DNA gel of psbA-FR primer. (b) Quality electrophoresis DNA gel of psbA-FR2 primer. (c) Quality electrophoresis DNA gel of psbA-F2R primer. Legend: CA: T. algeriensis. CC: C. compressa. CH: C. humilis. CS: T. sauvageauana. CST: C. amentacea var. stricta. Figure S7: Maximum likelihood phylogenetic tree obtained with Nearest Neighbour Interaction on mt23S sequences of samples from Sargassaceae family. Values of branches represent maximum likelihood bootstrap support values (>50%). *Cystoseira sl* samples sequenced in this study labelled with 1 or 2. Figure S8: Maximum likelihood phylogenetic tree obtained with Nearest Neighbour Interaction on tRNA-Lys spacer sequences of samples from *Sargassaceae* family. Values of branches represent maximum likelihood bootstrap support values (>50%). Samples sequenced in this study labelled with 1. Figure S9: Maximum likelihood phylogenetic tree obtained with Nearest Neighbour Interaction on psbA sequences of samples from *Sargassaceae* family. Values of branches represent maximum likelihood bootstrap support values (>50%). Samples sequenced in this study labelled with 1, 2 or 3.

REFERENCES

1. Cánovas, F.; Mota, C.; Serrão, E.; Pearson, G. Driving south: A multi-gene phylogeny of the brown algal family *Fucaceae* reveals relationships and recent drivers of a marine radiation. *Evol. Biol.* **2011**, *11*, 371. [CrossRef] [PubMed]
2. Draisma, S.; Ballesteros, E.; Florence, R.; Thibaut, T. DNA Sequence Data Demonstrate the Polyphily of the Genus *Cystoseira* and Other *Sargassaceae* Genera (*Phaeophyceae*). *J. Phycol.* **2010**, *46*, 1329–1345. [CrossRef]
3. García-Fernández, A.; Bárbara, I. Studies of *Cystoseira* assemblages in Northern Atlantic Iberia. *An. Jard. Bot. Madr.* **2016**, *73*, e0352016. [CrossRef]
4. Guiry, M.D.; Guiry, G.M.; World-Wide Electronic Publication, National University of Ireland, Galway. AlgaeBase. Available online: https://www.algaebase.org (accessed on 6 June 2020).
5. Thibaut, T.; Blanfuné, A.; Markovic, L.; Verlaque, M.; Boudouresque, C.F.; Perret-Boudouresque, M.; Macic, V.; Bottin, L. Unexpected abundance and long- term relative stability of the brown alga *Cystoseira amentacea*, hitherto regarded as a threatened species, in the north-western Mediterranean Sea. *Mar. Pollut. Bull.* **2014**, *89*, 305–323. [CrossRef]
6. Hsü, K.J. *The Mediterranean Was a Desert: A Voyage of the Glomar Challenger*; Princeton University Press: Princeton, NJ, USA, 1982; p. 197. ISBN -10.
7. Boudouresque, C. Marine biodiversity in the mediterranean status of species populations and communities. Sci. Rep. Port Cros Nat. Park **2004**, 20, 97–146.

8. Bermejo, R.; Chefaoui, R.M.; Engelen, A.H.; Buonomo, R.; Neiva, J.; Ferreira-Costa, J.; Pearson, G.A.; Marbà, N.; Duarte, C.M.; Airoldi, L.; et al. Marine forests of the Mediterranean-Atlantic *Cystoseira tamariscifolia* complex show a southern Iberian genetic hotspot and no reproductive isolation in parapatry. *Sci. Rep.* **2018**, *8*, 10427. [CrossRef]

9. Orellana, S.; Hernández, M.; Sansón, M. Diversity of *Cystoseira sensu lato* (Fucales, *Phaeophyceae*) in the eastern Atlantic and Mediterranean based on morphological and DNA evidence, including *Carpodesmia* gen. emend. and *Treptacantha* gen. emend. *Eur. J. Phycol.* **2019**, *54*, 447–465. [CrossRef]

10. Bruno de Sousa, C.; Cox, C.J.; Brito, L.; Pavão, M.M.; Pereira, H.; Ferreira, A.; Ginja, C.; Campino, L.; Bermejo, R.; Parente, M.; et al. Improved Phylogeny of Brown Algae *Cystoseira* (Fucales) from the Atlantic-Mediterranean Region Based on Mitochondrial Sequences. *PLoS ONE* **2019**, *14*, e0210143. [CrossRef]

11. Cheminée, A.; Sala, E.; Pastor, J.; Bodilis, P.; Thiriet, P.; Mangialajo, L.; Cottalorda, J.M.; Francour, P. Nursery value of *Cystoseira* forests for Mediterranean rocky reef fishes. *J. Exp. Mar. Biol. Ecol.* **2013**, *442*, 70–79. [CrossRef]

12. Bulleri, F.; Benedetti-Cecchi, L.; Acunto, S.; Cinelli, F.; Hawkins, S.J. The influence of canopy algae on vertical patterns of distribution of low-shore assemblages on rocky coasts in the northwest Mediterranean. *J. Exp. Mar. Biol. Ecol.* **2002**, *267*, 89–106. [CrossRef]

13. Ballesteros, E.; Torras, X.; Pinedo, S.; García, M.; Mangialajo, L.; De Torres, M. A new methodology based on littoral community cartography dominated by macroalgae for the implementation of the European Water Framework Directive. *Mar. Pollut. Bull.* **2007**, *55*, 172–180. [CrossRef]

14. Badreddine, A.; Abboud-Abi Saab, M.; Gianni, F.; Ballesteros, E.; Mangialajo, L. First assessment of the ecological status in the Levant Basin: Application of the CARLIT index along the Lebanese coastline. *Ecol. Indic.* **2018**, *85*, 37–47. [CrossRef]

15. Oceana: Protecting the World's Oceans. Available online: https://europe.oceana.org /en/home (accessed on 17 March 2019).

16. Roberts, M. Active speciation in the taxonomy of the genus *Cystoseira* C. Agard. In *Modern Approaches to the Taxonomy of Red and Brown Algae*; Irvine, D.E.G., Price, J.H., Eds.; Academic Press: London, UK; New York, NY, USA, 1978; Volume Systematics.

17. Ballesteros, E.; Catalán, J. Flora y vegetación marina y litoral del Cabo de Gata y el Puerto de Roquetas de Mar (Almería): Primera aproximación. *An. Univ. Murcia* **1981**, *42*, 237–277.

18. Ballesteros, E.; Pinedo, S. Los bosques de algas pardas y rojas. In *Praderas y Bosques Marinos de Andalucía*; Luque, A.A., Templado, J.C., Eds.; Consejería de

Medio Ambiente, Junta de Andalucía: Sevilla, Spain, 2004; Volume 336, pp. 199–222.

19. Mangialajo, L.; Chiantore, M.; Cattaneo-Vietti, R. Loss of fucoid algae along a gradient of urbanization, and structure of benthic assemblages. *Mar. Ecol. Prog. Ser.* **2008**, *358*, 63–74. [CrossRef]
20. Sales, M.; Cebrian, E.; Tomas, F.; Ballesteros, E. Pollution Impacts and Recovery Potential in Three Species of the Genus *Cystoseira* (*Fucales, Heterokontophyta*). *Estuar. Coast. Shelf Sci.* **2011**, *92*, 347–357. [CrossRef]
21. Thibaut, T.; Blanfuné, A.; Boudouresque, C.F.; Verlaque, M. Decline and local extinction of Fucales in the French Riviera: The harbinger of future extinctions? *Mediterr. Mar. Sci.* **2015**, *16*, 206–224. [CrossRef]
22. Susini, M.; Thibaut, T.; Meinesz, A.; Forcioli, D. A preliminary study of genetic diversity in *Cystoseira amentacea* (C. Agardh) Bory var. *stricta* Montagne (*Fucales, Phaeophyceae*) using random amplified polymorphic DNA. *Phycologia* **2007**, *46*, 605–611. [CrossRef]
23. Gianni, F.; Bartolini, F.; Airoldi, L.; Ballesteros, E.; Francour, P.; Guidetti, P.; Meinesz, A.; Thibaut, T.; Mangialajo, L. Conservation and restoration of marine forests in the Mediterranean Sea and the potential role of Marine Protected Areas. *Adv. Oceanogr. Limnol.* **2013**, *4*, 83–101. [CrossRef]
24. De La Fuente, G.; Asnaghia, V.; Chiantorea, M.; Thrushb, S.; Poveroa, P.; Vassalloa, P.; Petrilloa, M.; Paoli, C. The effect of Cystoseira canopy on the value of midlittoral habitats in NW Mediterranean, an emergy assessment. *Ecol. Model.* **2019**, *404*, 1–11. [CrossRef]
25. Zazo, C.; Goy, J.L.; Dabrio, C.J.; Bardají, T.; Hillaire-Marcel, C.; Ghaleb, B.; González-Delgado, J.A.; Soler, V. Pleistocene raised marine terraces of the Spanish Mediterranean and Atlantic coasts: Records of coastal uplift, sea-level high-stands and climate changes. *Mar. Geol.* **2003**, *194*, 103–133. [CrossRef]
26. Terradas-Fernández, M.; Botana Gómez, C.; Valverde Urrea, M.; Zubcoff, J.; Ramos-Esplá, A.A. The dynamics of phytobenthos and its main drivers on abrasion platforms with vermetids (Alicante, Southeastern Iberian Peninsula). *Mediterr. Mar. Sci.* **2018**, *0*, 58–68. [CrossRef]
27. ballesteros, E. Structure and dynamics of the *Cystoseira caespitosa* Sauvageau (*Fucales, Phaeophyceae*) community in the North-Western Mediterranean. *Sci. Mar.* **1990**, *54*, 155–168.
28. Pena-Martín, C.; Juan, A.; Crespo, J.M. Evaluación del estado de conservación de comunidades fitobentónicas en el litoral de Alicante (España). In *Oceanos III Milenio*, 2nd ed.; Editorial C.P.D.: Madrid, Spain, 2002; Volume 1, p. 142.

29. Hereu, B.; Garcia-Rubies, A.; Linares, C.; Navarro, L.; Bonaviri, C.; Cebrian, E.; Diaz, D.; Garrabou, J.; Teixidó, N.; Zabala, M. Impact of the Sant Esteve's storm (2008) on the algal cover in infralittoral rocky photophillic communities. In *Assessment of the Ecological Impact of the Extreme Storm of Sant Esteve's Day (26 December 2008) on the Littoral Ecosystems of the North Mediterranean Spanish Coasts*; Mateo, M.A., Garcia-Rubies, T., Eds.; Centro de Estudios Avanzados de Blanes, Consejo Superior de Investigaciones Científicas: Blanes, Spain, 2012; pp. 123–143, Final Report (PIEC, 200430E599).
30. Mariani, S.; Cefalì, M.E.; Chappuis, E.; Terradas-Fernández, M.; Pinedo, S.; Torras, X.; Jordana, E.; Medrano, A.; Verdura, J.; Ballesteros, E. Past and present of Fucales from shallow and sheltered shores in Catalonia. *Reg. Stud. Mar. Sci.* **2019**, *32*, 100824. [CrossRef]
31. Sales, M.; Ballesteros, E. Shallow *Cystoseira* (Fucales: *Ochrophyta*) assemblages thriving in sheltered areas from Menorca (NW Mediterranean): Relationships with environmental factors and anthropogenic pressures. *Est. Coast. Shelf Sci.* **2009**, *84*, 476–482. [CrossRef]
32. Medrano, A. Macroalgal Forests Ecology, Long-Term Monitoring, and Conservation in a Mediterranean Marine Protected Area. Ph.D. Thesis, Universitat de Barcelona, Barcelona, Spain, 2020.
33. Terradas-Fernández, M. Caracterización de las Fitocenosis de las Plataformas de Abrasión con Vermétidos del Sureste Ibérico. Master's Thesis, Universidad de Alicante-Universidad Miguel Hernández, Alicante, Spain, 2014.
34. Ballesteros, E. *Els Vegetals i la Zonació Litoral: Espècies, Comunitats i Factors que Influeixen en la Seva Distribució*; Arxius Secció Ciències; Institud de Estudis Catalans: Barcelona, Spain, 1992; Volume 101, p. 1.
35. Ballesteros, E. Production of seaweeds in Northwestern Mediterranean marine communities: Its relation with environmental factors. *Sci. Mar.* **1989**, *53*, 357–364.
36. Menconi, M.; Benedetti-Cecchi, L.; Cinelli, F. Spatial and temporal variability in the distribution of algae and invertebrates on rocky shores in the northwest Mediterranean. *J. Exp. Mar. Biol. Ecol.* **1999**, *233*, 1–23. [CrossRef]
37. Sangil, C.; Sansón, M.; Afonso-Carrillo, J. Spatial variation patterns of subtidal seaweed assemblages along a subtropical oceanic archipelago: Thermal gradient vs herbivore pressure. *Est. Coast. Shelf Sci.* **2011**, *94*, 322–333. [CrossRef]
38. Davis, T.A.; Volesky, B.; Mucci, A. A review of the biochemistry of heavy metal bio-absorption by brown algae. *Water Res.* **2003**, *37*, 4311–4330. [CrossRef]
39. Cho, G.Y.; Lee, S.H.; Boo, S.M. A new Brown algal order, *Ishigeales* (*Phaeophyceae*) established on the basis of plastid protein-coding rbcL, psaA, and psbA region comparisons. *J. Phycol.* **2004**, *40*, 921–936. [CrossRef]

40. Coyer, A.; Hoarau, G.; Le-Secq, M.; Stam, W.; Olsen, J.L. A mtDNA-based phylogeny of the brown algal genus *Fucus* (*Heterokontophyta; Phaeophyta*). *Mol. Phylogenet. Evol.* **2006**, *39*, 209–222. [CrossRef]

41. Ministerio Para la Transición Ecológica y el Reto Demográfico, Gobierno de España. Available online: https://www.miteco.gob.es/es/costas/temas/proteccion-medio-marino/biodiversidad-marina/espacio s-marinos-protegidos/red-natura-2000-ambito-marino/bm_emprot_rednat2000_marino_LIC.aspx (accessed on 17 July 2020).

42. European Commission: Natura 2000 Network. Available online: 000/SDF.aspx?site=ESZZ16008#6 (accessed on 5 August 2020).

43. Ministerio de Transportes, Movilidad y Agenda Urbana. Gobierno de España. Plan Nacional de Ortofotografía Aérea: Instituto Geográfico Nacional. Available online: http://www.ign.es/wms/pnoa-historico?request=GetCapabilities&service=WMS (accessed on 7 August 2019).

44. Anderson, M.J.; Gorley, R.N.; Clarke, K.R. *PERMANOVA+ for PRIMER: Guide to Software and Statistical Methods*; PRIMER-E: Plymouth, UK, 2008.

45. Gómez-Garreta, A. *Flora Phycologica Iberica. Vol. 1. Fucales*; Gómez-Garreta, A., Ed.; Universidad de Murcia: Murcia, Spain, 2000.

46. Cormaci, M.; Furnari, G.; Catra, M.; Alongi, G.; Giaccone, G. Flora marina bentonica del Mediterraneo: Phaeophyceae. Boll. Accad. Gioenia Sci. Nat. **2012**, 45, 1–508.

47. Lane, C.; Mayes, C.; Druehl, L.; Saunders, G. A multi-gene molecular investigation of the kelp (*laminariales, Phaeophyceae*) supports substantial taxonomic re-organization. *J. Phycol.* **2006**, *42*, 493–512. [CrossRef]

48. Rogers, S.; Bendich, A. Extraction of DNA from plant tissues. In *Plant Molecular Biology Manual.*; Gelvin, S.B., Schilperoort, R.A., Verma, D.P.S., Eds.; Springer: Dordrecht, Netherlands, 1989.

49. Stothard, P. The sequence manipulation suite: JavaScript programs for analyzing and formatting protein and DNA sequences. *BioTechniques* **2000**, *28*, 1102–1104. [CrossRef] [PubMed]

50. Madeira, F.; Park, Y.M.; Lee, J.; Buso, N.; Gur, T.; Madhusoodanan, N.; Basutkar, P.; Tivey, A.R.N.; Potter, S.C.; Finn, R.D.; et al. The EMBL-EBI search and sequence analysis tools APIs in 2019. *Nucleic Acids Res.* **2019**, *47*, 636–641. [CrossRef]

51. Kumar, S.; Stecher, G.; Tamura, K. MEGA7: Molecular Evolutionary Genetics Analysis Version 7.0 for Bigger Datasets. *Mol. Biol. Evol.* **2016**, *33*, 1870–1874. [CrossRef]

52. Edgar, R.C. MUSCLE: Multiple sequence alignment with high accuracy and high throughput. *Nucleic Acids Res.* **2004**, *32*, 1792–1797. [CrossRef]

53. Saitou, N.; Nei, M. The neighbor-joining method: A new method for reconstructing phylogenetic trees. *Mol. Biol. Evol.* **1987**, *4*, 406–425. [CrossRef]

54. Tamura, K.; Nei, M.; Kumar, S. Prospects for inferring very large phylogenies by using the neighbor-joining method. *Proc. Natl. Acad. Sci. USA* **2004**, *101*, 11030–11035. [CrossRef]

55. Felsenstein, J. Confidence Limits on Phylogenies: An Approach Using the Bootstrap. *Evolution* **1985**, *39*, 783–791. [CrossRef]

56. Celis-Pla, P.S.M.; Bouzon, Z.L.; Hall-Spencer, J.M.; Schmidt, E.C.; Korbee, N.; Figueroa, F.L. Seasonal biochemical and photophysiological responses in the intertidal macroalga *Cystoseira tamariscifolia* (*Ochrophyta*). *Mar. Environ. Res* **2016**, *115*, 89–97. [CrossRef]

57. Steneck, R.; Dethier, M. A Functional Group Approach to the Structure of Algal-Dominated Communities. *Oikos* **1994**, *69*, 476–498. [CrossRef]

58. Falace, A.; Zanelli, E.; Bressan, G. Morphological and reproductive phenology of *Cystoseira compressa* (Esper) Gerloff & Nizzamuddin (*Fucales*, *Fucophyceae*) in the Gulf of Trieste (North Adriatic Sea). *Ann. Hist. Sci. Soc.* **2005**, *15*, 71–78.

59. Mancuso, F.P.; Strain, E.M.A.; Piccioni, E.; De Clerck, O.; Sarà, G.; Airoldi, L. Status of vulnerable *Cystoseira* populations along the Italian infralittoral fringe, and relationships with environmental and anthropogenic variables. *Mar. Pollut. Bullet.* **2018**, *129*, 762–771. [CrossRef] [PubMed]

60. Thibaut, T.; Pinedo, S.; Torras, X.; Ballesteros, E. Long-term decline of the populations of *Fucales* (*Cystoseira* spp. and *Sargassum* spp.) in the Albères coast (France, North-western Mediterranean). *Mar. Pollut. Bullet.* **2005**, *50*, 1472–1489. [CrossRef]

61. Roberts, M. Studies on marine algae of the British Isles. 6. Cystoseira foeniculacea (Linnaeus) Greville. *Br. Phycol. Bull.* **1968**, *3*, 547–564. [CrossRef]

62. INPN: Inventaire National du Patrimoine Culturel. Muséum National d'Histoire Naturelle. Available online: https://inpn.mnhn.fr (accessed on 5 August 2019).

63. Devescovi, M. Effects of bottom topography and anthropogenic pressure on northern Adriatic *Cystoseira* spp. (*Phaeophyceae*, Fucales). *Aquat. Bot.* **2015**, *121*, 26–32. [CrossRef]

64. Pizzuto, F. On the structure, typology and periodism of a *Cystoseira brachycarpa* J. Agardh *emend.* Giaccone community and of a duby community from the eastern coast of Sicily (Mediterranean Sea). *Plant Biosyst.* **1999**, *133*, 15–35. [CrossRef]

65. Puertos del Estado. Predicción de viento y oleaje: Boyas y Mareógrafos. Available online: http://www.puertos.es/es-es/oceanografia/Paginas/portus.aspx (accessed on 22 August 2019).'

66. Jona Lasinio, G.; Tullio, M.A.; Ventura, D.; Ardizzone, G.; Abdelahad, N. Statistical analysis of the distribution of infralittoral *Cystoseira* populations on pristine coasts of four Tyrrhenian islands: Proposed adjustment to the CARLIT index. *Ecol. Ind.* **2017**, *73*, 293–301. [CrossRef]'
67. Milazzo, M.; Badalamenti, F.; Riggioa, S.; Chemello, R. Patterns of algal recovery and small-scale effects of canopy removal as a result of human trampling on a Mediterranean rocky shallow community. *Biol. Cons.* **2004**, *117*, 191–202. [CrossRef]
68. piattelli, M. Chemistry and taxonomy of Sicilian *Cystoseira* species. *New J. Chem* **1990**, *14*, 777–782.
69. Valls, R.; Piovetti, L.; Praud, A. The use of diterpenoids as chemotaxonomic markers in the genus *Cystoseira*. *Hydrobiol.* **1993**, *260*, 549–556. [CrossRef]
70. Gallardo, T.; Bárbara, I.; Afonso-Carrillo, J.; Bermejo, R.; Altamirano, M.; Gómez-Garreta, A.; Barceló Martí, C.; Rull-Lluch, J.; Ballesteros, E.; De la Rosa, J. Nueva lista crítica de las algas bentónicas marinas de España. *Bol. Soc. Esp. Ficol.* **2016**, *51*, 7–52.
71. Bosc, E.; Bricaud, A.; Antoine, D. Seasonal and inter-annual variability in algal biomass and primary production in the Mediterranean Sea, as derived from 4 years of SeaWiFS observations. *Global Biogeochem. Cycl.* **2004**, *18*. [CrossRef]
72. Amico, V.; Giaccone, G.; Colonna, P.; Mannino, A.M.; Randazzo, R. Un nuovo approccio allo studio della sistematica del genere Cystoseira C. Agardh (Phaeophyta, Fucales). *Bol. Sed. Accad. Gioenia Sci. Nat. Catania* **1985**, *18*, 887–985.
73. Pla, M.; Gómez-Garreta, A. Chorology of the genus Cystoseira C. Agardh (Phaeophyceae, Fucales). *An. Jard. Bot. Madr.* **1989**, *46*, 89–97.
74. Mangialajo, L.; Ruggieri, N.; Asnaghi, V.; Chiantore, M.; Povero, P.; Cattaneo-Vietti, R. Ecological status in the Ligurian Sea: The effect of coastline urbanization and the importance of proper reference sites. *Mar. Pollut. Bull.* **2007**, *55*, 30–41. [CrossRef]
75. Díaz-Valdés, M.; Abellán, E.; Izquierdo, A.; Ramos-Esplá, A.A. Estudio preliminar de comunidades de macroalgas de la franja litoral rocosa de la Comunidad Valenciana dentro de la Directiva Marco del Agua. In Proceedings of the XIV Simposio Ibérico de Estudios de Biología Marina, Barcelona, Spain, 12–15 September 2006.
76. Blanfuné, A.; Boudouresque, C.F.; Verlaque, M.; Thibaut, T. The fate of *Cystoseira crinita*, a forest-forming Fucale (*Phaeophyceae, Stramenopiles*): In France (North Western Mediterranean Sea). *Estuar. Coast. Shelf Sci.* **2016**, *181*, 196–208. [CrossRef]

CHAPTER-6

LABORATORY CULTURE-BASED CHARACTERIZATION OF THE RESTING STAGE CELLS OF THE BROWN-TIDE-CAUSING PELAGOPHYTE, *AUREOCOCCUS ANOPHAGEfFERENS*

1. INTRODUCTION

The non-motile, picoplanktonic (2–3 μm) pelagophyte *Aureococcus anophagefferens* Hargraves et Sieburth has caused numerous ecosystem disruptive algal blooms (EDABs), commonly known as "brown tides", in U.S. estuaries since 1985 [1,2]. This species lacked morphological features easily distinguishing it from other similar sized forms under light microscopy, but ultrastructural observations exhibited that each cell has a single chloroplast, nucleus, and mitochondrion and an unusual exocellular polysaccharide-like layer [2]. Although nontoxic to humans, *A. anophagefferens* blooms have significantly negative effects on the seagrass beds, shellfish industry, algal grazers, and zooplankton in the affected area [1]. In the summer of 1985, the first *A. anophagefferens* blooms have been reported in several estuaries of the north-eastern U.S. [2–4], and there has been an extra-large extension of the known range of *A. anophagefferens* along the U.S. East Coast since then, from Florida north to New Hampshire [5]. *Aureococcus anophagefferens* blooms have been reported for the first time in Saldanha Bay, South Africa in 1997 and several more during 1998–2003 [6,7]. Surprisingly, *A. anophagefferens* suddenly bloomed in the coastal waters of Qinhuangdao, China in early summer from 2009 to 2011, which caused significant negative impacts on the shellfish mariculture industry and large economic losses [8,9]. Its discontinuous global distribution and seemingly rapid geographic expansion, however, has been a vital but highly controversial issue [10]. It was hypothesized that *A. anophagefferens* was possibly introduced both within and outside (South Africa) the U.S. via ships' ballast water [11]. Previous studies have indeed shown that *A. anophagefferens* could endure the prolonged darkness for 30 days [12,13], a characteristic facilitating its anthropogenic transport by ships' ballast water. As a novel brown-tide-forming species in China, whether *A. anophagefferens* was an alien species recently introduced to China via anthropogenic transport processes or has been an indigenous species existing with a background abundance prior to the first reported bloom has thus become a question of ecological significance. A comparison of nearly the entire length of the 18S ribosomal RNA (rRNA) gene sequences of *A. anophagefferens* from Qinhuangdao, China with that from the USA has shown that there was relatively little genetic variability (0–6 bp differences) [9], which suggested that *A. anophagefferens* was possibly an alien species in China [10,14].

In nature, microalgae have evolved many survival strategies to withstand adverse environments such as forming resting cysts or spores [15–17], reduction in their metabolic rates [18,19], or reliance on the alternation of nutrition modes [20,21]. Inactive or resting stages are common in the life history of many microalgae (diatoms, dinoflagellates, haptophytes, green algae, cyanobacteria, raphidophytes, chrysophytes, euglenophytes, and cryptophytes) [22,23] and may provide tolerance to unfavourable conditions [17,22,23]. Resting stages refer to all types of cells that greatly reduce metabolic rate, cease cell division, but remain viable [24], and are often characterized by distinct morphological and compositional changes in cells such as thickened membranes and the formation of starch granules [25], lipid droplets [26], or red accumulation bodies [16]. Resting cells that undergo changes in morphology and physiology leading to a resting stage do not undergo major changes in cell surface or enclosing cell structures, which distinguishes them from resting cysts or spores [27]. The life history (life cycle) of *A. anophagefferens* had not been uncovered [2] until recently when a resting stage in another brown-tide-causing species, *Aureoumbra lagunensis*, was firstly described in pelagophytes from laboratory cultures [28], which provided insights into the life cycle of *A. anophagefferens*. One of the key questions regarding ballast water hypothesis of *A. anophagefferens* introduction is whether or not the species has a dormant stage in its life history, because it has been well established that all microalgal species that have been proved to be transported and introduced via ships' ballast tanks are cyst-forming species [10]. Therefore, answering whether or not *A. anophagefferens* has a resting stage will greatly help to elucidate the origin and seeding of *A. anophagefferens* found in China and its apparent geographic expansion around the world.

In our recently published work [29], we proved that *A. anophagefferens* has a resting stage in its life history via germination experiments of a sediment sample collected from the coast of Qinhuangdao, China, where blooms of this species occurred, and also found that this species has an extremely wide geographic distribution (a range of ~30° in latitude, ~15.7° in longitude) and a more than 1500-year presence in China and thus is not an alien species. With the support of mining the supplementary dataset in a recent work [30], we also found that *A. anophagefferens* in fact distributes globally [29]. As *A. anophagefferens* is a small-sized (2–3 μm) and morphologically simple alga, we were not able to recognize and isolate single resting stage cells (RSCs) from the marine sediment sample for morphological observations and also failed in establishing cultures from the germlings in germination experiments. A follow-up work using laboratory cultures is, therefore, highly desirable to characterize the morphology and physiological status of the RSCs and to observe the transformation process of resting stage cells into vegetative cells. The present study describes the basic morphological characteristics and physiological status of RSCs as well as their ability to revert from resting stage back to vegetative growth based on the long laboratory-raised culture of *A. anophagefferens* strain CCMP1984 (Provasoli-Guillard National Center for Culture of Marine Phytoplankton (West Boothbay Harbor, ME, USA), CCMP). We believe the knowledge of these aspects will be significantly important in understanding the basic biology and ecology of this notorious EDABs-causing pelagophyte.

2. MATERIALS AND METHODS

2.1 Culture Information and RSC Formation

The culture of *A. anophagefferens* strain CCMP1984 was obtained from Stony Brook University. The culture was routinely maintained in natural seawater-based f/2-Si medium [31] supplied with 10^{-8} M (final concentration) selenium (salinity 32) at 21°C in an incubator with a 12:12 h light: dark cycle and an irradiance of 100 μmol photons $m^{-2} \cdot s^{-1}$. An antibiotic solution (10,000 IU·mL^{-1} penicillin and 10,000 μg·mL^{-1} streptomycin, Solarbio, Beijing, China) was added into the medium immediately (final concentration 2%) before inoculation to discourage growth of bacteria. The RSCs were produced in the routinely maintained batch cultures of CCMP1984 at the late stationary growth stages (>20 days) and particularly in the cultures that were cultured to the late exponential phase and then placed in the dark for a few weeks or longer. The morphological observations on RSCs below were conducted with RSCs produced this way, except for that stated otherwise.

2.2 Basic Morphological Observation

To observe and contrast the general morphology of vegetative cells (VCs) and RSCs of\ *A. anophagefferens*, 5 mL VCs and RSCs were added into a 10 mL conical centrifuge tube, respectively. After being rinsed with filtered, sterilized seawater three times and diluted to appropriate concentration, the samples were transferred to clean slides and observed under an optical microscope (BX53, Olympus, Japan).

2.3 BODIPY 505/515 Fluorescence Staining

To detect lipid-rich structures in the RSCs of *A. anophagefferens*, an optimized BODIPY 505/515 fluorescence staining method was used by following the protocol described previously [32] with minor modifications, as described below: BODIPY 505/515 (4, 4-difluro-1, 3, 5, 7-tetramethyl-4-bora-3a, 4adiaza-s-indacene; Invitrogen Molecular Probes, USA) was dissolved in DMSO as a stock solution (0.5 mg·mL⁻1) and stored in a brown sample bottle away from light. RSCs of the *A. anophagefferens* culture were collected onto a polycarbonate filter (pore size 0.2 μm) and rinsed with 500 μL 0.22 μm-filtered, sterilized seawater three times. The cells were then stained with BODIPY 505/515 (final concentration 0.1 μg·mL⁻1) and incubated at 21 °C for 10 min in the dark. After incubation, the samples were centrifuged at 2000× *g* for 3 min and rinsed three times using filtered, sterilized seawater. The cells were then resuspended in 100 μL filtered, sterilized seawater and were transferred to clean slides for observation under an inverted microscope (IX73, Olympus, Japan) equipped with dichroic filters BP450–480 nm and BA >515 nm (Olympus, Japan).

2. 4 Adenosine Triphosphate (ATP) Assays

To characterize intracellular metabolic activity of RSCs and normal vegetative cells, the cellular content of ATP was quantified by using an Enhanced ATP assay Kit (S0027; Beyotime Biotechnology, Shanghai, China) following the manufacturer's protocol. Three 50 mL samples of RSCs and VCs of *A. anophagefferens* were collected onto a polycarbonate filter (pore size 0.2 μm) and rinsed with 500 μL filtered, sterilized seawater three times and then to be determined for the ATP content. The cellular ATP content was estimated according to an ATP standard curve and expressed as nmol/10^7 cells.

2.5 Germination Experiments

To evaluate the ability of RSCs to revert back to vegetative growth after being stored in a prolonged (3 months) darkness at 4 °C, which resembles winter environmental conditions of marine sediments, germination experiments were conducted using both culture plates and Erlenmeyer flasks as described below. RSCs in culture plates were first stored in darkness at 4 °C for 1, 2, and 3 months, with addition of antibiotics mixture to a final 2% concentration (10,000 IU·mL⁻1 penicillin and 10,000 μg·mL^{-1} streptomycin) in darkness every 10 d to discourage growth of bacteria [33]. Once the RSCs were stored for 1, 2, and 3 months, a subset of RSCs in plates was transferred to three Erlenmeyer flasks containing fresh medium, placed in the incubator for RSC germination with normal conditions used for culture maintenance, and monitored for cell density on day 0, 20, and 30, respectively. Moreover, daily observations on the germination of RSCs and subsequent vegetative growth were performed using culture plates. The RSCs were inoculated into wells of culture plates containing fresh culture medium, placed at normal culturing conditions, and examined daily with an Olympus microscope (IX73, Tokyo, Japan).

2.6 Statistical Analyses

For the germination experiments and intracellular metabolic activity of RSCs and normal vegetative cells, one-way ANOVA was used to assess the differences among cell densities of RSCs and vegetative cells at different incubation time, and G-test for ATP contents in RSCs and vegetative cells. In all cases, significance levels were set at $p < 0.05$.

3. RESULTS

3.1 Morphological Characteristics of Resting Cells

In the batch cultures of CCMP1984 that was routinely maintained as described above (salinity 32, 21 °C, 12:12 h photoperiod, an irradiance of 100 μmol photons m^{-2}·s^{-1}, with 2% antibiotics mixture), RSCs were generally formed at the late stationary growth stages (>20 days). Similar to normal vegetative cells, RSCs were spherical and of a relatively smooth surface (Figure 1b,d). Contrast to a size of 2–3 μm for vegetative cells, RSCs were characterized by an approximately cell diameter two-fold larger than vegetative cells (4–6 μm; Figure 1d). In RSCs, the granular cytoplasm became denser, the plastid became pale, fewer, and densely aggregated, and the vacuole space expanded (Figure 1b).

All RSCs contained a large and conspicuous red droplet-like organelle (Figure 1d), which was proven to be lipid-enriched "droplet" by BODIPY 505/515 fluorescence staining and appeared green under epi-fluorescence microscopy (Figure 2b). These morphological features are generally similar to RSCs of another pelagophyte, *A. lagunensis* [28].

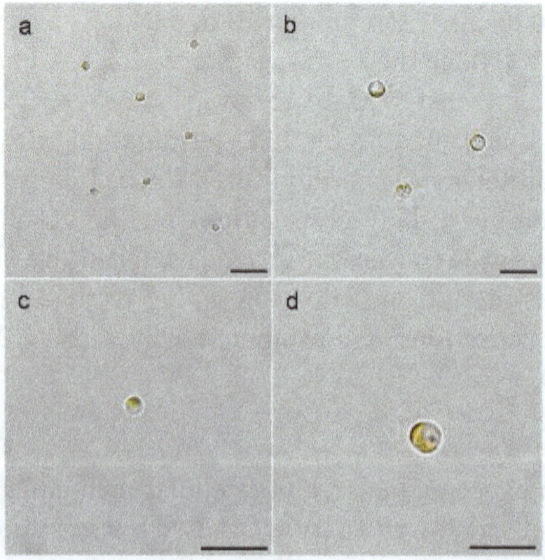

Figure 1. Light microscopic observations of the resting stage cells (RSCs) and normal vegetative cells (VCs) of *A. anophagefferens* strain CCMP1984. (a,c) VCs at magnifications of 400 (a) and 1000 (c); (b,d) RSCs at magnifications of 400 (b) and 1000 (d). Scale bars = 10 μm.

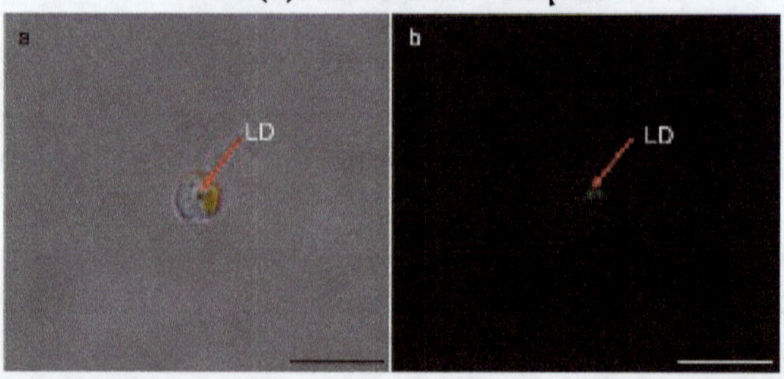

Figure 2. Lipid-enriched droplets (Ld) in the RSCs of *A. anophagefferens* strain CCMP1984 as indicated by the arrows. The RSCs of *A. anophagefferens* shown in (a) was observed under bright field microscopy, while (b) shows the same cell, but it was observed under epi-fluorescence microscopy (filters settings BP450–480 nm for excitation and BA > 515 nm for emission), as it was stained by BODIPY 505/515. Scale bars = 10 μm.

3.2 Metabolic Activity of RSCs as Indicated in Cellular ATP Content

Resting stage cells of *A. anophagefferens* were characterized by a significantly lowered cellular ATP content relative to that of normal vegetative cells, while the content of ATP in normal vegetative cells was 31.6 ± 0.51 nmol/10^7 cells, which was significantly higher than that in RSCs (2.9 ± 0.05 nmol/10^7 cells; Table 1; $p < 0.05$). This 9.9-fold drop in cellular ATP content from normal vegetative cells to RSCs indicated a significantly reduced metabolic activity in RSCs.

Table 1. Measurements of intracellular content of ATP in the vegetative cells and resting stage cells of *A. anophagefferens* strain CCMP1984. Standard deviation (± SD) of the data from three independent experiments. For each experiment, three technical replicates were performed.

Cell Type	ATP Content (nmol/107 Cells)
Resting stage cells	2.9 ± 0.5
Vegetative cells	31.6 ± 5.1

3.3 Resumption of Vegetative Growth from Resting State

To investigate the germination or transformation processes of the RSCs into vegetative cells, a time series observation was performed via culturing RSCs in fresh medium under 21 °C at 100 μmol photons m⁻2·s⁻1. While we did not observe the details of germination processes of RSCs into vegetative cells due to the small sizes of RSCs and a quick process, we observed that almost all RSCs transformed into vegetative cells within 10 days, resumed rapid vegetative growth, and reached a high cell density within 20 days (Figure 3). It was noteworthy that the germination of RSCs was a direct morphological transformation and quick resumption of cell division, similar to that in another pelagophyte *A. lagunensis* [28], but different from other species of Ochrophyta, e.g., three types of statospores (uninucleate, asexual; binucleate, asexual (potentially autogamic); binucleate, sexual (zygotic)) in the chrysophyte *Dinobryon cylindricum* [34,35], new germling escaped from the structure underneath the lid of germination pore in *Heterosigma akashiwo* (Raphidophyceae) [36], and dinoflagellates, in which a germination process within the cyst wall and a germling release through an archeopyle that has been numerously observed in the germination of resting cysts [33,37,38].

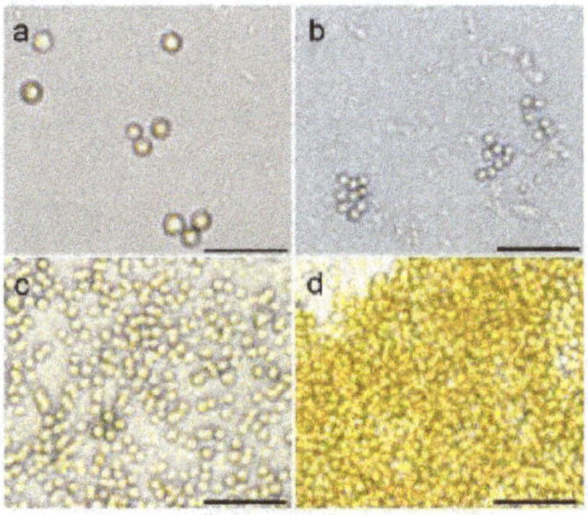

Figure 3. Germination of resting stage cells of *A. anophagefferens* strain CCMP1984 and subsequent rapid growth. The observations were conducted under the light microscope with the same magnification and sample volume (10 mL). Resting stage cells completely reverted to vegetative cells in about 10 days after being transferred to fresh f/2-Si medium and incubated under favorable condition (21

°C) and grew to a very high cell density in ~20 days. (a) 0 day, (b) 10th day, (c) 20th day, (d) 30th day. Scale bars = 10 μm.

3.4 The Ability of RSCs to Survive Prolonged Darkness and Coldness

The ability of RSCs of *A. anophagefferens* to survive prolonged darkness and coldness and to resume growth was assessed via storing RSCs in darkness at 4 °C for one to three months and incubating RSCs in fresh medium under routine culture maintenance conditions. Although the number of RSCs decreased significantly over time during the storage, which died during the storage because of light or nutrient shortage, or decomposed by bacteria, only about 28% of RSCs survived 3 months of darkness coldness (Figures 4 and 5). Under normal culture conditions, RSCs that survived 1, 2, and 3 months of storage could germinate and grow to a high cell density within 20 days (Figure 5). While RSCs declined from 7.8×10^5 cells·mL⁻1 to 2.2×10^5 cells·mL⁻1 after 3 months of storage ($p < 0.05$), the density of vegetative cells increased to 3.58×10^6 (~16 folds) and 8.31×10^6 (~38 folds) cells·mL⁻1 on day 20 and 30, respectively ($p < 0.05$), when RSCs were incubated in fresh medium under routine culturing conditions (Figure 5).

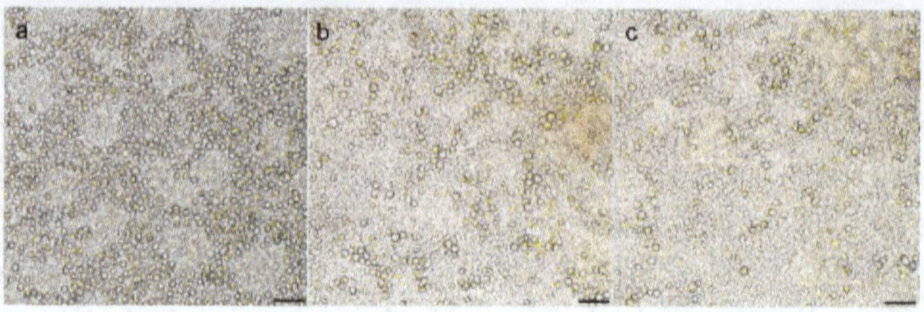

Figure 4. Appearance of resting stage cells of *A. anophagefferens* strain CCMP1984 after being stored in prolonged periods (a) 1 month, (b) 2 months, and (c) 3 months in darkness at 4 °C. Note that most of the cells died during the storage. Scale bars = 10 μm.

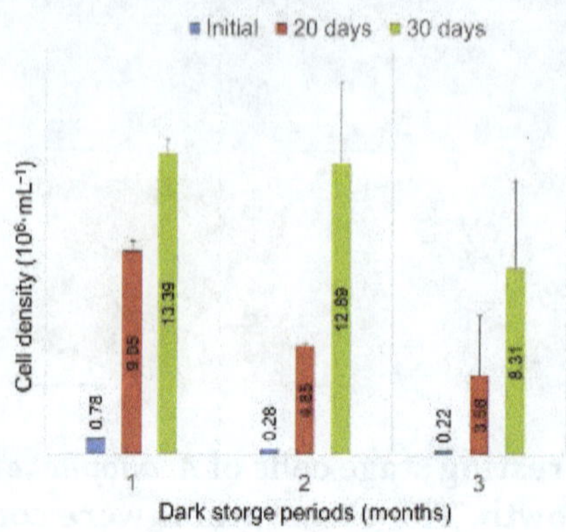

Figure 5. Cell density recovery of *A. anophagefferens* strain CCMP1984 after the cultures were stored in the dark at 4 °C for 1 to 3 months. Cell densities were

quantified microscopically at the time points of 0 (immediately after inoculation), 20, and 30 d after the dark-stored resting cells were inoculated to fresh f/2-Si medium and incubated at the conditions (12:12 h photoperiod, 21 °C, 100 μmol photons $m^{-2} \cdot s^{-1}$) that were used for the routine maintenance of cultures. Each data point was shown as mean ± SD (standard deviation) from triplicate samples.

4. DISCUSSION

Aureococcus anophagefferens blooms (also known as brown tides) have caused destructive environmental impacts and massive losses in aquaculture in the past three decades [39–42]. For a long period, however, brown tides were only reported in the U.S. East Coast until the first brown tides occurred at Saldanha Bay, South Africa in 1997 [6] and suddenly occurred in the coastal waters of Qinhuangdao, China in 2009 [9]. A hypothesis was put forward that ballast water vectoring could provide an explanation for the geographical spreading behavior of *A. anophageffrens* [10,11]. However, there have been many uncertainties regarding the hypothesis. Focal, unresolved questions include whether or not the species can produce a resting stage and/or has been existing at lower abundance in regions where brown tides seemed to occur suddenly [10]. It has been found that *A. anophageffrens* could survive prolonged darkness for more than 30 days in darkness, a characteristic facilitating its anthropogenic transport by ships [12]. A physiological change and adjustment of cells has been suggested to contribute to the capability [13]. It had been generally believed that pelagophytes may not form dormant cysts or resting stage cells [1] until the recent description of a resting stage in the pelagophyte *A. lagunensis* [28]. We recently proved the existence of a resting stage in *A. anophageffrens* via sediment germination experiments and also found that the alga in fact distributes globally [29]. Although sediment had been stored in darkness at 4 °C for more than one year prior to the germination experiments, the RSCs of *A. anophageffrens* in the sediment sample could recover (or "germinate") to normal vegetative growth within a reasonable time (a few weeks) when returned to favourable conditions [29]. However, it is noteworthy that our discovery of RSCs in *A. anophageffrens* did not support the "recent invasion of alien species" hypothesis for the sudden appearance of brown tides in Bohai Sea, because the sedimental record of *A. anophageffrens* demonstrated that the species has been existing in Bohai Sea for at least 1500 years [29]. On the other hand, this finding also does not disprove the possible roles of RSCs in the initiation and geographic expansion of brown tides and in the general ecology of this species. It is still possible that the RSCs provided the facility for *A. anophageffrens* to seed annual brown tides and to expand its geographical distribution via either artificial (e.g., ships' ballast water) or natural pathways up to a globally cosmopolitan distribution [29]. Therefore, following the above mentioned discovery of RSCs in *A. anophageffrens*, we characterized the general morphological features, physiological status, and the ability to resume vegetative growth of RSCs on the basis of observations on the laboratory-raised culture CCMP1984 in the present study.

Once the culture of *A. anophageffrens* that had been cultured to the late exponential phase was placed in the dark for a few weeks, almost all vegetative cells could transform into RSCs. Consumption of nutrients in the culture medium [34,43], growth of bacteria, and sustained darkness [26,44,45] may all be the cues driving the transformation of vegetative cells into RSCs. Whether or not a temperature stress (low or high and/or a rapid change) could drive the formation of RSCs also needs to be considered in future investigations [15,46]. In the investigation on *A. lagunensis*, vegetative cells were observed to quickly transform into resting cells upon exposure to higher temperatures [28].

RSCs of *A. anophageffrens* have a relatively smooth cell surface, which is different from the resting cysts or spores of many species that undergo more substantial changes in cell surface and/or surface structures including thickened walls or modified enclosing inorganic or organic thecae, as seen in dinoflagellates and some diatoms [27,47–51]. Similar to vegetative cells, RSCs of *A. anophageffrens* are spherical in shape, but are so different from vegetative cells in cell size: RSCs are significantly larger (doubled in cell diameter). RSCs of *A. anophageffrens* were also characterized by aggregated plastid and a large, red, and lipid-enriched droplet, both similar to the RSCs of *A. lagunensis* [28]. These red, lipid-enriched droplets may be biochemically similar to the red accumulation bodies found in the resting cysts of many dinoflagellates, although the latter has been hypothesized to be enriched with pigments [17,52,53]. It is likely that the red droplet in RSCs of *A. anophageffrens* is a lipid reserve used as energy storage for cellular metabolism during prolonged resting periods and germination [26]. Furthermore, lipid is an essential component of the membranes of microorganisms to control membrane fluidity and permeability [54–58]. Given that cell transition from resting stage to vegetative growth involves going from larger cells with a low ratio of surface to volume (cell membrane) to smaller cells with a high ratio, lipid may be utilized for membrane formation when vegetative growth resumes. Microalgae may increase lipid content in response to stress at the expense of protein and polysaccharides, which are considered as short-term energy storage [59]. Considering the findings mentioned above, we speculate that *A. anophageffrens* cells may accumulate photosynthetic products during normal growth stage and produce lipid-enriched droplets from the photosynthetic products under adverse conditions such as nutrient shortage and prolonged darkness as an adaptive mechanism for persistence as a resting cell. This process seems to take one or two weeks as indicated in the formation of RSCs. In this study, RSCs of *A. anophageffrens* readily reverted back to vegetative cells when optimal conditions provided after storage for 3 months (at 4°C in darkness).

Similarly, the RSCs of *A. anophageffrens* collected from marine sediment could also germinate and resume vegetative growth swiftly under favourable conditions even after having been stored in darkness and coldness for longer than one year [29]. These results indicated that the metabolic activity of RSCs was at a low level during the resting period, which made an extended resting stage possible in the field. The low metabolic activity of RSCs was also evidenced by an about 10-fold difference in the cellular content of ATP between the vegetative cells and RSC.

Almost all RSCs that were stored in darkness and coldness for 1–3 months could "germinate" (transform) into vegetative cells within a few to 10 days and then begin to divide and grow rapidly to a high cell density in ~20 days when they were placed at normal culturing conditions. This indicated that RSCs of *A. anophageffrens* are not in a deep dormancy and thus sensitive to alterations of environmental conditions. Since we did not capture the details of the cell transformation processes of RSCs when they were germinating into vegetative cells due to the resolution limitation of light microscopy for the small-sized RSCs and the swiftness of process, it is important that future studies investigate further the details of the cellular transformation during germination and whether the resting cells are formed via sexual or asexual processes, as has been conducted in dinoflagellates and many other microalgal groups [33,37,38,48,51].

5. CONCLUSIONS

Since *Aureococcus anophageffrens* is a globally cosmopolitan species [29], it is, therefore, changes in the environmental setting where the species forms blooms that account for the occurrences of *A. anophageffrens* brown tides. Our results indicate that the RSCs of *A. anophageffrens* are in a dormant state that differ both morphologically and physiologically from the vegetative cells. The ability of *A. anophageffrens* to form RSCs likely enables the species to survive unfavourable conditions and inoculate annual blooms and may have facilitated its geographic expansion and its globally cosmopolitan distribution that we recently documented. In conclusion, the novel knowledge of RSCs in the life history will be significantly important not only in obtaining insights into the basic biology of *A. anophagefferens*, but also the ecology of this brown-tide-causing pelagophyte.

REFERENCES

1. Gobler, C.J.; Sunda, W.G. Ecosystem disruptive algal blooms of the brown tide species, *Aureococcus anophagfferens* and *Aureoumbra lagunensis*. *Harmful Algae* **2012**, *14*, 36–45. [CrossRef]

2. Sieburth, J.M.; Johnson, P.W.; Hargraves, P.E. Ultrastructure and ecology of *Aureococcus anophageferens* gen. et sp. nov. (Chrysophyceae): The dominant picoplankter during a bloom in Narragansett Bay, Rhode Island, summer 1985. *J. Phycol.* **1988**, *24*, 416–425. [CrossRef]

3. Cosper, E.M.; Dennison, W.; Milligan, A.; Carpenter, E.J.; Lee, C.; Holzapfel, J.; Milanese, L. An examination of environmental factors important to initiating and sustaining "Brown Tide" blooms. In *Novel Phytoplankton Blooms: Causes and Impacts of Recurrent Brown Tides and Other Unusual Blooms*; Cosper, E.M., Bricelj, V.M., Carpenter, E.J., Eds.; Springer: Berlin/Heidelberg, Germany, 1989; pp. 317–340.
4. Olsen, P.S. Development and distribution of a brown-water algal bloom in Barnegat Bay, New Jersey with perspective on resources and other red tides in the region. In *Novel Phytoplankton Blooms: Causes and Impacts of Recurrent Brown Tides and Other Unusual Blooms*; Cosper, E.M., Bricelj, V.M., Carpenter, E.J., Eds.; Springer: Berlin/Heidelberg, Germany, 1989; pp. 189–212.
5. Popels, L.C.; Cary, S.C.; Hutchins, D.A.; Forbes, R.; Pustizzi, F.; Gobler, C.J.; Coyne, K.J. The use of quantitative polymerase chain reaction for the detection and enumeration of the harmful alga *Aureococcus anophagefferens* in environmental samples along the United States East Coast. *Limnol. Oceanogr. Methods* **2003**, *1*, 92–102. [CrossRef]
6. Probyn, T.; Pitcher, G.; Pienaar, R.; Nuzzi, R. Brown tides and mariculture in Saldanha Bay, South Africa. *Mar. Pollut. Bull.* **2001**, *42*, 405–408. [CrossRef]
7. Probyn, T.A.; Bernard, S.; Pitcher, G.C.; Pienaar, R.N. Ecophysiological studies on *Aureococcus anophagefferens* blooms in Saldanha Bay, South Africa. *Harmful Algae* **2010**, *9*, 123–133. [CrossRef]
8. Kong, F.; Yu, R.; Zhang, Q.; Yan, T.; Zhou, M. Pigment characterization for the 2011 bloom in Qinhuangdao implicated "brown tide" events in China. *Chin. J. Oceanol. Limnol.* **2012**, *30*, 361–370. [CrossRef]
9. Zhang, Q.-C.; Qiu, L.-M.; Yu, R.-C.; Kong, F.-Z.; Wang, Y.-F.; Yan, T.; Gobler, C.J.; Zhou, M.-J. Emergence of brown tides caused by *Aureococcus anophagefferens* Hargraves et Sieburth in China. *Harmful Algae* **2012**, *19*, 117–124. [CrossRef]
10. Smayda, T.J. Reflections on the ballast water dispersal—Harmful algal bloom paradigm. *Harmful Algae* **2007**, *6*, 601–622. [CrossRef]
11. Doblin, M.A.; Popels, L.C.; Coyne, K.J.; Hutchins, D.A.; Cary, S.C.; Dobbs, F.C. Transport of the harmful bloom alga *Aureococcus anophagefferens* by oceangoing ships and coastal boats. *Appl. Environ. Microbiol.* **2004**, *70*, 6495–6500. [CrossRef]
12. Popels, L.C.; Hutchins, D.A. Factors affecting dark survival of the brown tide alga *Aureococcus anophagefferens* (Pelagophyceae). *J. Phycol.* **2002**, *38*, 738–744. [CrossRef]
13. Popels, L.C.; MacIntyre, H.L.; Warner, M.E.; Zhang, Y.; Hutchins, D.A. Physiological responses during dark survival and recovery in *Aureococcus anophagefferens* (Pelagophyceae). *J. Phycol.* **2007**, *43*, 32–42. [CrossRef]

14. Bolch, C.J.S.; de Salas, M.F. A review of the molecular evidence for ballast water introduction of the toxic dinoflagellates *Gymnodinium catenatum* and the *Alexandrium* "*tamarensis complex*" to Australasia. *Harmful Algae* **2007**, *6*, 465–485. [CrossRef]

15. Anderson, D.M.; Coats, D.W.; Tyler, M.A. Encystment of the dinoflagellate *Gyrodinium uncatenum*: Temperature and nutrient effects. *J. Phycol.* **1985**, *21*, 200–206. [CrossRef]

16. Matsuoka, K.; Fukuyo, Y. *Technical Guide for Modern Dinoflagellate Cyst Study*; WESTPAC-HAB/WEATPAC/IOC; Japanese Society for the Promotion of Science: Tokyo, Japan, 2000; pp. 1–97.

17. Bravo, I.; Figueroa, R.I. Towards an ecological understanding of dinoflagellate cyst functions. *Microorganisms* **2014**, *2*, 11–32. [CrossRef]

18. Smayda, T.J.; Mitchell-Innes, B. Dark survival of autotrophic, planktonic marine diatoms. *Mar. Biol.* **1974**, *25*, 195–202. [CrossRef]

19. Dehning, I.; Tilzer, M.M. Survival of *Scenedesmus acuminatus* (Chlorophyceae) in darkness. *J. Phycol.* **1989**, *25*, 509–514. [CrossRef]

20. Deventer, B.; Heckman, C.W. Effects of prolonged darkness on the relative pigment content of cultured diatoms and green algae. *Aquat. Sci.* **1996**, *58*, 241–252. [CrossRef]

21. Wilken, S.; Huisman, J.; Naus-Wiezer, S.; Van Donk, E. Mixotrophic organisms become more heterotrophic with rising temperature. *Ecol. Lett.* **2013**, *16*, 225–233. [CrossRef] [PubMed]

22. Ellegaard, M.; Ribeiro, S. The long-term persistence of phytoplankton resting stages in aquatic 'seed banks'. *Biol. Rev.* **2018**, *93*, 166–183. [CrossRef]

23. Figueroa, R.I.; Estrada, M.; Garcés, E. Life histories of microalgal species causing harmful blooms: Haploids, diploids and the relevance of benthic stages. *Harmful Algae* **2018**, *73*, 44–57. [CrossRef] [PubMed]

24. von Dassow, P.; Montresor, M. Unveiling the mysteries of phytoplankton life cycles: Patterns and opportunities behind complexity. *J. Plankton Res.* **2011**, *33*, 3–12. [CrossRef]

25. Chapman, D.V.; Dodge, J.D.; Heaney, S.I. Cyst formation in the freshwater dinoflagellate *Ceratium hirundinella* (Dinophyceae). *J. Phycol.* **1982**, *18*, 121–129. [CrossRef]

26. Anderson, O.R. The ultrastructure and cytochemistry of resting cell formation in *Amphora coffaeformis* (Bacillariophyceae). *J. Phycol.* **1975**, *11*, 272–281. [CrossRef]

27. Sicko-Goad, L.; Stoermer, E.F.; Kociolek, J.P. Diatom resting cell rejuvenation and formation: Time course, species records and distribution. *J. Plankton Res.* **1989**, *11*, 375–389. [CrossRef]
28. Kang, Y.; Tang, Y.-Z.; Taylor, G.T.; Gobler, C.J. Discovery of a resting stage in the harmful, brown-tide-causing pelagophyte, *Aureoumbra lagunensis:* A mechanism potentially facilitating recurrent blooms and geographic expansion. *J. Phycol.* **2017**, *53*, 118–130. [CrossRef]
29. Tang, Y.Z.; Ma, Z.; Hu, Z.; Deng, Y.; Yang, A.; Lin, S.; Yi, L.; Chai, Z.; Gobler, C.J. 3000 km and 1500-year presence of *Aureococcus anophagefferens* reveals indigenous origin of brown tides in China. *Mol. Ecol.* **2019**, *28*, 4065–4076. [CrossRef]
30. de Vargas, C.; Audic, S.; Henry, N.; Decelle, J.; Mahé, F.; Logares, R.; Lara, E.; Berney, C.; Le Bescot, N.; Probert, I.; et al. Eukaryotic plankton diversity in the sunlit ocean. *Science* **2015**, *348*, 1261605. [CrossRef]
31. Guillard, R.R.L. Culture of Phytoplankton for Feeding Marine Invertebrates. In Culture of Marine Invertebrate Animals, Proceedings of the 1st Conference on Culture of Marine Invertebrate Animals Greenport, New York, October, 1972; Smith, W.L., Chanley, M.H., Eds.; Springer: Boston, MA, USA, 1975; pp. 29–60.
32. Wu, S.; Zhang, B.; Huang, A.; Huan, L.; He, L.; Lin, A.; Niu, J.; Wang, G. Detection of intracellular neutral lipid content in the marine microalgae *Prorocentrum micans* and *Phaeodactylum tricornutum* using Nile red and BODIPY 505/515. *J. Appl. Phycol.* **2014**, *26*, 1659–1668. [CrossRef]
33. Tang, Y.Z.; Gobler, C.J. The toxic dinoflagellate *Cochlodinium polykrikoides* (Dinophyceae) produces resting cysts. *Harmful Algae* **2012**, *20*, 71–80. [CrossRef]
34. Sandgren, C.D. Characteristics of sexual and asexual resting cyst (statospore) formation in *Dinobryon cylindricum* imhof (Chrysophyta). *J. Phycol.* **1981**, *17*, 199–210. [CrossRef]
35. Sandgren, C.D. Chrysophyte reproduction and resting cysts: A paleolimnologist's primer. *J. Paleolimnol.* **1991**, *5*, 1–9. [CrossRef]
36. Kim, J.-H.; Park, B.S.; Wang, P.; Kim, J.H.; Youn, S.H.; Han, M.-S. Cyst morphology and germination in *Heterosigma akashiwo* (Raphidophyceae). *Phycologia* **2015**, *54*, 435–439. [CrossRef]
37. Tang, Y.Z.; Gobler, C.J. Sexual resting cyst production by the dinoflagellate *Akashiwo sanguinea:* A potential mechanism contributing to the ubiquitous distribution of a harmful alga. *J. Phycol.* **2015**, *51*, 298–309. [CrossRef] [PubMed]
38. Chai, Z.Y.; Hu, Z.; Liu, Y.Y.; Tang, Y.Z. Proof of homothally of *Pheopolykrikos hartmannii* and details of cyst germination process. *J. Oceanol. Limnol.* **2020**, *38*, 114–123. [CrossRef]

39. Lomas, M.W.; Gobler, C.J. *Aureococcus anophagefferens* research: 20 years and counting. *Harmful Algae* **2004**, *3*, 273–277. [CrossRef]

40. Wazniak, C.E.; Glibert, P.M. Potential impacts of brown tide, *Aureococcus anophagefferens*, on juvenile hard clams, *Mercenaria mercenaria*, in the Coastal Bays of Maryland, USA. *Harmful Algae* **2004**, *3*, 321–329. [CrossRef]

41. Gobler, C.J. Harmful Algal Species Fact Sheet: *Aureococcus anophagefferens* Hargraves et Sieburth & Aureoumbra lagunensis De Yoe et Stockwell—Brown Tides. In *Harmful Algal Blooms: A Compendium Desk Reference*; Shumway, S.E., Burkholder, J.M., Morton, S.L., Eds.; John Wiley & Sons, Ltd.: Hoboken, NJ, USA, 2018; pp. 583–584.

42. Huang, B.; Liang, Y.; Pan, H.; Xie, L.; Jiang, T.; Jiang, T. Hemolytic and cytotoxic activity from cultures of *Aureococcus anophagefferens*—A causative species of brown tides in the north-western Bohai Sea, China. *Chemosphere* **2020**, *247*, 125819. [CrossRef]

43. Figueroa, R.I.; Bravo, I.; Garcés, E. Effects of nutritional factors and different parental crosses on the encystment and excystment of *Alexandrium catenella* (Dinophyceae) in culture. *Phycologia* **2005**, *44*, 658–670. [CrossRef]

44. Anderson, D.M.; Taylor, C.D.; Armbrust, E.V. The effects of darkness and anaerobiosis on dinoflagellate cyst germination. *Limnol. Oceanogr.* **1987**, *32*, 340–351. [CrossRef]

45. Itakura, S.; Nagasaki, K.; Yamaguchi, M.; Imai, I. Cyst formation in the red tide flagellate *Heterosigma akashiwo* (Raphidophyceae). *J. Plankton Res.* **1996**, *18*, 1975–1979. [CrossRef]

46. Anderson, D.M. Effects of temperature conditioning on development and germination of *Gonyaulax tamarensis* (Dinophyceae). *J. Phycol.* **1980**, *16*, 166–172. [CrossRef]

47. Mertens, K.N.; Gu, H.; Gurdebeke, P.R.; Takano, Y.; Clarke, D.; Aydin, H.; Li, Z.; Pospelova, V.; Shin, H.H.; Li, Z.; et al. A review of rare, poorly known, and morphologically problematic extant marine organic-walled dinoflagellate cyst taxa of the orders Gymnodiniales and Peridiniales from the Northern Hemisphere. *Mar. Micropaleontol.* **2020**, *159*, 101773. [CrossRef]

48. Liu, Y.; Hu, Z.; Deng, Y.; Tang, Y.Z. Evidence for resting cyst production in the cosmopolitan toxic dinoflagellate *Karlodinium veneficum* and the cyst distribution in the China seas. *Harmful Algae* **2020**, *93*, 101788. [CrossRef] [PubMed]

49. Van Nieuwenhove, N.; Head, M.J.; Limoges, A.; Pospelova, V.; Mertens, K.N.; Matthiessen, J.; De Schepper, S.; de Vernal, A.; Eynaud, F.; Londeix, L.; et al. An overview and brief description of common marine organic-walled

dinoflagellate cyst taxa occurring in surface sediments of the Northern Hemisphere. *Mar. Micropaleontol.* **2020**, *159*, 101814. [CrossRef]

50. Limoges, A.; Van Nieuwenhove, N.; Head, M.J.; Mertens, K.N.; Pospelova, V.; Rochon, A. A review of rare and less well known extant marine organic-walled dinoflagellate cyst taxa of the orders Gonyaulacales and Suessiales from the Northern Hemisphere. *Mar. Micropaleontol.* **2020**, *159*, 101801. [CrossRef]
51. Liu, Y.; Hu, Z.; Deng, Y.; Tang, Y.Z. Evidence for production of sexual resting cysts by the toxic dinoflagellate *Karenia mikimotoi* in clonal cultures and marine sediments. *J. Phycol.* **2020**, *56*, 121–134. [CrossRef]
52. Bibby, B.T.; Dodge, J.D. The encystment of a freshwater dinoflagellate: A light and electron-microscopical study. *Br. Phycol. Bull.* **1972**, *7*, 85–100. [CrossRef]
53. Loeblich, A.R.; Loeblich, L.A. 13—Dinoflagellate Cysts. In *Dinoflagellates*; Spector, D.L., Ed.; Academic Press: San Diego, CA, USA, 1984; pp. 443–480.
54. Bloch, K. Sterol molecule: Structure, biosynthesis, and function. *Steroids* **1992**, *57*, 378–383. [CrossRef]
55. Volkman, J. Sterols in microorganisms. *Appl. Microbiol. Biotechnol.* **2003**, *60*, 495–506. [CrossRef]
56. Pichrtová, M.; Arc, E.; Stöggl, W.; Kranner, I.; Hájek, T.; Hackl, H.; Holzinger, A. Formation of lipid bodies and changes in fatty acid composition upon pre-akinete formation in Arctic and Antarctic Zygnema (Zygnematophyceae, Streptophyta) strains. *FEMS Microbiol. Ecol.* **2016**, *92*. [CrossRef]
57. Barati, B.; Gan, S.Y.; Lim, P.E.; Beardall, J.; Phang, S.M. Green algal molecular responses to temperature stress. Acta Physiol. Plant. **2019**, 41, 26. [CrossRef]
58. Da Costa, E.; Silva, J.; Mendonça, S.H.; Abreu, M.H.; Domingues, M.R. Lipidomic approaches towards deciphering glycolipids from microalgae as a reservoir of bioactive lipids. *Mar. Drugs* **2016**, *14*, 101. [CrossRef] [PubMed]
59. Smith, A.E.; Morris, I. Synthesis of lipid during photosynthesis by phytoplankton of the Southern Ocean. *Science* **1980**, *207*, 197–199. [CrossRef]

CHAPTER-7

MORPHOLOGY AND PHYLOGENY OF *SCRIPPSIELLA PRECARIA MONTRESOR* & ZINGONE (THORACOSPHAERALES, DINOPHYCEAE) FROM KOREAN COASTAL WATERS

1. INTRODUCTION

The dinoflagellate genus *Scrippsiella* Balech ex A.R.Loeblich belongs to the family Thoracosphaeraceae and was established by Balech, with *Scrippsiella sweeneyae* Balech as the type species [1]. *Scrippsiella* species share the consistent plate pattern of Po, X, $4'$, 3a, $7''$, 6C, 5S, $5'''$, $2''''$ [1–5], and based on morpho-molecular features, including the plate pattern and phylogenetic position [6], approximately 30 species are currently assigned to the genus *Scrippsiella*, which includes some previously recorded *Calcigonellum* Deflandre species [6–11]. Most of the *Scrippsiella* species produce resting cysts with distinctive calcareous ornaments [12], but some *Scrippsiella* species, such as *Scrippsiella donghaiensis* H.Gu and *Scrippsiella enormis* H.Gu, produce noncalcareous resting cysts [13,14]. Recently, the morphological features of calcareous ornaments have been used to classify *Scrippsiella* species [13,15,16].

In previous studies, *Scrippsiella* sensu lato (s.l.) was identified based on internal transcribed spacer (ITS) sequences and large subunit ribosomal DNA (LSU rDNA) [13,15,17,18]. As *Scrippsiella* s.l. includes not only common *Scrippsiella* species but also several cyst- based genera, such as *Calciodinellum*, *Pernambugia* Janofske & Karwath, and *Naiadinium polonicum* (Woloszynskia) S. Carty, which is a freshwater species [15,18], *Scrippsiella* s.l. is not fully resolved, either phylogenetically or taxonomically. Consequently, the morphological and genetic diversity of *Scrippsiella* species from many coastal areas requires additional exploration. However, most studies on the classification of *Scrippsiella* species using morphological and genetic approaches have focused on the Mediterranean and Atlantic diversities [18,19]. Gu et al. [13,20] and Luo et al. [15] recently reported the diversity of *Scrippsiella* species from Chinese coastal areas, and the morphology and phylogeny of several *Scrippsiella* species from Korean coastal areas, such as *Scrippsiella lachrymosa* and *Scrippsiella masanensis*, have been described [21,22].

S. precaria Montresor & Zingone was originally described from established cultures in coastal waters of the Gulf of Naples [23], where cyst formation of this species was also reported [24]. Gu et al. [19] later provided the morphological details and elucidated the phylogenetic relationships of *S. precaria* collected from Chinese coastal waters. Cysts of *S. precaria* have also been reported in Japanese coastal waters [25,26]; however, vegetative cells and resting cysts of *S. precaria* have not yet been reported from Korean coastal areas. According to Montresor and Zingone [23] and Gu et al. [20], *S. precaria* share three anterior intercalary plates with *Scrippsiella ramonii* and *Scrippsiella irregularis*, and these species form a clade phylogenetically. However, the clade also includes *N. polonicum*, which has two anterior intercalary plates. This indicates that the arrangements of adjacent plates, possibly including precingular plates, can vary depending on the number and arrangement of anterior intercalary plates in the group, and thus, this character may be useful for the classification of species that are members of this clade.

During a study of marine dinoflagellates and resting cysts from Korean coastal areas, a *Scrippsiella*-like species and calcareous resting cysts were found. The cultures were success- fully established based on the incubation of the collected vegetative cell and germination experiments. These cultures were examined using light and scanning electron microscopy, and ITS and LSU rDNA were sequenced. The results indicated that the cultures belonged to *S. precaria*. In the present study, we established the morphological details and phylogenetic positions of the Korean cultures of *S. precaria*, and then we compared them with those previously reported in other studies.

2. MATERIALS AND METHODS

2. 1 Sample Collection and Culture

Sediment samples were collected from a station at Jinhae-Masan Bay (34°59′36″ N, 128°0′33″ E) in July 2013 using a gravity corer. The top 2 cm of the core samples was sliced and then stored in dark and cool conditions at 4°C prior to further analysis. Sample analysis was conducted using the panning method of Matsuoka and Fukuyo [27]; approximately 2 g of each sample was placed in a beaker, rinsed with filtered seawater, and sonicated for approximately 30 s. The sediment suspension was sieved using stainless steel screens with 125 and 10 μm mesh opening sizes. The residue on the 10 μm mesh was washed into a watch glass. By panning on the watch glass, cysts and light particles were separated from heavier sand grains and were sieved again onto the 10 μm sieve and concentrated into a final sample volume of 10 mL. To observe the cysts, 1 mL of the sample was placed in a Sedgewick Rafter chamber (Pyser-SGI, Edenbridge, Kent, UK), and the cysts of *Scrippsiella* species characterized by numerous calcareous spines were photographed using a digital camera and then isolated under an inverted microscope.

The isolated cysts were inoculated into individual wells of 96-well tissue culture plates filled with f/2-Si culture medium (Marine Water Enrichment Solution, Sigma-Aldrich, Saint Louis, MO, USA) via micropipetting using a capillary pipette and cultured at 20°C and ≈100 μmol photons m^{-2} s^{-1} cool-white illumination under a 14L:10D photocycle. Germinated cells were transferred into the wells of six-well tissue plates. After sufficient growth, the cells in the six-well tissue plates were transferred to a 30 mL culture flask containing 25 mL sterile f/2-Si medium. A monoclonal culture of the germinated cells was successfully established and deposited as strain LMBE-C28 in the Library of Marine Samples, Korea Institute of Ocean Science and Technology, Republic of Korea; however, the strain is currently not available.

On 15 March 2019, plankton samples were collected at a station in the Jinhae-Masan Bay, Korea (35°00′38″ N, 128°34′03″ E) using a 20 μm mesh plankton net. In the laboratory, a single cell of *Scrippsiella* species was isolated from the samples using a capillary pipette. The isolated cell was inoculated into a well of a 48-well tissue culture plate filled with f/2-Si culture medium and cultured at a temperature of 20°C and ≈100 μmol photons m^{-2} s^{-1} cool-white illumination under a 12L:12D photocycle. The cultured cell was transferred into the wells of six-well tissue plates. After sufficient growth, the cells in the six-well tissue plates were transferred to a 30 mL culture flask containing 25 mL sterile f/2-Si medium. A monoclonal culture of *Scrippsiella* species was successfully established and deposited as strain LIMS-PS-2799 in the Library of Marine Samples, Korea Institute of Ocean Science and Technology, Republic of Korea.

2.2 Light Microscopy

Vegetative cells were examined and photographed at ×400 magnification using an ultra-high-resolution digital camera (DS-Ri2, Nikon, Tokyo, Japan) on an upright micro- scope (ECLIPSE Ni, Nikon, Tokyo, Japan). For fluorescence microscopy, approximately 1 mL of culture strain LIMS-PS-2799 was transferred to a 1.5 mL microcentrifuge tube and SYTOX® Green Nucleic Acid Stain (Molecular Probes, Eugene, OR, USA) was added at a final concentration of 1.0 μM. The cells were incubated in the dark at room temperature for 30 min and then observed using a Zeiss Filterset (emission: BP 450–490; beam splitter: FT 510) and photographed using an AxioCam MRc digital camera on an upright microscope (Axio Imager 2, Zeiss, Jena, Germany).

To observe changes in the morphology of the resting cyst, a round resting cyst produced from the culture strain LIMS-C28 was isolated and inoculated into Petri dishes (50 × 15 mm; SPL, Pocheon, South Korea) containing 10 mL of f/2-Si culture medium with a salinity of 32. The cyst was incubated on a live-cell observation system (Zeiss) at 20°C under 100 μmol photons m^{-2} s^{-1} cool-white illumination in a 24L:0D photocycle. Time-lapse sequences were captured at ×400 magnification with an AxioCam MRm digital camera on an Axio Imager 2 upright microscope (Zeiss) using bright field illumination.

2.3 Scanning Electron Microscopy (SEM)

For the SEM, the cells were fixed with Lugol's solution for 4 h at room temperature. The fixed cells were deposited on polycarbonate membrane filters (2.0 μm pore size; Millipore, Billerica, MA, USA). The filters were rinsed twice with deionized water and dehydrated in a graded ethanol series (10–99.9% in eight steps) for 10 min per step. Filters were then critical point dried (CPD) using a critical point drying apparatus (Spi-Dry^{TM} Regular Critical Point Dryer, SPI Supplies, West Chester, PA, USA) and liquid CO_2. After they were CPD, the filters were mounted on stubs. Finally, the samples were coated with platinum-palladium and examined using a field emission scanning electron microscope equipped with X-ray energy dispersive spectroscopy (EDS) (JSM 7600F, JEOL, Tokyo, Japan).

2.4 DNA Extraction and Sequencing

Genomic DNA was extracted from 1 mL of exponentially growing cultures using the DNeasy® Plant Mini Kit (QIAGEN Inc., Valencia, CA, USA). The ITS1-5.8S-ITS2 sequences for two strains LIMS-PS-2799 and LMBE-C28 were amplified using the primer pairs ITSFor and ITSRev [28]. Part of the LSU region sequence for strain LIMS-PS-2799 was amplified using the primer pairs 25F1 and R2 [29]. Polymerase chain reaction (PCR) was conducted using a thermo block (T100^{TM} Thermal Cycler; Bio-Rad, Hercules, CA, USA) using the following protocol: 95°C for 4 min; 30 cycles of denaturation at 95°C for 10 s, annealing at 52°C for 40 s, extension at 72°C for 1 min, and a final elongation at 72°C for 5 min. The PCR-amplified products were confirmed using 1% agarose gel electrophoresis. The PCR products were purified with the QIAquick PCR Purification kit (Qiagen, Valencia, CA, USA). A cycle-sequencing reaction was performed using the ABI PRISM® Big DyeTM Terminator Cycle Sequencing Ready Reaction Kit (Applied Biosystems, Foster City, CA, USA).

2.5 Alignment and Phylogenetic Analyses

Sequences were viewed and assembled in DNABaser version 4.36 (http://www.dnabaser.com). Contigs were aligned using Mafft v6.624b online version 7 (http://mafft. cbrc.jp/alignment/server/) [30] and the Q-INS-I option was used to consider rRNA sec- ondary structures. The GTR+I+G substitution model was selected using the Akaike in- formation criterion, as implemented in jModelTest version 2.1.4 [31]. For the analysis of ITS1-5.8S-ITS2 sequences, the dataset contained 51 taxa and consisted of 621 characters (including gaps inserted for alignment). The armored dinoflagellate *Ensiculifera* aff. *loeblichii* (HQ845328) and *Pentaphasodinium dalei* (JX262496) were used as outgroup taxa. For the analysis of LSU sequences, the dataset contained 44 taxa and consisted of 1024 characters (including gaps inserted for alignment). The armored dinoflagellate *Prorocentrum cordadum* (AF260379) was used as the outgroup taxon.

Phylogenetic trees for both datasets were independently constructed using maximum likelihood (ML) analyses and Bayesian inference. ML analyses were performed using PhyML ver. 3.1 [32]. The starting tree was generated using BIONJ with optimization of the topology, branch lengths, and selected rate parameters. Six different substitution rates were selected. Bootstrap analyses for both datasets were carried out using ML with 1000 replicates to evaluate the statistical reliability. Bayesian inference analyses were conducted on both datasets using the MrBayes program version 3.2 [33]. The evolutionary model used in the Bayesian inference analyses was the TVM+I+G model in the ITS region and the GTR+I+G model for the LSU region with a gamma-distributed rate variation across sites. Five Markov chain Monte Carlo (MCMC) chains were run for 10 million generations, sampling every 100 generations.

3.RESULTS

3.1 Morphology of the Resting Cysts of Scrippsiella precaria

Cysts of *Scrippsiella precaria* were observed in sediment samples (Figure 1a,b), as well as in clonal cultures (Figure 1c–i). Cysts were yellowish or grayish, with granular cell contents (Figure 1). The cyst bodies were spherical to ovoid and the cyst diameters ranged from 18.8 to 25.6 μm (mean = 22.3 μm, *n* = 30). One or two red accumulation bodies were located toward the apical or central part of the cyst bodies (Figure 1a,c–e). The archeopyle was mesoepicystal, representing the loss of 2–4$^{\prime}$ and 1–3a paraplates (Figure 1b). The cysts were covered with numerous narrow spines ranging in length from 0.7 to 6.6 μm (mean = 3.9 μm, *n* = 30), and the EDS indicated that the spines were composed of calcium carbonate (CaCo3) (Figure 1i). The SEM observations revealed that the calcareous spines that emerged directly from the cyst bodies were pentagonal and became narrower toward the apex (Figure 1f–h). However, in the clonal culture, cysts without spiny processes were occasionally observed (Figure 2a,b). Time-lapse sequences obtained using a live-cell observation system showed the formation and changes in the length of the spiny processes of the cyst (Figure 2). Resting cysts had a rounded shape without spines until 2 days of incubation, whereas very short spines emerged from the cyst bodies on day 3, where the spine lengths subsequently increased until day 8 of incubation (Figure 2).

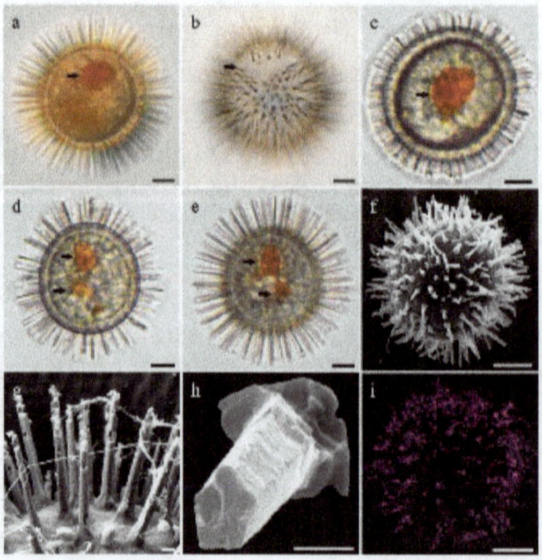

Figure 1. Resting cysts of *Scrippsiella precaria* from the sediment and cultures: (a) resting cyst from the sediment showing a red pigment body (arrow), (b) empty cyst showing the archeopyle (arrow), (c–e) resting cyst from the culture showing one or two red pigment bodies, (f) scanning electron microscopy (SEM) micrograph of a resting cyst, (g,h) details of calcareous spines, and (i) energy dispersive spectroscopy (EDS) result showing the detection of calcium carbonate (CaCo3) on the surface of a resting cyst. Scale bars: 5 µm.

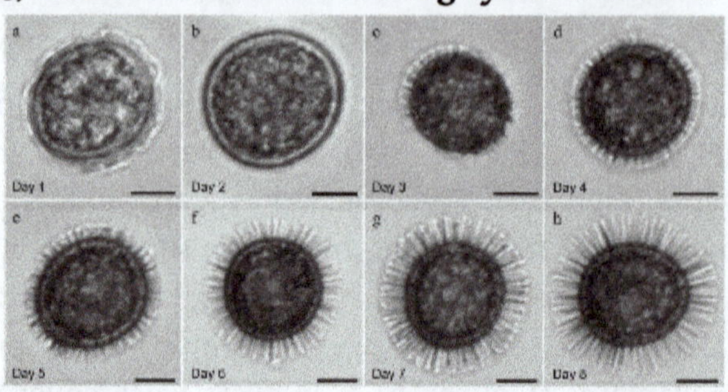

Figure 2. Time series of the formation and changes in the length of calcareous spiny processes of a round *Scrippsiella precaria* cyst observed in culture LMBE-C28: (a,b) round form of the resting cyst, (c–g) showing an increase in the length of the spiny processes over time, and (h) complete formation of a cyst. Scale bars: 10 µm.

3.2 Morphology of Vegetative Cells of Scrippsiella precaria

There were no significant differences in the morphology between motile cells germinated from resting cysts and those collected from water samples. Motile cells of *S. precaria* were oval and slightly dorso-ventrally compressed (Figures 3 and 4), and were usually solitary and rarely found in pairs (Figure 3e). Cells were yellowish or grayish and filled with pale white and grayish granules (Figure 3a–e), and were small, ranging from 20.4–28.5 μm in length (mean = 24.7 μm, *n* = 36) and 18.1–24.4 μm in width (mean = 21.5 μm, *n* = 36). The epitheca of the cell was conical and had a rounded apex, the cingulum was wide and the hypotheca was hemispherical (Figure 3a,c–e and Figure 4a,b). The left half of the hypotheca was sometimes prominent in the ventral view (Figure 3a,c). The epitheca was slightly longer than the hypotheca, occasionally showing a red pigment in the epitheca (Figure 3a,c–e and Figure 4a–d). A potential red eyespot was visible in the sulcal area (Figure 3c,d). In the ventral view, the spherical nucleus was visible and located in the anterior part of the cell (Figure 3f). The chloroplast was distributed toward the marginal regions of the cell but its shape was unclear (Figure 3f).

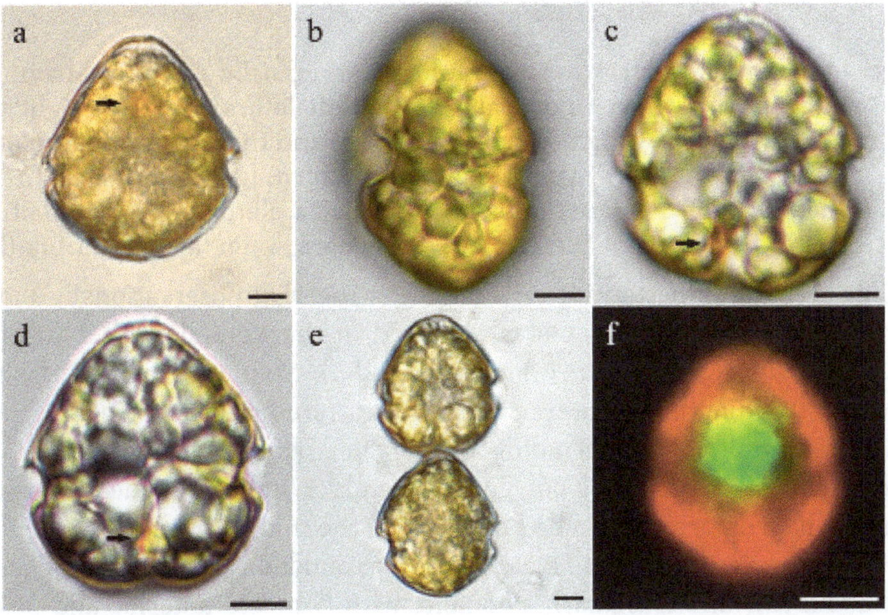

Figure 3. Light and fluorescence micrographs of *Scrippsiella precaria* (strains LMBE-C28 and LIMS-PS-2799): (a) ventral view showing a red pigment body in the epitheca, (b) left lateral view, (c) ventral view showing a red pigment body in the hypotheca, (d) dorsal view showing a red pigment body in the hypotheca, (e) two-celled chain, and (f) ventral view showing the position of the nucleus (green) and chloroplast (red). Scale bars: 5 μm.

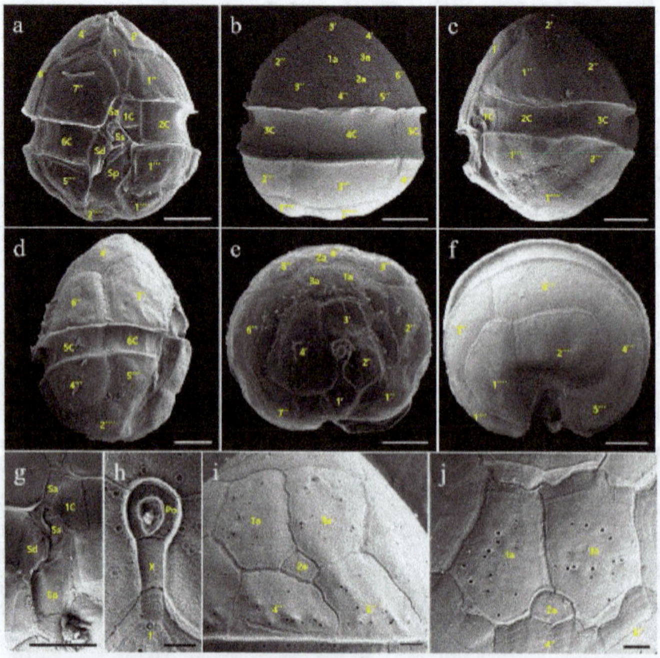

Figure 4. Scanning electron micrographs of *Scrippsiella precaria* (strain LIMS-PS-2799): (a) ventral view, (b) dorsal view, (c) left lateral view, (d) right lateral view, (e) apical view showing the apical pore (Po) and three intercalary plates, (f) antapical view showing the hypothecal plate pattern, (g) sulcal plates, (h) detail of Po, (i) detail of the $5''$ plate that contacted the 2a plate, and (j) detail of the $5''$ plate that did not contact the 2a plate. Scale bars: 5 µm (a–g) and 1 µm (h–j).

The thecal plate pattern of cells based on the SEM observations is shown in Figure 5. The cells had thin plates arranged in the Kofoidian plate formula of Po, X, $4'$, 3a, $7''$, 6C, 4S, $5'''$, $2''''$ (Figures 4 and 5). Numerous small pores (mean diameter = 0.13 µm, n = 30) were randomly scattered on the thecal surface. The apical pore complex (APC), including an elongated pentagonal canal plate (X), was keyhole-shaped, comprising a polygonal pore plate (Po) surrounded by a rim and a comma-shaped pore surrounded by a rim in the middle of the APC (Figure 4h). The X plate contacted plate $1'$. Plate $1'$ was surrounded by plates $2'$, $4'$, $1''$, and $7''$, and was narrow and irregularly rhombic, with truncate anterior and posterior ends (Figure 4a,e). Plate $2'$ was pentagonal, whereas plates $3'$ and $4'$ were hexagonal and heptagonal, respectively (Figure 4e). Plate $4'$ was larger than plates $2'$ and $3'$. There were three anterior intercalary plates in the dorsal part of the epitheca and the plates contacted each other (Figure 4b,e,i,j and Figure 6). Plates 1a and 3a were hexagonal and similar in size, while plate 2a was much smaller, rhombic but occasionally triangular or rectangular, and either contacted plate $5''$ or not (Figure 4b,i,j and Figure 6). Plates $3''$, $4''$, and $5''$ were much smaller than plates $1''$, $2''$, $6''$, and $7''$ due to the presence of intercalary plates (Figure 4b,e). Plates $4''$ and

$5^{!!}$ contacted the intercalary plates. The cingulum, comprising six plates, was wide and deeply excavated and descended to about its own width (Figure 4b,i). Among the cingulum plates, the first cingulum plate (1C) was the smallest, while the other six plates were similar in size (Figures 4a–d and 5). There were four sulcal plates. The anterior sulcal plate (Sa) was covered by the 7” plate, contacted 1C and the right sulcal plate (Sd), and had a sickle-like extension. Sd was narrow, with a nib-shaped posterior end, and was wing-shaped on its left side (Figure 4a,g). The left sulcal plate (Ss) was covered by Sa and Sd (Figure 4a,g). The posterior sulcal plate (Sp) was wide and long and extended into the hypotheca without reaching the antapex (Figure 4a).

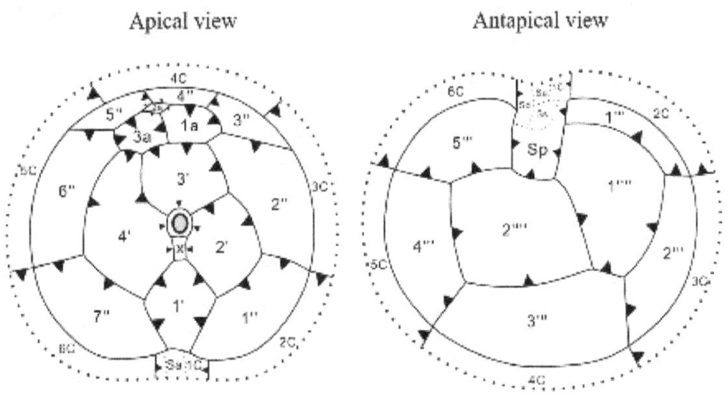

Figure 5. Schematic drawings of thecal plate patterns of *Scrippsiella precaria*. Arrowheads indicate the plate overlap pattern.

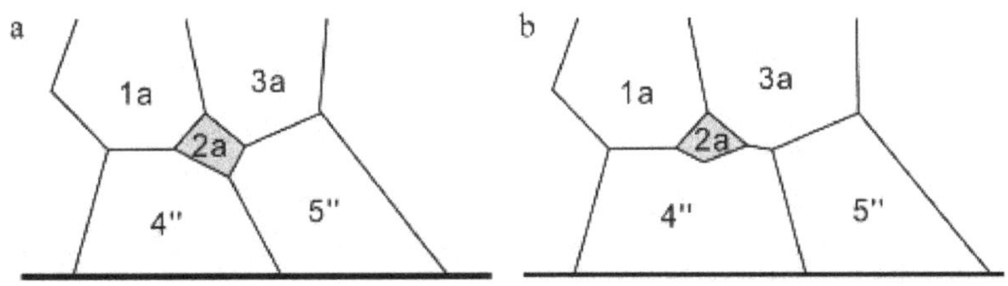

Figure 6. Schematic drawings of the intercalary plates and two location types of the fifth precingular plate ($5^{!!}$) of *Scrippsiella precaria*: (a) $5^{!!}$ plate that contacted the 2a plate and (b) $5^{!!}$ plate that did not contact the 2a plate.

In the postcingular series, plate $1^{!!!}$, which contacted plates Sp, 1C, 2C, $1^{!!!!}$, and $2^{!!!!}$, was tetragonal and wider than the other four plates (Figure 4a). Plate $3^{!!!}$ was the largest postcingular plate and contacted the antapical plates ($1^{!!!!}$ and $2^{!!!!}$). Plate $2^{!!!}$ was the second-largest plate and only contacted plate $1^{!!!!}$. Plates $4^{!!!}$ and $5^{!!!}$ contacted plate $2^{!!!!}$, whereas plate $5^{!!!}$ also contacted plates Sp and Sd. The antapical plates were pentagonal and similar in size and contacted plate Sp.

The plate overlap patterns of the epithecal, cingular, and hypothecal plate series followed two general gradients: from dorsal to ventral and from cingulum to the two poles (Figure 5). The fourth precingular ($4''$) and third postcingular ($3'''$) plates and the fourth cingular (4C) plate were identified as keystone plates that overlapped all adjacent plates. In the sulcal plate series, the Sp plate was overlapped by the cingulum and hypothecal plates.

3.3 Molecular Phylogeny of Scrippsiella precaria

The inferred phylogenies from the ML and Bayesian inference (BI) analysis based on ITS region (ITS1, ITS2, and 5.8S rDNA) and LSU rDNA sequences are shown in Figures 7 and 8. The sequence of one strain of *Scrippsiella precaria* (LIMS-PS-2799) for LSU rDNA, and the sequences of two strains of *S. precaria* (LIMS-PS-2799 and LMBE-C28) for the ITS region, with other related species available from GenBank, were used for the phylogenetic analysis.

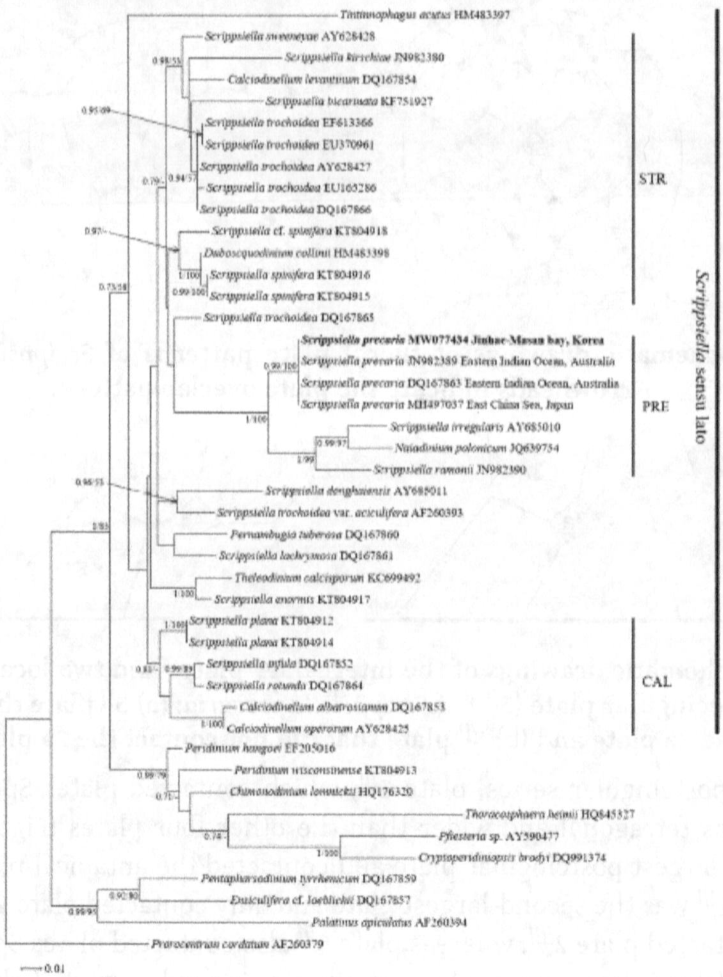

Figure 7. Phylogenetic positions of Korean isolates of *Scrippsiella precaria* that were inferred from the LSU rDNA sequences based on maximum likelihood (ML).

The numbers on each node are the Bayesian posterior probability (PP) followed by the bootstrap values (%). Only bootstrap values above 50% and PP above 0.7 are shown. CAL: clade of *Calciodinellum* and its relatives, STR: clade of *S. trochoidea* and its relatives, PRE: clade of *S. precaria* and its relatives. These clades were based on the results of Gottschling et al. [18] and Luo et al. [15]. Scale bar: 0.01 nucleotide substitutions per site.

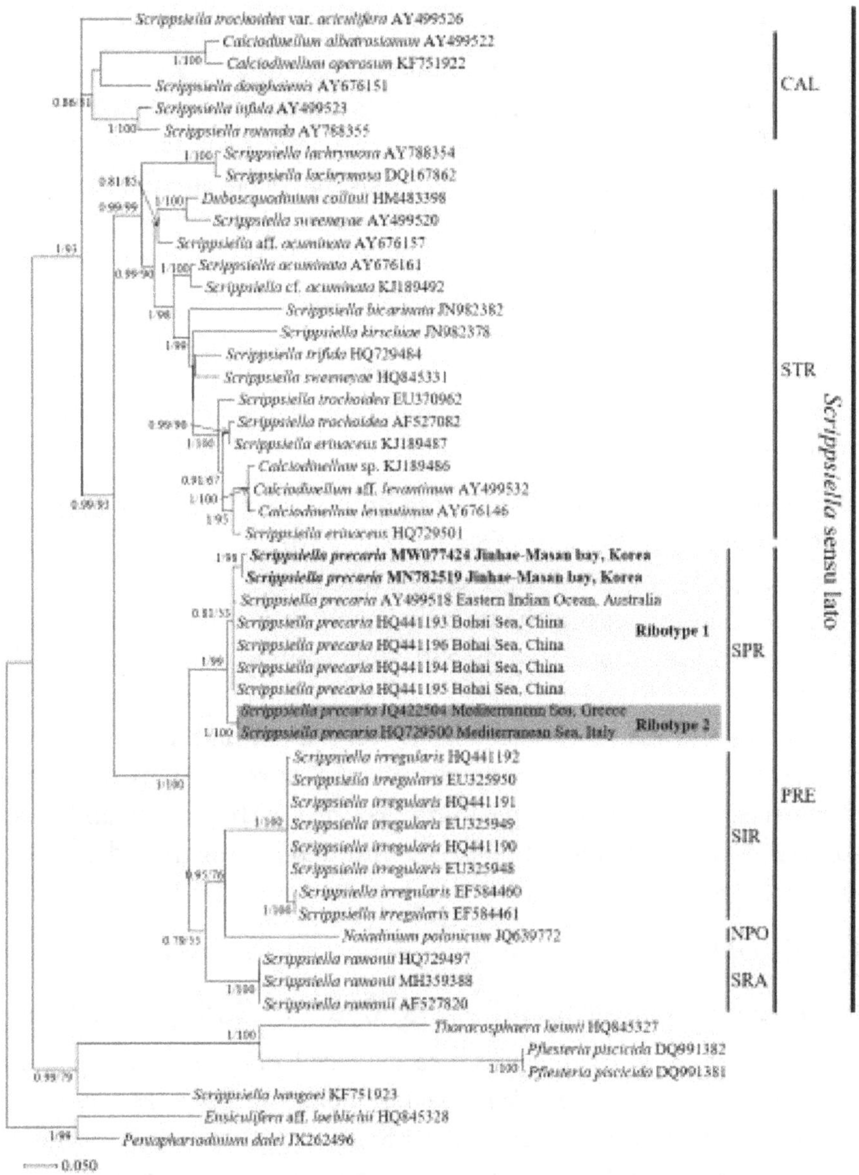

Figure 8. Phylogenetic positions of Korean isolates of *Scrippsiella precaria* that were inferred from the ITS and 5.8S sequences based on ML. *Ensiculifera* aff. *loeblichii* and *Pentapharsodinium dalei* were used as outgroup taxa. The numbers on each

node are the Bayesian PP followed by the bootstrap values (%). Only bootstrap values above 50% and PP above 0.7 are shown. STR: clade of *Scrippsiella trochoidea* species and its relatives, PRE: clade of *Scrippsiella precaria* and its relatives; CAL: clade of *Calciodinellum* and its relatives; SPR: subclade of PRE, consisting of *S. precaria*; SIR: subclade of PRE, consisting of *S. irregularis*; NPO: subclade of PRE, consisting of *Naiadinium polonicum*; SRA: subclade of PRE, consisting of *S. ramonii*. These clades were based on the results of Gottschling et al. [18] and Luo et al. [15]. Scale bar: 0.05 nucleotide substitutions per site.

ML and BI based on ITS region and LSU rDNA generated similar phylogenetic trees that differed by only a few topological features. The molecular tree based on LSU rDNA formed a clade (*Scrippsiella* sensu lato) that mainly consisted of *Scrippsiella* species and those of genera *Calciodinellum*, *Duboscquodinium*, *Naiadinium*, *Thelodinium*, and *Tintinnoph- agus* (Figure 7); however, this clade had weak support (posterior probability/ML boot- strap = 0.73/58). Three subclades, namely, CAL (*Calciodinellum* and related species), STR (*Scrippsiella trochoidea* and related species), and PRE (*S. precaria* and related species), were identified (Figure 7), and the Korean strain of *S. precaria* was nested within the PRE clade and shared an identical sequence with the Australian (DQ167863 and JN982389) and Japanese (MH497037) strains of *S. precaria* (Figure 7).

The molecular tree based on the ITS region had a similar topology to the LSU rDNA gene tree; however, the clade of *Scrippsiella* sensu lato received strong support (1/93) (Figure 8) and comprised three clades (CAL, STR, and PRE). Korean strains of *S. precaria* were nested within the PRE clade, which consisted of four subclades: the SPR clade containing *S. precaria*, the SIR clade containing *S. irregularis*, the NPO clade consisting of *Naiadinium polonicum*, and the SRA clade composed of *S. ramonii* (Figure 8). In the SPR clade, two ribotypes were identified among the Chinese, Korean, Australian, Italian, and Greek strains of *S. precaria*. Korean strains of *S. precaria* were nested within ribotype 1 and grouped with Chinese and Australian strains (0.81/53), whereas ribotype 2 was composed of Italian and Greek strains (1/100) of *S. precaria*.

4. DISCUSSION

4. 1 Morphological Features of Resting Cysts of Scrippsiella precaria

According to Gu et al. [13], the presence of a red body and the shape of the archeopyle in resting cysts of *Scripsiella* species, including *S. precaria,* are potentially useful taxonomic characters for the identification of the species because these characteristics are constant among strains and from one generation to the next generation of cysts. Montresor and Zingone [23] also described resting cysts of *S. precaria* with a red body. However, resting cysts of *S. precaria* in this study had one or two prominent red bodies. Variability in the number of red bodies has been observed in *Scrippsiella spinifera* [15]. This indicates that the number of red bodies should not be used as a taxonomic character to distinguish between vegetative cells of *Scrippsiella* species and other related species. By contrast, the shape of the archeopyle appears to be a stable character because *Scrippsiella* species, including *S. precaria,* usually have a cap-shaped operculum that includes apical (2'–4') and/or intercalary (1a–3a) paraplates (e.g., [4]). Such an archeopyle was found in *S. precaria* and has also been reported in *Scrippsiella regalis*, *S. trochoidea*, *Scrippsiella rotuda*, and *S. donghaiensis* [12,13].

Cysts of *Scrippsiella* species are usually surrounded by calcareous ornaments [2,3,34,35], whereas some species can produce noncalcareous cysts [15,22]. Previous studies on cyst–theca relationships in *S. precaria* revealed that resting cysts are characterized by calcareous spines with capitate or pointed ends [20,23,24]. Calcareous spines of resting cysts collected from Korean sediments and those formed in clonal cultures also had pointed spines. However, the capitate ends recorded by Monstresor and Zingone [23,24] were not observed in this study, and the spine ends of resting cysts of *S. precaria* from Chinese and Japanese coastal areas were pointed as well (e.g., [20,26]). This indicates that *S. precaria* has two types of spines. According to Montresor et al. [19], the morphology of calcareous ornamentations can be influenced by environmental conditions, such as macro- and micro-nutrient concentrations, pH, and light availability. We were not able to address this issue based on the data obtained in our study. Differences in spine type may be related to differences in environmental conditions based on the geographic origins of *S. precaria* because strains of *S. precaria* from Asian coastal areas, such as the Bohai Sea of Chinese and Korean coast, have identical ITS and LSU rDNA sequences (Figures 7 and 8).

Several studies reported resting cysts of *Scrippsiella* species without calcareous ornaments in culture, despite the observation of spiny cysts in the field (e.g., [13,24,36,37]). Wang et al. [36] and Shin et al. [37] reported two morphotypes of *S. trochoidea* cysts from surface sediments and cultures: a typical type with short calcareous spines and a transparent type without calcareous spines (naked-type cyst). Montresor and Marino [24] observed the formation of spines from the rounded form of resting cysts of *S. precaria* that were identical to our observations. These observations indicate that true resting cysts of calcareous *Scrippsiella* species can have a rounded form. Shin et al. [37] also observed that in acidic environments, the lengths of calcareous spines were shorter than those observed for typical cysts and the shape of the spines was abnormal, indicating that the length and shape of calcareous spines of *Scrippsiella* species are plastic and therefore unreliable characteristics for classification [13].

According to Shin et al. [37], the formation of calcareous spines can help to protect the cyst body from aerobic and anaerobic decay or mechanical damage during ingestion and digestion by predators. If so, the rapid formation of calcareous spines may ensure cyst survival. In this study, it took 8 days for the complete formation of calcareous spines of *S. precaria*. This was the first record of the time required to form calcareous spines in *Scrippsiella* species; however, because the relationship between cyst survival and the formation time of calcareous spines remains unknown, further studies are needed.

4. 2 Morphological Comparisons of Korean Isolates of Scrippsiella precaria with Previously Described Species

According to Montresor and Zingone [23] and Gu et al. [20], *Scrippsiella precaria* specimens from the Gulf of Naples and the Bohai Sea have the plate formula of Po, X, $4'$, 3a, 7”, 6C, 5S, $5'''$, $2''''$. They described a median sulcal plate (Sm) in the sulcal region, whereas no Sm plate was observed in the Korean isolates. This was possibly because the Sm was almost hidden by the wing of the right sulcal plate [23]. *S. precaria* described from previous studies was characterized as being slightly compressed dorso-ventrally, with the presence of a red pigment body and spherical nucleus, where these morphological features are consistent with those observed in Korean isolates of *S. precaria*. A red pigment body in the lower sulcal area was observed in *S. precaria* specimens collected from the Gulf of Naples [23,24], which is consistent with our findings in Korean isolates. However, reported cell sizes of isolates of *S. precaria* differed between studies: isolates from the Gulf of Naples and the Bohai Sea, China, have similar sizes (15–25 μm in length and 13.5–20 μm in width) [20,23], but isolates from Korean and Japanese coastal waters are slightly larger (e.g., [25,26]).

Scrippsiella species usually have three intercalary plates, with plates 1a and 3a separated by plate 2a (e.g., [13,15,18,19,22]). However, in some *Scrippsiella* species, such as *S. irregularis*, *S. ramonii*, and *S. precaria*, the plates 1a and 3a contact each other, and thus the 2a plate contacts not only the 1a and 3a plates, but also two precingular plates ($4''$ and $5''$) [23,34,38]. Gu et al. [20] described the 2a plate of *S. precaria* collected from the Bohai Sea of China as touching the 4” and 5” precingular plates. However, Korean isolates of *S. precaria* had two types of the $5''$ plate: one type that contacted the 2a plate and the other that did not. It is possible that Gu et al. [20] did not observe a specimen in which the $5''$ plate did not contact the 2a plate because in most *S. precaria* cells, the 2a plate contacts the $4''$ and $5''$ plates. According to Attaran-Fariman and Bloch [38] and Gu et al. [20], differences in the size of the 2a plate and the shape of the hypotheca are key characteristics that can be used to separate *S. precaria* from *S. irregularis* and *S. ramonii*. Consequently, the smaller size of the 2a plate, the shape of the hypotheca, and the locations of the

5'' plate may be distinct features that distinguish *S. precaria* from other *Scrippsiella* species.

The plate overlap pattern is considered to be conserved at higher taxonomic levels [39]. The epithecal plate overlap pattern of *S. precaria* is the same as that found in most peridinioid dinophytes [3,40], with the fourth precingular plate forming the keystone. The 4C plate is the keystone plate of the cingular series of *S. precaria*, identical to *S. acuminata* (heterotypic synonym of *S. trochoidea*) [3].

4.3 Phylogenetic Position of the Korean Isolates of Scrippsiella precaria

The molecular phylogeny of *Scrippsiella precaria* based on the ITS and LSU sequences in the present study was consistent with those reported in previous studies (e.g., [15,18,20]); in a *Scrippsiella* sensu lato clade consisting of mainly *Scrippsiella* species, *S. precaria* was nested within the PRE clade. Gu et al. [20] found that only one *S. precaria* isolate from the Bohai Sea of China was a member of the PRE clade; however, Luo et al. [15] found that an isolate from Australia was grouped with the Chinese isolate in the PRE clade. This result agrees with the results of the current study. In addition, the PRE clade was divided into four subclades (SPR, SIR, NPO, and SRA) in our phylogeny based on the ITS region, and the Korean isolates were nested within the SPR clade (Figure 8). The SPR clade comprised two ribotypes of *S. precaria*: ribotype 1 consisting of isolates from Korea, China, and Australia, and ribotype 2 consisting of isolates from Italy and Greece. According to Shin et al. [41,42] and Hallegraeff [43], the dispersal of harmful microalgae, such as *Alexandrium* species in Asian coastal areas and Australia, is related to ocean currents and ballast waters via international shipping, and because of this, the harmful species from Jinhae-Masan Bay, Korea, are phylogenetically grouped with those from China and Australia. This suggests that the lineages between isolates of the ribotype 1 could also be due to dispersal. In the SPR clade, Korean isolates are characterized by cysts with pointed spines, as was also reported for the Chinese and Japanese strains [20,26], while the Italian cysts are spiny and mostly have a capitate end [23]. In addition, the 5" plate either contacts or does not contact the 2a plate, which may be a distinguishing feature of ribotype 1, although 5" plates that do not contact the 2a plate have not been found in the isolates from China and Australia. Consequently, further studies are needed to clarify the morphological differences between the ribotypes of *S. precaria*.

REFERENCES

1. Balech, E. Two new genera of dinoflagellates from California. *Biol. Bull.* **1959**, *116*, 195–203. [CrossRef]

2. Kretschmann, J.; Zinssmeister, C.; Gottschling, M. Taxonomic clarification of the dinophyte *Rhodosphaera erinaceus* Kamptner, *Scrippsiella erinaceus* comb. nov. (Thoracosphaeraceae, Peridiniales). *Syst. Biodivers.* **2014**, *12*, 393–404. [CrossRef]

3. Kretschmann,J.;Elbrächter,M.;Zinbmeister,C.;So″hner,S.;Kirsch,M.;Kusber, W.-H.; Gottschling, M. Taxonomic clarification of the dinophyte *Peridinium acuminatum* Ehrenb., *Scrippsiella acuminata* comb. Nov. (Thoracosphaeraceae, Peridiniales). *Phytotaxa* **2015**, *220*, 239–256. [CrossRef]

4. Lewis, J. Cyst-theca relationships in *Scrippsiella* (DInophyceae) and related Orthoperidinioid Genera. *Bot. Mar.* **1991**, *34*, 91–106. [CrossRef]

5. Zinssmeister, C.; Soehner, S.; Facher, E.; Kirsch, M.; Meier, K.J.S.; Gottschling, M. Catch me if you can: The taxonomic identity of *Scrippsiella trochoidea* (F.S TEIN) A.R.LOEBL. (Thoracosphaeraceae, Dinophyceae). *Syst. Biodivers.* **2011**, *9*, 145–157. [CrossRef]

6. Guiry, M.D.; Andersen, R.A. Validation of the generic name *Symbiodinium* (Dinophyceae, Suessiaceae) revisited and the reinstatement of *Zooxanthella* K.Brandt. *Not. Algarum* **2018**, *58*, 1–5.

7. Craveiro, S.C.; Daugbjerg, N.; Moestrup, O.; Calado, A.J. Studies on Peridinium aciculiferum and Peridinium malmogiense(=Scrippsiella hangoei): Comparison with Chimonodinium lomnickiiand description of Apocalathium gen. nov. (Dinophyceae). *Phycologia* **2016**, *56*, 21–35. [CrossRef]

8. Elbrächter, M.; Gottschling, M.; Hildebrand-Habel, T.; Keupp, H.; Kohring, R.; Lewis,J.;Meier,K.S.;Montresor,M.;Streng,M.;Versteegh,G.J.;et al. Establishing an agenda for calcareous dinoflagellate research (Thoracosphaeraceae, Dinophyceae) including a nomenclatural synopsis of generic names. *Taxon* **2008**, *57*, 1289–1303. [CrossRef]

9. Horiguchi, T.; Chihara, M. *Scrippsiella hexapraecingula* sp. nov. (Dinophyceae), a Tide Pool Dinoflagellate from the Northwest Pacific. *Bot. Mag.* **1983**, *96*, 351–358. [CrossRef]

10. Horiguchi, T.; Pienaar, R.N. Ultrastructure of a new sand-dwelling dinoflagellate, *Scrippsiella arenicola* sp. nov. *J. Phycol.* **1988**, *24*, 426–438.

11. Horiguchi, T.; Pienaar, R.N. Validation of *Bysmatrum arenicola* Horicuchi et pienaar, sp. nov. (Dinophyceae). *J. Phycol.* **2000**, *36*, 237. [CrossRef]

12. Janofske, D. *Scrippsiella trochoidea* and *Scrippsiella regalis*, nov. comb. (Peridiniales, Dinophyceae): A comparison. *J. Phycol.* **2000**, *36*, 178–189. [CrossRef]

13. Gu, H.; Sun, J.; Kooistra, W.H.; Zeng, R. Phylogenetic position and morphology of thecae and cysts of *Scrippsiella* (Dinophyceae) species in the East China Sea. *J. Phycoloy* **2008**, *44*, 478–494. [CrossRef]

14. Gu, H.; Luo, Z.; Liu, T.T.; Lan, D. Morphology and phylogeny of *Scrippsiella enormis* sp. nov. and *S.* cf. *spinifera* (Peridiniales, Dinophyceae) from the China Sea. *Phycologia* **2013**, *52*, 182–190. [CrossRef]

15. Luo, Z.; Mertens, K.N.; Bagheri, S.; Aydin, H.; Takano, Y.; Matsuoka, K.; McCarthy, F.M.G.; Gu, H. Cyst–theca relationship and phylogenetic positions of *Scrippsiella plana* sp. nov. and *S. spinifera* (Peridiniales, Dinophyceae). *Eur. J. Phycol.* **2016**, *51*, 188–202. [CrossRef]

16. Matsuoka, K.; Head, M.J. Clarifying cyst–motile stage relationships in dinoflagellates. In *Biological and Geological Perspectives of Dinoflagellates*; Geological Society: London, UK, 2013; pp. 320–350.

17. D'Onofrio, G.; Marino, D.; Bianco, L.; Busico, E.; Montresor, M. Toward an assessment on the taxonomy of dinoflagellates that produce calcareous cysts (Calciodinelloideae, Dinophyceae): A morphological and molecular approach. *J. Phycol.* **1999**, *35*, 1063–1078. [CrossRef]

18. Gottschling, M.; Knop, R.; Plotner, J.; Kirsch, M.; Willens, H.; Keupp, H. A molecular phylogeny of *Scrippsiella* sensu lato (Calciodinellaceae, Dinophyta) with interpretations on morphology and distribution. *Eur. J. Phycol.* **2005**, *40*, 207–220. [CrossRef]

19. Montresor, M.; Sgrosso, S.; Procaccini, G.; Kooistra, W.H. Intraspecific diversity in *Scrippsiella trochoidea* (Dinopbyceae): Evidence for cryptic species. *Phycologia* **2003**, *42*, 56–70. [CrossRef]

20. Gu, H.; Luo, Z.; Wang, Y.; Lan, D.Z. Diversity, Distribution, and new phylogenetic information of calcareous dinoflagellates from China Sea. *J. Syst. Evol.* **2011**, *49*, 126–137. [CrossRef]

21. Lee, S.Y.; Jeong, H.J.; Yun, J.H.; Kim, S.J. Morphological and genetic characterization and the nationwide distribution of the phototrophic dinoflagellate *Scrippsiella lachrymose* in the Korean waters. *Algae* **2018**, *33*, 21–35. [CrossRef]

22. Lee, S.Y.; Jeong, H.J.; Kim, S.J.; Lee, K.H.; Jang, S.H. *Scrippsiella masanensis* sp. nov. (Thoracosphaerales, Dinophyceae), a phototrophic dinoflagellate from the coastal waters of southern Korea. *Phycologia* **2019**, *58*, 287–299. [CrossRef]

23. Montresor, M.; Zingone, A. *Scrippsiella precaria* sp. nov. (Dinophyceae), a marine dinoflagellate from the Gulf of Naples. *Phycologia* **1988**, *27*, 387–394. [CrossRef]

24. Montresor, M.; Marino, D. Reproduction and cyst formation in *Scrippsiella precaria* (Dinophyceae). *G. Bot. Ital.* **1989**, *123*, 157–167. [CrossRef]
25. Ishikawa, A.; Taniguchi, A. Some cysts of the genus *Scrippsiella* (Dinophyceae) newly found in Japanese waters. *Bull. Plankton Soc. Jpn.* **1993**, *40*, 1–7.
26. Kobayashi, S.; Itakura, S.; IMai, I. *Scrippsiella precaria* Montresor and Zingone (Dinophyceae) newly found from Hiroshima bay, west Japan. *Bull. Plankton Soc. Jpn.* **1994**, *40*, 169–173.
27. Matsuoka, K.; Fukuyo, Y. *Technical Guide for Modern Dinoflagellate Cyst Study*; Asian Natural Environmental Science Center, The University of Tokyo, Westpac-Hab/Westpac/Ic: Tokyo, Japan, 2000; p. 29.
28. Murray, S.A.; Garby, T.; Hoppenrath, M.; Neilan, B.A. Genetic diversity, morphological uniformity and polyketide production in dinoflagellates (*Amphidinium*, Dinoflagellata). *PLoS ONE* **2012**, *7*, e38253. [CrossRef]
29. Yamaguchi, A.; Horiguchi, T. Molecular phylogenetic study of the heterotophic dinoflagellate genus *Protoperidinium* (Dinophyceae) inferred from small subunit rRNA gene sequences. *Phycol. Res.* **2005**, *53*, 30–42. [CrossRef]
30. Katoh, K.; Rozewicki, J.; Yamada, K.D. MAFFT online service: Multiple sequence alignment, interactive sequence choice and visualization. *Brief Bioinform.* **2019**, *20*, 1160–1166. [CrossRef]
31. Darriba, D.; Taboada, G.L.; Doallo, R.; Posada, D. jMod-elTest 2: More models, new heuristics and parallel computing. *Nat. Methods* **2012**, *9*, 772. [CrossRef]
32. Guindon, S.; Dufayard, J.F.; Lefort, V.; Anisimova, M.; Hordijk, W.; Gascuel, O. New algorithms and methods to estimate maximum-likelihood phylogenies: Assessing the performance of PhyML 3.0. *Syst. Biol.* **2010**, *59*, 307–321. [CrossRef]
33. Ronquist, F.; Teslenko, M.; van der Mark, P.; Ayres, D.L.; Darling, A.; Hohna, S.; Larget, B.; Liu, L.; Suchard, M.A.; Huelsenbeck, J.P. MrBayes 3.2: Efficient Bayesian phylogenetic inference and model choice across a large model space. *Syst. Biol.* **2012**, *61*, 539–542. [CrossRef] [PubMed]
34. Montresor, M. *Scrippsiella ramonii* sp. nov. (Peridiniales, Dinophyceae), a marine dinoflagellate producing a calcareous resting cyst. *Phycologia* **1995**, *34*, 87–91. [CrossRef]
35. Head, M.J.; Lewis, J.; de Vernal, A. The cyst of the calcareous dinoflagellate *Scrippsiella trifida*: Resolving the fossil record of its organic wall with that of *Alexandrium tamarense*. *J. Paleontol.* **2006**, *80*, 1–18. [CrossRef]
36. Wang, Z.H.; Qi, Y.-Z.; Yang, Y.-F. Cyst formation; an important mechanism for the termination of *Scrippsiella trochoidea* (Dino- phyceae) bloom. *J. Plankton Res.* **2007**, *29*, 209–218. [CrossRef]

37. Shin, H.H.; Jung, S.W.; Jang, M.C.; Kim, Y.O. Effect of pH on the morphology and viability of *Scrippsiella trochoidea* cysts in the hypoxic zone of a eutrophied area. *Harmful Algae* **2013**, *28*, 37–45. [CrossRef]

38. Attaran-Fariman, G.; Bolch, C.J.S. *Scrippsiella irregularis* sp. nov. (Dinophyceae), a new dinoflagellate from the southeast coast of Iran. *Phycologia* **2007**, *46*, 572–582. [CrossRef]

39. Netzel, H.; Dürr, G. Dinoflagellate cell Cortex. In *Dinoflagellates*; Spector, D.L., Ed.; Academic Press Inc. (London) Ltd.: Orlando, FL, USA, 1984; pp. 43–105.

40. Kretschmann,J.;ŽerdonerCˇalasan,A.;Gottschling,M.Molecularphylogenetics of dinophytes harboring diatoms as endosym-bionts (Kryptoperidiniaceae, Peridiniales), with evolutionaty interpretations and a focus on the identity of *Durinskia oculata* from Prague. *Mol. Phylogenetics Evol.* **2018**, *118*, 392–402. [CrossRef]

41. Shin, H.H.; Matsuoka, K.; Yoon, Y.H.; Kim, Y.O. Response of dinoflagellate cyst assemblages to salinity changes in Yeoja Bay, Korea. *Mar. Micropaleontol.* **2010**, *77*, 15–24. [CrossRef]

42. Shin, H.H.; Li, Z.; Kim, E.S.; Park, J.W.; Lim, W.A. Which species, *Alexandrium catenella* (Group I) or *A. pacificum* (Group IV), is really responsible for past paralytic shellfish poisoning outbreaks in Jinhae-Masan Bay, Korea? *Harmful Algae* **2017**, *68*, 31–39. [CrossRef]

43. Hallegraff, G.M. Transport of toxin dinoflagellates via ships' ballast water: Bioeconomic risk assessment and efficacy of possible ballast water management strategies. *Mar. Ecol. Prog. Ser.* **1998**, *168*, 297–309. [CrossRef]

CHAPTER-8

CHANGES IN FREE-LIVING AND PARTICLE-ASSOCIATED BACTERIAL COMMUNITIES DEPENDING ON THE GROWTH PHASES OF MARINE GREEN ALGAE, *TETRASELMIS SUECICA*

1. INTRODUCTION

Marine green microalga genus *Tetraselmis* is well-known to have a high lipid content and fast growth [1,2], and thus it has been widely used in multiple industries, for example, a source of nutrition for invertebrates in aquaculture [3], feedstock of biofuel production [4], and cosmetic applications [5]. In order to save the cost for algal production, there have been attempts to advance the algal-culture technique which enables gaining a higher biomass of this green algae.

Algal-culture techniques have been developed based on adjusting physiochemical factors (e.g., light intensity, nutrient limitation, temperature, pH, CO_2 concentration, and salinity) which are well known to have an intimate association with algal growth [6,7]. Recently, bacteria which have a symbiotic relationship to algae have been considered as a new factor to advance the algal-culture technique, allowing to gain a higher algal biomass. Bacteria can affect the growth of algae in various ways which ranged from mutualism to parasitism [8–12]. Interestingly, the maximum algal cell density is obviously enhanced when symbiotic bacteria are added to the algal culture, compared to that in the optimum culture condition (e.g., culture media, temperature, salinity, pH, etc.) without bacteria, for example, the biomass productivity of *T. striata* was enhanced up to two-fold by the addition of two bacterial strains (*Pelgibacter bermudensis* and *Stappia* sp.) [4]. These findings show the possibility that bacteria can be an important factor to advance a culture technique for algae.

Due to the high economic value of *T. suecica*, studies on developing the mass culture technique for this algae have been carried out, and in order to advance this culture technique, bacteria which enhanced the algal growth were isolated more recently [13,14]. Thus, understanding mutualism between *T. suecica* and co-existing bacteria is thought to be important, but there is a significant knowledge gap for this ecological relationship due to a lack of evidence. The two bacterial groups exist in the algal culture. These include free-living (FLB) and particle-associated (PAB) bacteria which are phylogenetically distinct. Based on previous findings, these two groups are closely associated with the growth of algae even though PAB showed stronger species-specific association to the host algae [8,15–17]. In addition, their community compositions are distinctly varied de- pending on the growth phase of algae due to the change in composition and quantity of dissolve organic matters (DOM) released from the host algae. However, in previous studies, only the FLB composition was identified, and the previous analysis method (terminal restriction fragment length polymorphism) may not be adequate to gain a high resolution for the characterization of bacterial community composition in *T. suecica* culture due to the methodological limitation [13].

The present study aims to elucidate bacterial taxa which were associated with the growth of *T. suecica*. To address this, we investigated the community composition of both FLB and PAB and a variation in the two bacterial groups depending on the growth phase of *T. suecica* with the next generation sequencing (NGS) approach, allowing a high taxonomic resolution.

2. MATERIALS AND METHODS

2.1 Algal Culture and Cell Growth Analysis

A culture of *T. suecica* P039 (cell size: Average 16.2 μm length and 10.6 μm width in 33 cells) which was isolated from coastal water in Deukryang Bay, Korea, was obtained from the Korea Marine Microalgae Culture Center (Pukyong National University, Busan, Korea). It has been maintained in an autoclaved f/2 medium [18] at 20°C in a 12:12 h light:dark cycle with a photon flux density of approximately 65 μmol photons m^{-2} s^{-1}. The growth stage of *T. suecica* was determined based on the growth curve (Figure S1).

Cell growth curves were analyzed at every 2-day intervals using cell numbers measured with the Sedgwick-Rafter counting chamber. At the same time, bacterial cells were counted by using the 4′,6-diamidino-2-phenylindole (DAPI) (D9542, Sigma-Aldrich, Darmstadt, Germany) staining method [19]. Briefly, all glass wares and reagents were sterilized with 10% nitric acid treatments and filtration with a 0.22 μm nuclepore membrane (Millipore, Cork, Ireland), respectively. As for bacteria counting, 20 mL of cultures was preserved with 1 mL of formaldehyde (37% formaldehyde), and then stored at 4°C and a dark condition. For DAPI staining, we mixed 1 mL of the preserved cultures and 100 μL of DAPI (0.1 μg mL^{-1}), and incubated it for 5 min. Then, the cells were filtered through the pore size 0.2 μm black filter (Isopore membrane; Millipore, Bedford, MA, USA) using a hand pump with less than 178 mmHg in pressure. The filters were mounted on an objective slide, and observed using a fluorescent microscope (Axioskop, Carl Zeiss, Oberkochen, Germany).

2.2 Sample Collection and DNA Extraction

To investigate a change in the bacteria community composition depending on the growth stages, the *T. suecica* culture was harvested in lag (day 2), exponential (day 8), stationary (day 20), and decline (day 32) growth phases. In addition, we used the size- fractionated filtration method for collecting the separate bacteria of both PAB and FLB from the culture. One hundred mL of *T. suecica* culture were filtered sequentially through a 10 μm (PAB) and 0.2 μm (FLB) pore size membrane filter (diameter: 47 mm; Millipore, Cork, Ireland). Each loaded filter was immediately placed in a 2-mL microtube (Axygen Sciences, CA) that contained 800 μL extraction buffer (100 mM Tris-HCl, 100 mM Na_2-EDTA, 100 mM sodium phosphate, 1.5 M NaCl, 1% CTAB) and was then stored at -80°C until the DNA was extracted.

The DNA was extracted using a modified CTAB protocol [20]. Briefly, 2 mL of micro- tubes containing membrane filters and the extraction buffer (100 mM Tris-HCl, 100 mM Na_2-EDTA, 100 mM sodium phosphate, 1.5 M NaCl, 1% CTAB) were subjected to three cycles of immersion in liquid N_2 until completely frozen and then thawed in a 65°C water bath. After 8 μL of proteinase K was added (10 mg mL^{-1} in TE buffer), the samples were incubated at 37°C for 30 min. Then, after the addition of 80 μL of 20% sodium dodecyl sulfate (SDS), the samples were incubated at 65°C for 2 h, shaken gently with an equal volume of chloroform-isoamyl alcohol (24:1), and then centrifuged at 10,000× *g* for 5 min. The aqueous phase was transferred to a new 2 mL tube containing 88.8 μL of 3 M sodium acetate (pH 5.1), and 587 μL of isopropanol (≥99%) was added. Following centrifugation at 14,000× *g* for 20 min, the supernatant was decanted, 1 mL of cold 70% ethanol was added, and the samples were centrifuged at 14,000× *g* for 15 min. The pellets were air dried at room temperature before being dissolved in 100 μL of TE buffer (10 mM Tris-HCl, 1 mM EDTA; pH 8).

2.3 Pyrosequencing

Metagenomic sequencing was performed using the 454 GS FLX Titanium Sequencer System (Roche, Basel, Switzerland). Briefly, target rDNA retrieved from the cultured samples was amplified using the polymerase chain reaction (PCR) performed with two universal bacterial primers: 27F, 5′-GAG TTT GAT CMT GGC TCA G-3′ and 518R, 5′-TTA CCG CGG CTG CTG G-3′. Each primer was tagged using multiplex identifier (MID) adaptors according to the manufacturer's instructions (Roche, Mannheim, Germany), which allowed for the automatic sorting of the pyrosequencing-derived sequencing reads based on the MID adaptors. In addition, MID-linked 27F and 518R were linked to the pyrosequencing primers 5′-CGT ATC GCC TCC CTC GCG CCA TCA G-3′ and 5′-CTA TGC GCC TTG CCA GCC CGC TCAG-3′, respectively, according to the manufacturer's instructions (Roche, Mannheim, Germany).

Metagenomic PCR was performed with 20 μL reaction mixtures containing 2 μL 10× Ex Taq buffer (TaKaRa, Kyoto, Japan), 2 μL of a dNTP mixture (4 mM), 1 μL of each primer (10 pM), 0.2 μL Ex Taq polymerase (2.5 U), and 0.1 μg of the environmental DNA template. PCR cycling was performed in an iCycler (Bio-Rad, Hercules, CA, USA) at 94°C for 5 min, followed by 35 cycles at 94°C for 20 s, 52°C for 40 s, and 72°C for 1 min, and a final extension at 72°C for 10 min. The resulting PCR products were electrophoresed in 1.0% agarose gel, stained with ethidium bromide, and viewed under ultraviolet transillumination.

Prior to pyrosequencing, amplified PCR products were individually purified using a Dual PCR Purification Kit (Bionics, Seoul, Korea) and, subsequently, equal volumes of each purified PCR product were mixed together. Pyrosequencing of the MID-tagged PCR amplicons was performed using a 454 GS FLX Titanium system (Roche, Mannheim, Germany) with a commercial service at Macrogen, Inc. (Seoul, Korea).

2.4 Pyrosequencing Data Analysis

After each sequencing procedure had been completed, a quality check was per- formed to remove short sequence reads (e.g., less than 150 bp), low-quality sequences, sequence artifacts and chloroplast sequences, and any non-bacterial ribosome sequences and chimeras [21,22]. Using the basic local alignment search tool (BLAST), all of the se- quence reads were compared to the Silva rRNA database [23]. Sequence reads which were similar with an E-value of less than 0.01 were admitted as partial 16S rDNA sequences. The taxonomy of the sequence with the highest similarity was assigned to the sequence read (genus and class level). To analyze operational taxonomic units (OTUs), the CD-HIT-OTU software was used for clustering [24] and the Mothur software was used for Shannon- Weaver diversity and Chao richness estimation [25]. All data from the 454 pyrosequences were deposited into the Genebank database.

To examine the similarities and differences in the community composition (at the OTU level) among the samples (FLB and PAB communities depending on the growth phase of *T. suecica*), PRIMER ver. 7 [26] and R studio ver. 5.3 [27] were used to generate the heatmap and dendrogram for the hierarchical cluster analysis, together with similarity profiles (SIMPROF, $p = 0.05$). All the analyses were performed based on the Bray-Curtis dissimilarity index.

3. RESULTS AND DISCUSSION

3.1 Cell Growth of Tetraselmis suecica and Bacteria in Cultures

The green algae *T. suecica* and FLB in the culture appeared as a common, typical S-shaped growth curve (Figure 1), but the growth of FLB was faster than that of *T. suecica*. The highest density (3,774,000 cells mL⁻1) of bacterial cells was shown at day 10, whereas

T. suecica reached the maximum cell density (827,830 cells mL⁻1) at day 18. Community compositions of FLB and PAB were investigated using the samples which were collected at day 2, 8, 20, and 32, referring to the lag, exponential, stationary, and death growth phase of *T. suecica*.

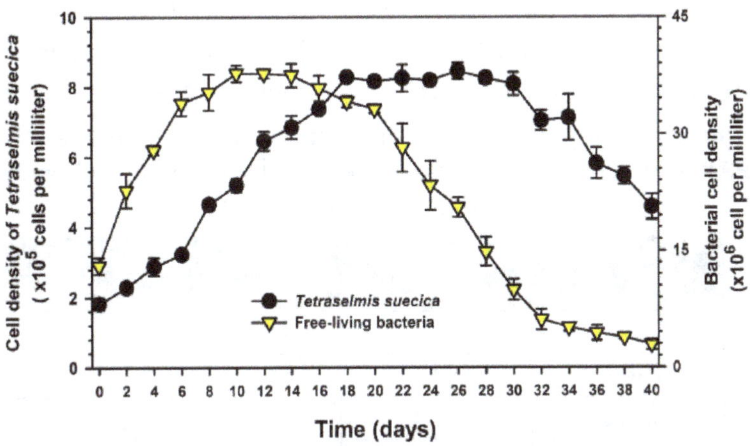

Figure 1. The growth curves of *Tetraselmis suecica* and the associated bacteria. Arrows represent each sampling point. Abbreviations: FLB: Free-living bacteria; PAB: Particle-associated bacteria; Exp: Exponential; Stn: Stationary; Dth: Death.

3.2 Pyrosequencing Data and Comparison of FLB and PAB in Tetraselmis suecica

The pyrosequencing analysis generated 3742–9409 nucleotide fragments of FLBs and 8078–13,725 nucleotide fragments of PAB. All of the samples reached saturation in the rarefaction curve indicating that a sufficient amount of sequences was identified by the pyrosequencing used in this study (Figure S1). The number of bacterial OTUs (≥97% similarity) in FLB and PAB were 35 and 30, respectively, and all bacterial OTUs in PAB were present in FLB (Figure 2A).

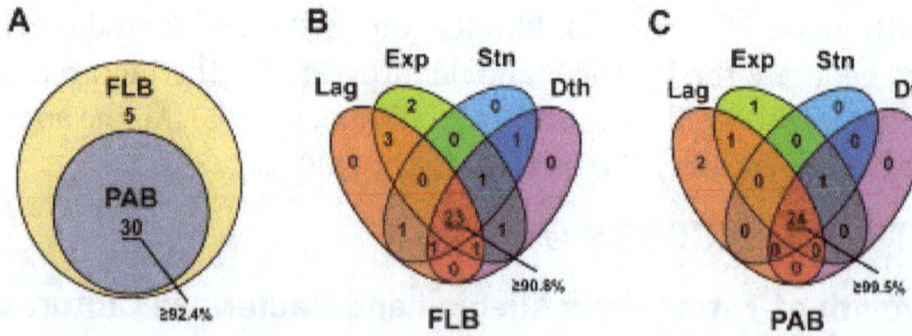

Figure 2. Bacterial operational taxonomic units (OTUs) Venn diagram. (A): Venn diagram enumerates shared bacterial OTUs between free-living (FLB) and particle-associated (PAB); (B,C): Venn diagrams enumerate FLB and PAB OTUs, shared by and exclusive to lag, exponential (Exp), stationary (Stn), and death (Dth) growth phases of *T. suecica*. The percentage number (lower right corner of each Venn-diagram) indicates the proportion of shared bacterial OTUs.

Based on alpha diversity indices (Shannon and Simpson), the level of bacterial diversity in FLB was clearly higher than that in PAB (Table 1). The averages of diversity indices of FLB were 3.39 (Shannon) and 0.84 (Simpson), but those of PAB were 1.99 (Shannon) and 0.57 (Simpson), respectively. Together with these findings, the bacterial community in FLB was more diverse than that in PAB.

Table 1. Number of sequence reads and alpha diversity indices (Shannon and Simpson) PAB and FLB samples, which were isolated from each growth phase of *T. suecica* culture (P-039).

	Read Count	Diversity Indices Shannon Simpson	
Lag	10,683	2.023	0.569
Exponential	13,725	1.908	0.543
Stationary	9433	1.961	0.565
Death	8078	2.066	0.594
Lag	7233	3.105	0.792
Exponential	3742	3.540	0.850
Stationary	9409	3.418	0.844
Death	7854	3.490	0.863

The proportion of shared OTUs between FLB and PAB communities was remarkably high. It was 100% of PAB and ≥92.4% of FLB (Figure 2A). According to previous findings, this proportion is determined by the concentration of particles [28–30], and the proportion of shared bacterial OTUs between FLB and PAB communities was relatively high in certain marine environments where a higher concentration of particles existed [31–34]. In this study, we used the culture which contained a high concentration of cells. Therefore, this might result in a high proportion of shared bacterial OTUs between the two bacterial communities in this study. Interestingly, we have obtained similar results (i.e., high proportion of shared bacterial OTUs) in previous studies where dinoflagellate cultures were used [28,35].

3.3 Difference between FLB and PAB Communities in Tetraselmis suecica Culture

Interestingly, there was a clear difference in the bacterial community composition between FLB and PAB communities even though a proportion of shared bacterial OTUs was remarkably high. The genus *Roseobacter*, belonging to the class Rhodobacteraceae (56.7–60.1%), was most dominant in the FLB community (Figure 3). The second most dominant groups were genera *Oceanicaulis* (family Hyphomonadaceae, 6.3–9.3%), *Loktanella* (Rhodobacteraceae, 2.9–8.6%), *Muricauda* (Flavobacteriaceae, 4.8–6.5%), *Roseibacterium* (Rhodobacteraceae, 3.5–6.3%), and *Marinobacter* (Alteromonadaceae, 1.9–7.1%). Whereas, in the PAB community, the family Flavobacteriaceae (19.5–36.9%, unidentified at the genus level) and the genus *Marinobacter* (10.2–30.1%) were predominant groups, with *Oceanicaulis* (8.2–14.3%), *Loktanella* (4.8–5.8%), and *Muricauda* (5.5–8.1%) the second most dominant genera (Figure 3). These two bacterial communities at the OTU level were significantly distinct (77.5% of similarity level, $p = 0.05$) in the cluster analysis and the SIMPROF test (Figure 4). Organic matters released from phytoplankton play an important role in the determination of abundance and community composition of heterotrophic bacteria in aquatic environments [8]. There are two types of organic matters which originated from phytoplankton [8], one is the low molecular weight (LMW) molecule (e.g., amino acids, organic acids, and carbohydrate, etc.) and the other is the high molecular weight (HMW) macromolecule (e.g., polysaccharides, proteins, nucleic acids, and lipids, etc.). The environments where FLB and PAB communities distribute are clearly distinct in terms of the ratio between LMW and HMW. Thus, this may lead to a significant difference in the community composition between FLB and PAB. For example, the most dominant taxa in FLB and PAB were Rhodobacteraceae (genus *Roseobacter*) and Flavobacteriaceae, respectively. Based on previous findings, the *Roseobacter* clade is capable of utilizing various LMW compounds that originated from phytoplankton as a carbon source [36–38]. Whereas, Flavobacteriaceae prefer to use HMW compounds and can convert these into LMW compounds [39].

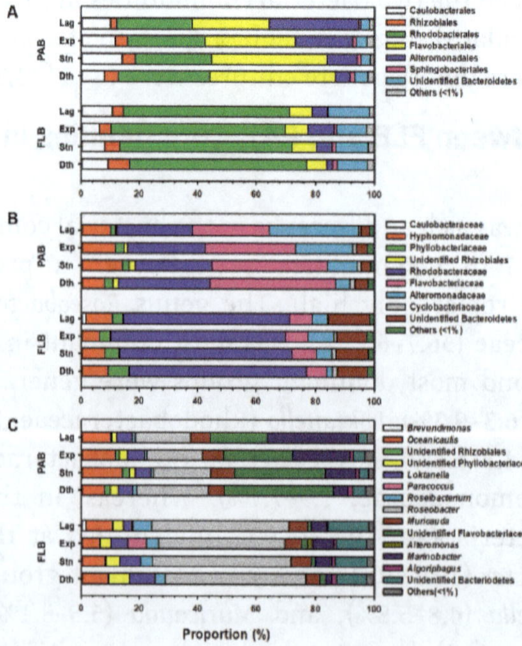

Figure 3. Composition of the FLB and PAB communities in lag, exponential (Exp), stationary (Stn), and death (Dth) growth phases of *T. suecica*. (A): Order; (B): Family; and (C): Genus level.

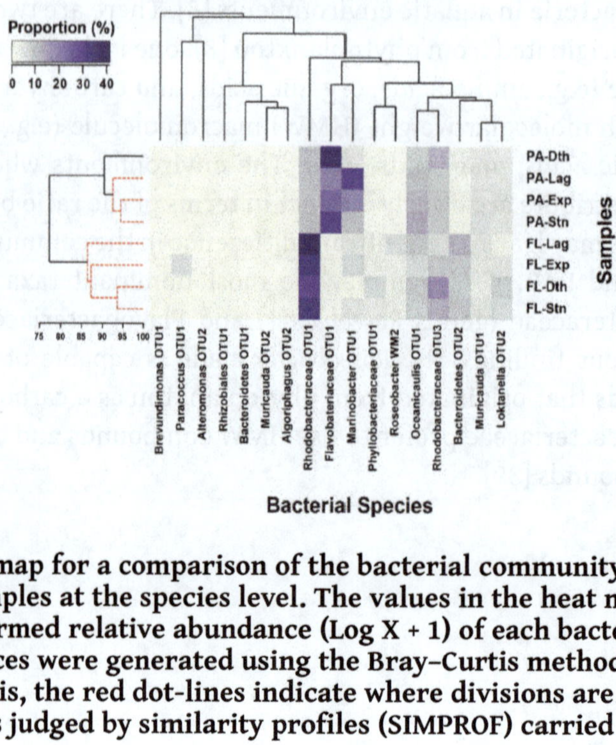

Figure 4. Heatmap for a comparison of the bacterial community composition among the samples at the species level. The values in the heat map represent the log-transformed relative abundance (Log X + 1) of each bacterial OTU, and similarity matrices were generated using the Bray–Curtis method. In the cluster analysis on Y-axis, the red dot-lines indicate where divisions are not statistically significant, as judged by similarity profiles (SIMPROF) carried out at p = 0.05.

Variation in Bacterial Community Composition Depending on the Growth Stage of Tetraselmis suecica The number of bacterial OTUs observed in the lag, exponential, stationary, and death growth phases of *T. suecica* was 29, 31, 27, and 28, respectively for FLB (Figure 3B) and 27, 27, 25, 25, respectively for PAB (Figure 2C). There was no significant difference in bacterial diversity depending on the growth phase of host green algae (Table 1). The range of the Shannon index was 3.105–3.540 (FLB) and 1.908–2.066 (PAB), and the range of the Simpson index was 0.792–0.863 (FLB) and 0.543–0.594 (PAB). Whereas, depending on the growth phase of *T. suecica*, the community composition of FLB and PAB was varied, and, interestingly, the PAB community was more variable than the FLB community (Figures 3 and 4). In the hierarchical cluster analysis, the PAB community composition in the death phase of *T. suecica* was significantly ($p = 0.05$) distinct to that in the other growth phases (lag, exponential, and stationary phases) at the OTU level, where there was no significantly different composition of FLB communities in each growth phase of green algae (Figure 4). To our knowledge, PAB has a stronger species-specific association to the growth and physiological condition of host algae [15–17,28,35], since the microhabitat for PAB is provided by algae [40,41].

Rhodobacteraceae OTU1, belonging to the *Roseobacter* clade, was generally predominant in both FLB and PAB communities even though its proportion in FLB was higher than that in PAB. This suggests that this bacterial OTU may be intimately associated with *T. suecica*. The *Roseobacter* clade can provide growth-promoting compounds (e.g., vitamin B12) to various algal species, including *Tetraselmis* [12,42,43]. Based on a recent study [13], the growth yield of *T. suecica* F and M33 was increased when three isolates from the *Roseobacter* clade were co-cultivated. Interestingly, the bacterial isolate, belonging to genus *Muricauda*, has also shown the capability to enhance the growth of *T. suecica* F and M33 [13].

This genus was one of the dominant taxa in both FLB and PAB communities in our study. Given these findings, dominant bacterial groups of PAB and FLB communities established in *T. suecica* culture may positively contribute to the growth of host green algae. However, the actual impact of these bacterial taxa on the growth of *T. suecica* is thought to be unclear. For example, a cell number of *T. suecica* was largely decreased in the death phase, but the proportion of these bacterial taxa was not significantly decreased (Figures 1 and 3). To our knowledge, inorganic nutrients which are essential to algal growth were generally depleted in the death phase of algae due to the consumption by the algae. Thus, it is more likely that the decrease in *T. suecica* cells in the death phase may be caused by the depletion of inorganic nutrients even though growth promoting bacteria (e.g., *Roseobacter* clade and *Muricauda*) were present.

In the PAB community, the proportion of *Marinobacter* was gradually decreased in accordance with the increase in cell number of *T. suecica*. Its proportion in the lag and exponential growth phase was clearly higher than that in the stationary and death phase, in which the algal growth rate was relatively low (Figures 1 and 3). Several clades of this bacterial genus can secrete a siderophore called vibrioferrin, allowing the promotion of the algal assimilation of iron [44]. Therefore, a lower proportion of *Marinobacter* in the stationary and death phases may adversely affect the growth of *T. suecica*.

Flavobacteriaceae can primarily consume HMW compounds and convert them into LMW compounds [39]. Thus, an abundance of this bacteria is generally elevated when a number of HMW compounds were released from the algae, as a result of cell lysis [45–48]. Similarly, in this study, the proportion of Flavobacteriaceae in the PAB community gradually increased in accordance with an increase in the cell number of *T. suecica* (Figures 1 and 3), resulting in an elevation of the HMW compound concentration.

4. CONCLUSIONS AND REMARKS

The present study is the first to characterize FLB and PAB communities in *T. suecica* culture using the NGS approach. Based on our findings, FLB and PAB communities were significantly distinct ($p = 0.05$), the PAB community was more affected depending on the growth phases of *T. suecica* than FLB community. In addition, bacterial taxa (e.g., *Roseobacter* clade and *Muricauda*) which are capable of enhancing the growth of host green algae were dominant in both FLB and PAB communities, irrespective of growth phase of *T. suecica*. These findings suggest that bacterial community in *T. suecica* culture may positively affect the growth of host algae. However, to evaluate the actual impact of bacterial communities on the growth of *T. suecica*, further extensive studies, such as growth promoting mechanism of bacteria in algal culture, are needed.

REFERENCES

1. Chisti, Y. Biodiesel from microalgae. *Biotechnol. Adv.* **2007**, *25*, 294–306. [CrossRef]
2. Teo, C.L.; Jamaluddin, H.; Zain, N.A.M.; Idris, A. Biodiesel production via lipase catalysed transesterification of microalgae lipids from *Tetraselmis* sp. *Renew. Energy* **2014**, *68*, 1–5. [CrossRef]
3. Peña, M.R.; Villegas, C.T. Cell growth, effect of filtrate and nutritive value of the tropical Prasinophyte *Tetraselmis tetrathele* (Butcher) at different phases of culture. *Aquac. Res.* **2005**, *36*, 1500–1508. [CrossRef]

4. Park, J.; Park, B.S.; Wang, P.; Patidar, S.K.; Kim, J.H.; Kim, S.H.; Han, M.S. Phycospheric native bacteria *Pelagibaca bermudensis* and *Stappia* sp. ameliorate biomass productivity of *Tetraselmis striata* (KCTC1432BP) in co-cultivation system through mutualistic interaction. *Front. Plant Sci.* **2017**, *8*, 289. [CrossRef]

5. Pertile, P.; Zanella, L.; Herrmann, M.; Joppe, H.; Gaebler, S. Extracts of *Tetraselmis* sp. U.S. Patent US2010/0143267 A1, 10 June 2010.

6. Mitra, M.; Patidar, S.K.; George, B.; Shah, F.; Mishra, S. A euryhaline *Nannochloropsis gaditana* with potential for nutraceutical (EPA) and biodiesel production. *Algal Res.* **2015**, *8*, 161–167. [CrossRef]

7. Patidar, S.K.; Mitra, M.; George, B.; Soundarya, R.; Mishra, S. Potential of *Monoraphidium minutum* for carbon sequestration and lipid production in response to varying growth mode. *Bioresour. Technol.* **2014**, *172*, 32–40. [CrossRef]

8. Buchan, A.; LeCleir, G.R.; Gulvik, C.A.; González, J.M. Master recyclers: Features and functions of bacteria associated with phytoplankton blooms. *Nat. Rev. Microbiol.* **2014**, *12*, 686–698. [CrossRef]

9. Geng, H.; Belas, R. Molecular mechanisms underlying *Roseobacter*–phytoplankton symbioses. *Curr. Opin. Biotechnol.* **2010**, *21*, 332–338. [CrossRef]

10. Park, B.S.; Joo, J.H.; Baek, K.D.; Han, M.S. A mutualistic interaction between the bacterium *Pseudomonas asplenii* and the harmful algal species *Chattonella marina* (Raphidophyceae). *Harmful Algae* **2016**, *56*, 29–36. [CrossRef]

11. Park, B.S.; Erdner, D.L.; Bacosa, H.P.; Liu, Z.; Buskey, E.J. Potential effects of bacterial communities on the formation of blooms of the harmful dinoflagellate *Prorocentrum* after the 2014 Texas City "Y" oil spill (USA). *Harmful Algae* **2020**, *95*, 101802. [CrossRef] [PubMed]

12. Seyedsayamdost, M.R.; Case, R.J.; Kolter, R.; Clardy, J. The Jekyll-and-Hyde chemistry of *Phaeobacter gallaeciensis*. *Nat. Chem.* **2011**, *3*, 331–335. [CrossRef] [PubMed]

13. Biondi, N.; Cheloni, G.; Rodolfi, L.; Viti, C.; Giovannetti, L.; Tredici, M.R. *Tetraselmis suecica* F&M-M33 growth is influenced by its associated bacteria. *Microb. Biotechnol.* **2018**, *11*, 211–223.

14. Piampiano, E.; Pini, F.; Biondi, N.; Garcia, C.J.; Decorosi, F.; Tomàs-Barberàn, F.A.; Giovannetti, L.; Viti, C. *Tetraselmis suecica* F&M-M33 phycosphere: Associated bacteria and exo-metabolome characterisation. *Eur. J. Phycol.* **2020**, 1–11. [CrossRef]

15. Crump, B.C.; Armbrust, E.V.; Baross, J.A. Phylogenetic analysis of particle-attached and free-living bacterial communities in the Columbia River, its

estuary, and the adjacent coastal ocean. *Appl. Environ. Microbiol.* **1999**, *65*, 3192–3204. [CrossRef] [PubMed]

16. Jasti, S.; Sieracki, M.E.; Poulton, N.J.; Giewat, M.W.; Rooney-Varga, J.N. Phylogenetic diversity and specificity of bacteria closely associated with *Alexandrium* spp. and other phytoplankton. *Appl. Environ. Microbiol.* **2005**, *71*, 3483–3494. [CrossRef]
17. Rooney-Varga, J.N.; Giewat, M.W.; Savin, M.C.; Sood, S.; LeGresley, M.; Martin, J.L. Links between phytoplankton and bacterial community dynamics in a coastal marine environment. *Microb. Ecol.* **2005**, *49*, 163–175. [CrossRef] [PubMed]
18. Guillard, R.R.; Ryther, J.H. Studies of marine planktonic diatoms: I. Cyclotella nana Hustedt, and *Detonula confervacea* (Cleve) Gran. *Can. J. Microbiol.* **1962**, *8*, 229–239. [CrossRef]
19. Turley, C.M. Direct estimates of bacterial numbers in seawater samples without incurring cell loss due to sample storage. In *Handbook of Methods in Aquatic Microbial Ecology*; CRC Press: Boca Raton, FL, USA, 1993; pp. 143–147.
20. Park, B.S.; Wang, P.; Kim, J.H.; Kim, J.H.; Gobler, C.J.; Han, M.S. Resolving the intra-specific succession within *Cochlodinium polykrikoides* populations in southern Korean coastal waters via use of quantitative PCR assays. *Harmful Algae* **2014**, *37*, 133–141. [CrossRef]
21. Gontcharova, V.; Youn, E.; Wolcott, R.D.; Hollister, E.B.; Gentry, T.J.; Dowd, S.E. Black box chimera check (B2C2): A windows-based software for batch depletion of chimeras from bacterial 16S rRNA gene datasets. *Open Microbiol. J.* **2010**, *4*, 47. [CrossRef]
22. Huse, S.M.; Huber, J.A.; Morrison, H.G.; Sogin, M.L.; Welch, D.M. Accuracy and quality of massively parallel DNA pyrosequenc- ing. *Genome Biol.* **2007**, *8*, R143. [CrossRef] [PubMed]
23. Quast, C.; Pruesse, E.; Yilmaz, P.; Gerken, J.; Schweer, T.; Yarza, P.; Peplies, J.; Glöckner, F.O. The SILVA ribosomal RNA gene database project: Improved data processing and web-based tools. *Nucleic Acids Res.* **2012**, *41*, D590–D596. [CrossRef]
24. Li, W.; Fu, L.; Niu, B.; Wu, S.; Wooley, J. Ultrafast clustering algorithms for metagenomic sequence analysis. *Brief Bioinform.* **2012**, *13*, 656–668. [CrossRef] [PubMed]
25. Schloss, P.D.; Westcott, S.L.; Ryabin, T.; Hall, J.R.; Hartmann, M.; Hollister, E.B.; Lesniewski, R.A.; Oakley, B.B.; Parks, D.H.; Robinson, C.J.; et al. Introducing mothur: Open-source, platform-independent, community-supported software for describing and comparing microbial communities. *Appl. Environ. Microbiol.* **2009**, *75*, 7537–7541. [CrossRef]

26. Clarke, K.; Gorley, R. *PRIMER v7: User Manual/Tutorial*; Primer-E: Plymouth, UK, 2015; 296p.

27. Team, R. *RStudio: Integrated Development for R*; RStudio Inc.: Boston, MA, USA, 2015; 700p.

28. Park, B.S.; Guo, R.; Lim, W.A.; Ki, J.S. Importance of free-living and particle-associated bacteria for the growth of the harmful dinoflagellate *Prorocentrum minimum*: Evidence in culture stages. *Mar. Freshw. Res.* **2018**, *69*, 290–299. [CrossRef]

29. Parveen, B.; Mary, I.; Vellet, A.; Ravet, V.; Debroas, D. Temporal dynamics and phylogenetic diversity of free-living and particle- associated Verrucomicrobia communities in relation to environmental variables in a mesotrophic lake. *FEMS Microbiol. Ecol.* **2013**, *83*, 189–201. [CrossRef]

30. Rösel, S.; Grossart, H.P. Contrasting dynamics in activity and community composition of free-living and particle-associated bacteria in spring. *Aquat. Microb. Ecol.* **2012**, *66*, 169–181. [CrossRef]

31. Noble,P.A.;Bidle,K.D.;Fletcher,M.Natural Microbial Community Compositions Compared by a Back-Propagating Neural Network and Cluster Analysis of 5S rRNA. *Appl. Environ. Microbiol.* **1997**, *63*, 1762–1770. [CrossRef]

32. Riemann, L.; Winding, A. Community dynamics of free-living and particle-associated bacterial assemblages during a freshwater phytoplankton bloom. *Microb. Ecol.* **2001**, *42*, 274–285. [CrossRef] [PubMed]

33. Tang, X.; Li, L.; Shao, K.; Wang, B.; Cai, X.; Zhang, L.; Chao, J.; Gao, G. Pyrosequencing analysis of free-living and attached bacterial communities in Meiliang Bay, Lake Taihu, a large eutrophic shallow lake in China. *Can. J. Microbiol.* **2015**, *61*, 22–31. [CrossRef] [PubMed]

34. Zhang, R.; Liu, B.; Lau, S.C.; Ki, J.S.; Qian, P.Y. Particle-attached and free-living bacterial communities in a contrasting marine environment: Victoria Harbor, Hong Kong. *FEMS Microbiol. Ecol.* **2007**, *61*, 496–508. [CrossRef]

35. Park, B.S.; Guo, R.; Lim, W.A.; Ki, J.S. Pyrosequencing reveals specific associations of bacterial clades *Roseobacter* and Flavobac- terium with the harmful dinoflagellate *Cochlodinium polykrikoides* growing in culture. *Mar. Ecol.* **2017**, *38*, e12474. [CrossRef]

36. Buchan, A.; González, J.M.; Moran, M.A. Overview of the marine *Roseobacter* lineage. *Appl. Environ. Microbiol.* **2005**, *71*, 5665–5677. [CrossRef]

37. Merzouk, A.; Levasseur, M.; Scarratt, M.; Michaud, S.; Lizotte, M.; Rivkin, R.B.; Kiene, R.P. Bacterial DMSP metabolism during the senescence of the spring diatom bloom in the Northwest Atlantic. *Mar. Ecol. Prog. Ser.* **2008**, *369*, 1–11. [CrossRef]

38. Wagner-Döbler, I.; Biebl, H. Environmental biology of the marine *Roseobacter* lineage. *Annu. Rev. Microbiol.* **2006**, *60*, 255–280. [CrossRef] [PubMed]

39. Teeling, H.; Fuchs, B.M.; Becher, D.; Klockow, C.; Gardebrecht, A.; Bennke, C.M.; Kassabgy, M.; Huang, S.; Mann, A.J.; Waldmann, J.; et al. Substrate-controlled succession of marine bacterioplankton populations induced by a phytoplankton bloom. *Science* **2012**, *336*, 608–611. [CrossRef]

40. Bagatini, I.L.; Eiler, A.; Bertilsson, S.; Klaveness, D.; Tessarolli, L.P.; Vieira, A.A.H. Host-specificity and dynamics in bacterial communities associated with bloom-forming freshwater phytoplankton. *PLoS ONE* **2014**, *9*, e85950. [CrossRef]

41. Worm, J.; Søndergaard, M. Dynamics of heterotrophic bacteria attached to *Microcystis* spp. (Cyanobacteria). *Aquat. Microb. Ecol.* **1998**, *14*, 19–28. [CrossRef]

42. Helliwell, K.E.; Wheeler, G.L.; Leptos, K.C.; Goldstein, R.E.; Smith, A.G. Insights into the evolution of vitamin B12 auxotrophy from sequenced algal genomes. *Mol. Biol. Evol.* **2011**, *28*, 2921–2933. [CrossRef] [PubMed]

43. Wagner-Döbler, I.; Ballhausen, B.; Berger, M.; Brinkhoff, T.; Buchholz, I.; Bunk, B.; Cypionka, H.; Daniel, R.; Drepper, T.; Gerdts, G.; et al. The complete genome sequence of the algal symbiont *Dinoroseobacter shibae*: A hitchhiker's guide to life in the sea. *ISME J.* **2010**, *4*, 61–77. [CrossRef]

44. Amin, S.A.; Green, D.H.; Hart, M.C.; Küpper, F.C.; Sunda, W.G.; Carrano, C.J. Photolysis of iron–siderophore chelates promotes bacterial–algal mutualism. *Proc. Natl. Acad. Sci. USA* **2009**, *106*, 17071–17076. [CrossRef]

45. Azam, F.; Malfatti, F. Microbial structuring of marine ecosystems. *Nat. Rev. Microbiol.* **2007**, *5*, 782–791. [CrossRef]

46. Pinhassi, J.; Sala, M.M.; Havskum, H.; Peters, F.; Guadayol, O.; Malits, A.; Marrasé, C. Changes in bacterioplankton composition under different phytoplankton regimens. *Appl. Environ. Microbiol.* **2004**, *70*, 6753–6766. [CrossRef] [PubMed]

47. Riemann, L.; Steward, G.F.; Azam, F. Dynamics of bacterial community composition and activity during a mesocosm diatom bloom. *Appl. Environ. Microbiol.* **2000**, *66*, 578–587. [CrossRef] [PubMed]

48. Simon, M.; Glöckner, F.O.; Amann, R. Different community structure and temperature optima of heterotrophic picoplankton in various regions of the Southern Ocean. *Aquat. Microb. Ecol.* **1999**, *18*, 275–284. [CrossRef]

CHAPTER-9

COAGULANT PLUS *BACILLUS NITRATIREDUCENS* FERMENTATION BROTH TECHNIQUE PROVIDES A RAPID ALGICIDAL EFFECT OF TOXIC RED TIDE DINOFLAGELLATE

1. INTRODUCTION

At present, red tides have become one of the ocean's most catastrophic global disasters. The occurrence frequency, outbreak intensity, and impact range of red tides worldwide have been increasing and causing various degrees of harm to many countries' coastal regions. It is conservatively estimated that the annual loss of fishery and tourism caused by harmful algae blooms (HABs) in Europe is up to EUR 862 million, while the annual loss caused by HAB in the United States is up to USD 82 million. Between 1995 and 2004, an average of USD 1.31 million was lost annually in South Korea's fisheries due to harmful algal blooms. Red tide refers to the rapid proliferation or accumulation of various dinoflagellates and diatoms under external environmental conditions, mainly in marine environments such as coastlines or estuaries [1].The algal cell density reaches a certain level, causing water discoloration and affecting coastal areas and aquatic ecosystems, which, in turn, cause severe effects and hinder tourism [2]. The dinoflagellate *Gymnodinium catenatum* is a dominant harmful algae bloom (HAB)-forming species along coasts worldwide [3,4]. For example, *G. catenatum* is a bloom-forming species that forms HABs in China's Fujian coastal waters almost every year [5]. In 2017, a red tide of *G. catenatum* broke out in the coastal waters near Quanzhou and the Zhangzhou Sea. The continuous eruption of *Gymnodinium* species and cystic settlements formed by sexual reproduction in sediments poses a significant threat to the aquaculture industry and people's health. Therefore, dinoflagellate cyst deposits in coastal areas, the distribution patterns, and the abundance of species are an urgent need to control the harmful effects generated by HABs [6].

Different methods have been developed to prevent and eliminate red tides, including physical, chemical, and biological methods in recent decades. These three methods have their advantages and disadvantages in red tide management [7]. Among these methods, physical methods will not harm the original ecosystem, but they are also inefficient, expensive, not suitable for large-scale red tides, and can only suppress red tides quickly. Chemical methods of algae removal are very efficient and can destroy algae cells. Regrettably, these methods are quite costly and may lead to secondary pollution [8–10]. Biological methods, however, are economical, effective, and environmentally friendly. In particular, microbial algae suppression methods offer many advantages such as simple operation, complete algae killing, and no secondary pollution to the environment [11]. The management of current red tide hotspot areas is of great significance, and the prospect of controlling HAB algae-killing microorganisms has also been in rapid development. In particular, these algaecide bacteria can lyse algae by directly or indirectly attacking cells [12–15]. Certain substances secreted by microorganisms will cause cell lysis and death by invading and contacting algae cells, although the action time is longer.

Nowadays, related research has discovered that the combined algae removal effect of multiple methods is more effective than that of a single method. However, the inhibitory effect of combined methods on coastal HAB species is much less studied. In the present study, chemical algae removal can reduce algae cells' density for a temporary period during a red tide outbreak. Among these particular methods, the coagulation method has relatively high safety [16]. Polyaluminum chloride (PAC), aluminum sulfate [$Al_2(SO_4)_3$], ferric sulfate [$Fe_2(SO_4)_3$], and ferric chloride ($FeCl_3$) can also increase HAB removal efficiency. Besides biological algae removal methods, algae suppression bacteria and fermentation broth algae suppression methods have become new research directions in recent years. Few studies have demonstrated that *Bacillus* sp. can suppress the growth of harmful algal bloom species [17]. However, the effect of the combined approach on micro-algae is little known. To the best of our knowledge, there have been no reports so far. In this study, we first compared four coagulants, namely PAC, $Al_2(SO_4)_3$, $Fe_2(SO_4)_3$, and $FeCl_3$, against *G. catenatum*. The coagulant we selected has the best flocculation effect and the best algae inhibition effect; on this basis, we explored the algaecide effect of *Bacillus* fermentation broth (Ba3) and further studied the inhibitory effect of the combination of coagulant and Ba3 on *G. catenatum*. Furthermore, we used this method to inhibit the germination of algae cysts in the sediments. We planned a systematic study of the comprehensive effect of the coagulant Ba3 on the red tide of algae and its germination as controlled from the source red tide outbreak, which may have crucial effective management strategies. The framework of the study design is shown schematically in Figure S1.

2. MATERIAL AND METHODS

2.1 Cultivation of the Dinoflagellate and Bacteria

Gymnodinium catenatum was obtained from the State Key Laboratory of Marine Environmental Science at Xiamen University, China. We maintained the axenic algal culture at 20 ± 2°C in a sterile L1 medium prepared with natural seawater filtered to 0.45 μm and maintained under a 12:12 h light/dark cycle. We also counted cell numbers under a microscope. The algal culture was transferred once a week to a fresh, sterilized medium, which ensured that experiments were always conducted with cultures during the exponential growth phase.

We previously identified the algicidal bacteria preserved in the College of Life Sciences, Fujian Normal University, as *Bacillus nitratireducens*. The strain was initially cultivated with a modified Bacillus Medium (Peptone 10 g/L, sodium chloride 5 g/L, beef paste 5 g/L, pH value 7.2~7.4, dissolved in deionized water) in a rotary shaker (30°C, 180 rpm), and the resultant mixture kept in our laboratory was thoroughly stirred for subsequent experiments.

2.2 Cultivation of Gymnodinium catenatum Cysts

Dinoflagellate has a high cyst formation rate under low nitrogen and phosphorus environments [18]. Usually, the limitation of nutrients in the water phase is an effective way to induce vegetative cells to form cysts. Therefore, an L1 medium with low phosphate and nitrate was used to prepare *G. catenatum* cysts. The dilution ratio of phosphate and nitrate was 1:15 (named L15). In brief, we added 20 mL of 10^6 cells/L of *G. catenatum* in a 50-mL centrifuge at 3000 rpm for 10 min, discarded the supernatant, and then slowly added 20 mL of L15 medium. After adding the medium, the algae cells in the centrifuge tube were mixed with the medium and then transferred to a sterile Erlenmeyer flask for cultivation.

2.3 Selecting Coagulants and Preparing Concentration

Four agents, including PAC, $Al_2(SO_4)_3$, $Fe_2(SO_4)_3$, and $FeCl_3$, were used in the experiments. The coagulant concentration was set to 0, 10, 20, 30, 50, 70, and 90 mg/L; the first group (0 mg/L) was used as the control, and the following six were used as the treatment groups. Each treatment was established in triplicate.

2.4 Cell Inhibition Efficiency

2.4.1 First Experiment

The efficiency of the cell removal experiment was tested in 25-mL sterile test tubes. A 20-mL aliquot of the algal culture was placed in each tube with a density of 2×10^6~ 4×10^6 cells/L; then, 0, 10, 20, 30, 50, 70, or 90 mg/L of the four agents' stock solution was evenly added to the tube. After adding the agent, each tube was mixed thoroughly, and each concentration was tested in triplicate on the alga cells. After three periods (3, 24, and 48 h), three replicate samples from each group were pipetted from the upper-middle region of the collected liquid surface, and each sample's concentration of algal cell was determined under the microscope; photographs were taken for the bottom flocs of each treatment. In addition, the supernatant's pH value was measured to determine the total removal rate of zeta potential, turbidity, and UV254; then the positive effects and the best dose of the coagulant were noted for further experiments.

2.4.2 Second Experiment

B. nitratireducens bacteria were inoculated into the culture medium and grown to the stationary phase (30°C at 180 rpm for 24 h). The extraction of bacterial fermentation broth (Ba3) was collected using centrifugation (10,000, 15 min) over three days (1, 2, and 3 d). The supernatants were filtrated through 0.22-µm Millipore membrane filters and then used. Ba3 volume ratios of 0.3, 0.7, 1.0, and 2.0% were added to the 30-mL exponential phase *G. catenatum* algae. Each group was placed in a light incubator for culturing, shaken twice a day, and sampled once every two days until the end of the experiment. The group with no Ba3 served as a control for the experiments, and all experiments were repeated in triplicate.

2.4.3 Third Experiment

In the first and second experiments, a more effective dosage of mixed coagulant and Ba3 fermentation broth was selected. This experiment determined whether it was possible to improve the previous experiments results using combinations of more effective agents and Ba3. The above experiments' dosage of the various coagulants and the fermentation broth dosage for two days were 0.3, 0.7, 1.0, and 2.0%. The combined coagulants and fermentation broth were added to 20 mL *G. catenatum* alga in triplicate samples for each group for better cell inhibition efficiency. Each group was placed in a light incubator for cultivation, and samples were assessed at 3, 24, 48, and 96 h for density determination to ensure that they had significant inhibition efficiency.

2.5 Calculation Method of Algae Removal Rate

The algae removal rate was monitored by estimating cell numbers utilizing a micro- scope and calculated according to the following formula.

$$RE\ (\%) = (1 - N_t/N_0) \times 100\%$$

where RE is the removal rate of algal cells; N_0: vegetative cell density before adding algae inhibitor; and N_t is the vegetative cell density after adding the algae inhibitor.

2.6 Analytical Methods

An optical microscope was used to observe the effect of fermentation broth on the morphological characteristics of algal cells. After the addition of algal inhibitor, 1 mL of the supernatant was removed, placed on a slide, and observed under a microscope. The size of the coagulant and the flocs as well as the compactness of the flocs were photographed for further characterization of the coagulation effect of different coagulants on the cells of dinoflagellates. UV_{254} was determined by using the spectrometer as an indicator to reflect the content of organic pollutants in water. The spectrometer reading does not represent a specific organic substance, but rather the total amount of multiple organic substances; therefore, the spectrometer reading refers to the many soluble fine particles and various inorganic compounds present in the experiment. In this experiment, after stirring, the supernatant was taken from 2 cm below the liquid level after standing for 24 h. The supernatant was filtered using a 0.45-μm acetate fiber membrane, and then, pure water was used as the reference solution to determine the results of each experiment group. A spectrophotometer analyzed turbidity with a wavelength of 660 nm. Zeta potential is one of the indexes that reflect the stability of suspended matter or colloid in water. The zeta potential was monitored using Zetasizer software. In this experiment, 2 mL of algal sedimentation floc was taken and injected into the sample pool with an injection needle, and the sample pool was put into the card slot to ensure that the algal sedimentation floc in the sample pool did not exceed the electrodes at both ends of the sample pool. Zetasizer software was used to measure and record the results. Dissolved inorganic phosphorus (DIP) and dissolved inorganic nitrogen (DIN) were determined using a spectro-photometer at a wavelength of 882 nm.

2.7 Provenance Control Experiment

This experiment was divided into two groups; one group was marine sediments—PAC + Ba3 + sediment (marked as I-1, I-2, and I-3)—and the other group was marine sediments + pure-breed *Gymnodinium* cysts 25 ± 2 cysts/g DW—PAC + Ba3 + sediments + pure-breed (marked as II-1, II-2, and II-3). The framework of the study design is shown schematically in Table S1. The dosages of the treatment used in the study were as follows: PAC 50 mg/L, and PAC + Ba3 50 mg/L + 0.3%.

2.8 A Calculation Method of Algae Cell Abundance

We calculated the algae cell abundance rate after germination in the sediment were calculated by visual observation method using a microscope and calculated according to the following formula.

$$N = (40/30) \times (V_0/V) \times n$$

where N is the abundance of algal cells after germination, with the unit of cells/mL; V is the count volume taken in each observation, with the unit of ml; V_0 is the volume of the sample after concentration, with the unit of mL; n is the number of algae in each sample.

2.9 Data Processing

The experimental data were analyzed using SPSS 22.0 software, and the difference between the data ($p < 0.05$) was significant ($p < 0.01$).

3. RESULTS

3. 1 The Effect of Different Coagulants on the Algae Inhibition

The four kinds of agents have noticeable removal effects on *G. catenatum*, but different degrees of the "back-dissolving" phenomenon appeared over time. Figure 1 shows that when the reagent dosage was low (10 mg/L) at 3 h, the PAC, $Fe_2(SO_4)_3$, $Al_2(SO_4)_3$, and $FeCl_3$ algae removal rates achieved were 80, 68.5, 77.5, and 80%, respectively. When the dosage was increased, the difference between PAC, $Fe_2(SO_4)_3$, $Al_2(SO_4)_3$, and $FeCl_3$ in algae removal efficiency rate gradually narrowed, while that of the $FeCl_3$ group was relatively low. The removal rate of *G. catenatum* in the PAC group increased when the dosage was increased. When the dosage was 50 mg/L, the removal rate reached more than 95.1%; when the dosage was 70 and 90 mg/L, the removal rate reached 100%.

After 24 h of dosing, the removal rates of different dosing amounts were between 65.0 and 92.9%. When the dosage was 50 mg/L, the removal rate of *G. catenatum* decreased to 80.3%. After 48 h of dosing, the resolution phenomenon was more marked when the dosage was 10 mg/L, which reduced the removal rate to 52.9%. After 96 h of dosing, the removal effect of *G. catenatum* increased compared with that at 48 h, and the removal rate was 69.4% at 10 mg/L. However, when the dosage was greater than 50 mg/L, the removal rate of *G. catenatum* was still higher than 80% over time. After 3 h of dosing, the different dosages' removal rates differed significantly from those between 24 and 96 h ($p < 0.05$). When the dosage of $Fe_2(SO_4)_3$ was higher than 20 mg/L, the removal rate was higher than 85%. After 48 h of administration, the removal rate decreased; after 96 h of administration, the removal rates of the 10, 20, and 30 mg/L groups increased, but when the dosage of ferric sulfate increased to 90 mg/L, the removal rate decreased to 69.4%. The removal rate was not altered between 24 and 96 h. When the $Al_2(SO_4)_3$ dosage was 30 mg/L, the removal rate reached 85% or more; above this dosage, the removal rate did not change significantly ($p > 0.05$). When the dosage was 90 mg/L, the removal rate of *G. catenatum* between 24 and 96 h was slightly higher than that after 3 h. The $FeCl_3$ group activity was relatively low with the increase in the dosage, and the removal rate did not increase significantly ($p > 0.05$); when the dosage was 90 mg/L, the removal rate was 90.2%. With time, the removal rate decreased significantly at a low dosage (10~50 mg/L) ($p < 0.05$). When the dosage was 70 mg/L, the removal rate was stable (Figure 1). After adding different dosages, the removal rate did not change significantly between 24 and 96 h.

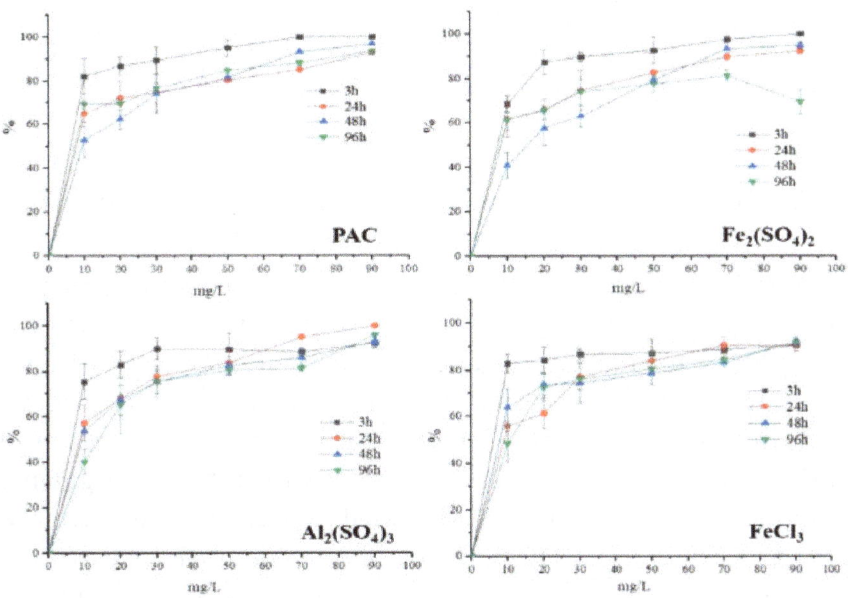

Figure 1. Removal effect of four different coagulants on *G. catenatum*.

Figure 2A shows the four coagulants' influence the pH range, and the control group's pH value was 8.1. With the addition of coagulants, the pH values of the four groups showed a downward trend. $FeCl_3$ had the smallest effect on pH at 90 mg/L, and the pH value was the lowest, which was 7.6. The pH range of the $Al_2(SO_4)_3$ group was 8.1~6.8. The influence trend of the PAC and $Fe_2(SO_4)_3$ groups affected *G. catenatum* in almost the same way, and the variation range was 8.1~6.8. The pH range between 6.8 and 8.1 exhibited a good algal removal effect at both a higher dosage of coagulant and a lower dosage of coagulant, a pH decline, and the efficiency was slightly less reactive (Figure 2A).

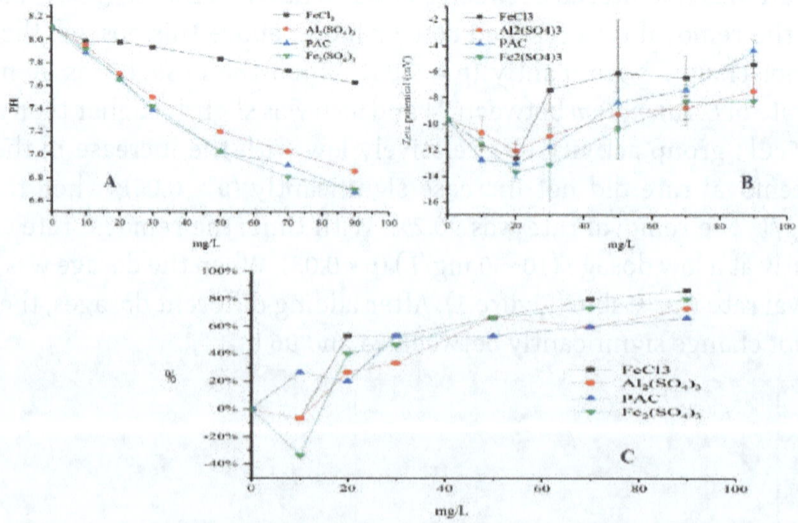

Figure 2. (A) Effect of coagulants on pH; (B) changes in zeta potential; (C) removal effects of coagulants on turbidity.

Zeta potential is one of the evaluation indices of the coagulation method's water treatment effect, showing the stability of colloids or suspended solids in a solution system. Coagulation and algae removal usually use the positively charged aluminum or iron hydrolyzed cations formed by the coagulant in water and the negative charge on the algae cells' surface to attract each other. Figure 2B shows the potential change in the surface of *G. catenatum* with different coagulants added. The potential of the four coagulants decreased first and then increased with the increase in the dosage. The lowest point of the potential appeared when the dosage was 20 mg/L. Among them, the rising trend of FeCl3 was the most obvious. When the dosage was 50 mg/L, the potential was the highest at -6.1 mV, but the fluctuation range was also higher. The Al2(SO4)3 group fluctuated slowly with the dosage increase, and the dosage potential at 20 mg/L was higher than other coagulant groups. The Fe2(SO4)3 group's potential decreased the most, and at a high dose (90 mg/L), the potential exceeded the other coagulant groups. In the PAC group, after the dosage was higher than 20 mg/L, and with the increase in the dosage, the surface potential of the algae cells increased rapidly. The highest point appeared at 90 mg/L, with a potential of -5.3 mV (Figure 2B). The turbidity of the system reflects the sedimentation effect of algae cells, and different dosages of different coagulants have different effects on the removal of turbidity.

Figure 2C illustrates the dosage at 10 mg/L. The turbidity of the four retardants showed significant differences, among which the turbidity of Fe2(SO4)3 was the lowest, followed by the turbidity of Al2(SO4)3, FeCl3, and PAC. That said, the removal rate was higher except for PAC; the other three coagulant groups' turbidity removal rates were all negative, thus increasing the system's turbidity. When the dosage was greater than 30 mg/L, the turbidity's changing trend was stable (Figure 2C). Comprehensive analysis of the removal effect of the density, turbidity, and organic matter of *G. catenatum* when using the four coagulants PAC, $FeCl_3$, $Fe_2(SO_4)_3$, and $Al_2(SO_4)_3$ showed that PAC had the most robust ability to remove algae on *G. catenatum* and had a low re-solubilization rate. Large flocs made of a large number of flocculated algal cells were formed, and PAC proved to have a good removal effect on their turbidity and organic matter.

An optical microscope was used to observe the settled flocs to observe coagulant and dinoflagellate coagulation effects. Figure 3a–d show the coagulation effects of the four groups of coagulants on *G. catenatum*. The micrograph suggests that the algae cells in the flocs, formed by the coagulation of aluminum sulfate and algae cells, were not tight enough, and there were few algae cells fixed in the flocs; however, the cell morphology of *G. catenatum* was also observed, and the cell shape was still active. The floc's micrograph that settled on the bottom after the $FeCl_3$ group was added for 96 h (Figure 3b). It is also evident that ferric chloride has a good coagulation effect on *G. catenatum* because the flocs were large, and few algal cells were swimming outside the flocs. The algae cells in the flocs were many and dense, but the flocs were not tight. The swimming algae cells could easily break away from the flocs and return to the water body. The morphology of algae cells in the flocs was almost unchanged at 400×. A micrograph of the floc that settled to the bottom after the $Fe_2(SO_4)_3$ group, which was added for 96 h, is shown in Figure 3c.

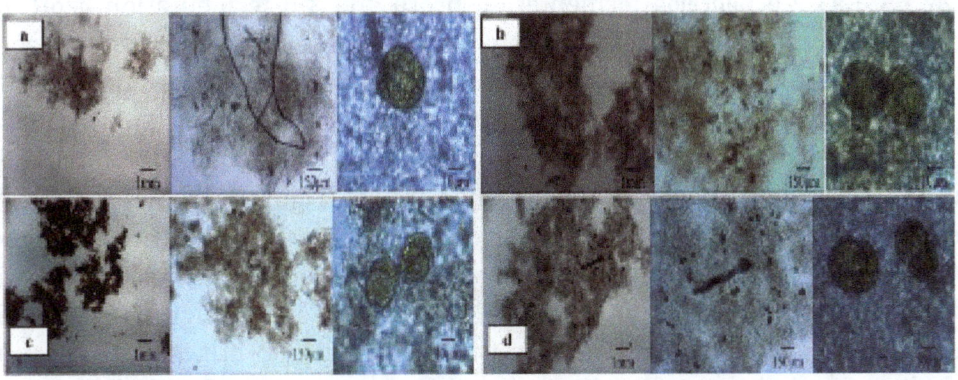

Figure 3. (a–d) Micrographs of flocs after sedimentation of (a) $Al_2(SO_4)_3$, (b) $FeCl_3$, (c) $Fe_2(SO_4)_3$, and (d) polyaluminum chloride (PAC) group for 96 h.

Clearly, the flocs were small and scattered, and algae cells were swimming outside the flocs. As the floc is small, it distributes the algae cells on the edge of the floc. As the algae cells swim, the algal cells may break free from the flocs, causing a "back-dissolution" phenomenon. A micrograph of a floc that settled to the bottom after PAC administration for 96 h is shown in Figure 3d. At 40×, we could see that the flocs formed by the flocculation and *G. catenatum* cells were more extensive, and there were fewer algal cells outside the flocs. Under the 100× microscope, we observed that the algae cells in the flocs were relatively dense, there were few algae cells at the edges of the flocs, and the coagulation effect was better. PAC has a minor effect on the cell morphology of *G. catenatum*, and the four coagulants have different effects on the removal of algae. Although the surface potential of the algal cells increased after PAC addition, the fluctuations were larger.

3.2 The Effect of Bacterial Fermentation Broth (Ba3) on the Algae Inhibition

Ba3 fermented broth from 1, 2, and 3 d was added to an algal concentration of 1200~1500 cells/mL at four dosage levels of 0.3, 0.7, 1.0, and 2.0% (*v*/*v*). The experimental results are displayed in Figure 4A. With the increase in the action time, the concentration of *G. catenatum* continued to decrease. After the second day of dosing, the concentration of algal cells decreased to below 500 cells/mL. The removal rates of all groups were at least 68.1%, and the highest one was 92.9%. On days 2–6, the concentration was stable. After adding Ba3 bacteria fermentation broth, compared with the control group, all concentrations of fermentation broth had a more obvious removal effect. From the results of adding Ba3 bacteria 1-d fermentation broth, we observed that with the increase in the addition, the concentration of *G. catenatum* showed a downward trend. When the addition was 2.0%, the cell number reached the lowest value. The removal rate was as high as 82.1%. Ba3 2-d bacteria fermentation broth and 3-d fermentation broth also increased the algaecide effect with the dosage. The fermentation broth usage at different fermentation times also has a certain impact on the algaecide effect. The figure clearly shows that the effect of Ba3 bacteria 2-d fermentation broth was higher than that of the 1-d fermentation broth and 3-d fermentation broth but was not affected by the dosage or influence of time. Figure 4B shows the effect of removing algae from fermentation broth on turbidity. We designed this experiment with two groups—with and without algae—to observe the turbidity changes by adding bacteria fermentation liquid. The turbidity of the algal and algal-free groups increased with the addition of the fermentation broth. Among them, the turbidity of each dosage of Ba3 bacteria 2-d fermentation broth was slightly lower than that of the 1-d and 3-d fermentation broths. Except for the control group, the algae-containing group's turbidity and the non-algae-containing group in each experimental group differed, showing that after the bacterial fermentation broth destroyed the algal cells, a large amount of content was released, leading to a significant increase in turbidity.

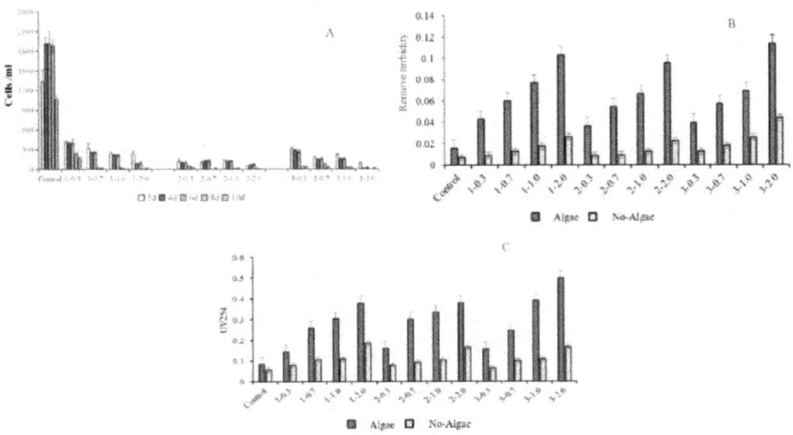

Figure 4. (A) Effect of fermentation broth on the growth of *G. catenatum;* (B) effect of algal removal from fermentation broth on turbidity; (C) effect of algal removal from fermentation broth on UV254.

The bacterial fermentation broth consisted of a variety of active products secreted by the bacterial body. Therefore, the various organic substances present in the bacterial fermentation broth not only increased the turbidity of the water body but also increased the UV254 value of the water body. The algae cells ruptured, and the organics inside the cells flowed out. In short, the fermentation broth can destroy the integrity of algal cells, but the action time is longer. After the algae cells dissolve, the body's organic matter will be released, and the bacterial fermentation broth will maintain certain turbidity. As a result, the water phase's turbidity and organic content will increase after processing the bacterial fermentation broth (Figure 4C).

Figure 5 shows the influence of algal cell morphology. The surface of the algae cells of *G. catenatum* was smooth and complete before the bacterial fermentation broth treatment, with prominent horizontal grooves and nuclei (Figure 5A). After 12 h of Ba3 treatment (Figure 5B), the upper shell of algal cells had become transparent, and the nucleus was visible. The cell wall showed damage to a certain extent, but the cell membrane was still intact and no content flowed out. Over 24 h (Figure 5C), the algae cells' surfaces were loose and ruptured, and the cell membrane was damaged. At 36 h (Figure 5D), the cell membrane was revealed to be severely damaged. Many granular materials of different sizes appeared around the cells. It is possible that after the cell membrane of the cell wall is ruptured, the contents of the cell overflow, and the algal cell's morphological structure becomes relatively blurred. After 96 h of observation, *G. catenatum* had lost its morphology entirely, and the algal cells gradually decomposed and ruptured into small particles that were almost unrecognizable (Figure 5E,F). From the analysis, the algicidal effect of

Ba3 on *G. catenatum* is triggered via indirect algaecide activity. The bacteria secrete some active substances for algae cells to lyse the algae cells, achieving the algae-killing effect.

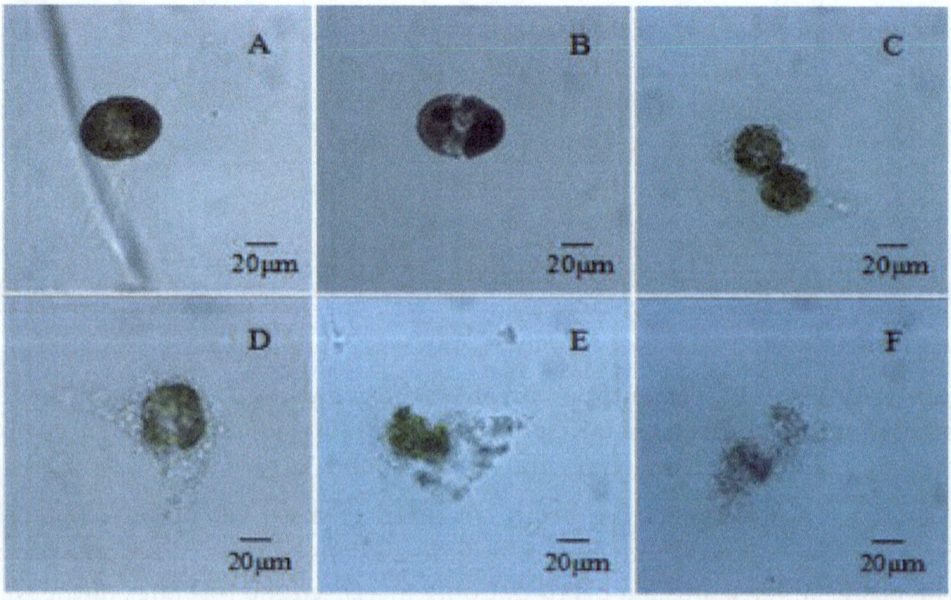

Figure 5. Effect of fermentation broth on *G. catenatum* ((A)—0 h; (B)—12 h; (C)—24 h; (D)—36 h; (E) and (F)—96 h).

3.3 The Effect of Combined Coagulant and Ba3 on the Algae Inhibition

In this study, the coagulant and Ba3 fermentation broth were combined to eliminate algal cells in order to increase the removal effects on *G. catenatum*. As shown in Figure 6A, the removal effect of the four coagulants combined with four concentrations of Ba3 broth on *G. catenatum* occurred at different times. The PAC and Ba3 fermented broth group had the best algae removal effect. The removal rate reached 100%, and the number of algal cells in the overlying water did not increase over time. The effect of other coagulants and algae cells combined with the algae inhibition method was lower than that of the PAC and Ba3 fermented broth group. Figure 6B shows that the group of $Al_2(SO_4)_3$, combined with the Ba3 fermented broth, caused the algae density to reach the highest value at 24 h; the removal rate of *G. catenatum* was 82.3%. Figure 6C shows that with the $Fe_2(SO_4)_3$ and Ba3 broth group, the removal of algal cells reached the highest rate at 80.7%. Figure 6D shows the $FeCl_3$ and Ba3 fermented broth group and the changing trend; all combinations reach the highest value at 24 h. Overall, the combination of PAC and Ba3 broth has the best algae inhibition effect, and when the volume ratio of bacterial fermentation broth is 0.3%, the removal rate can reach 100%. The combination method of algae suppression can improve algae suppression efficiency quickly and reduce the effect of algae cell re-dissolution. This method can be combined with the addition of the coagulant to act on the algae cells to form flocs and precipitate to the bottom; the bacterial fermentation broth also acts on the algae cells, which are gradually lysed under the stimulation of the active substance, and the exercise ability gradually decreases until the cells rupture.

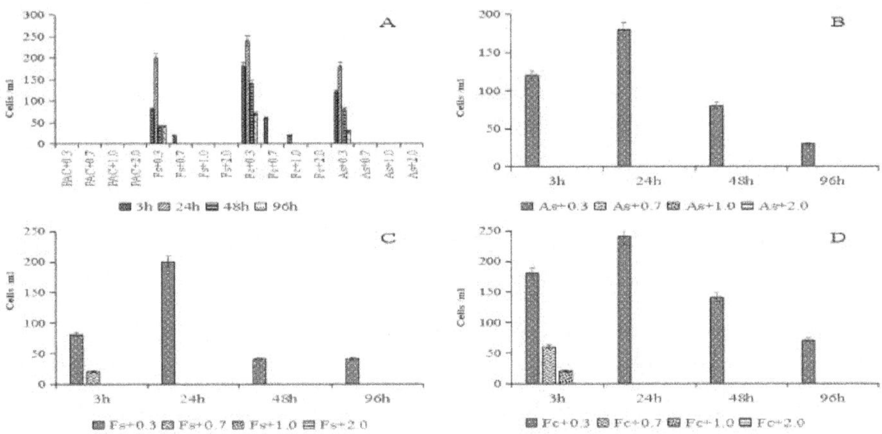

Figure 6. Effects of coagulant and fermentation broth on growth of *G. catenatum*. (A) Combination of four coagulants and fermentation broth to inhibit algae; (B) $Al_2(SO_4)_3$ + Ba3 broth; (C) $Fe_2(SO_4)_3$ + Ba3 broth; and (D) $FeCl_3$ + Ba3 broth.

Figure 7 shows the effect of the combination method of coagulant and Ba3 fermented broth on UV254. With the increase in Ba3 broth dosage, each group's organic removal effect showed a downward trend. In the PAC-combined Ba3 broth group, when the dosage of the bacterial fermentation broth was 0.3% (*v/v*), the maximum removal rate was 41.0%, and the lowest removal rate was 1.0%. This removal rate confirms that when the dosage of Ba3 broth is higher than 1.0%, the content of organic matter in water is higher than the flocculation effect of PAC. The removal rate of $Al_2(SO_4)_3$-combined Ba3 broth group was negative when the dosage was higher than 0.3%, demonstrating that the removal effect of $Al_2(SO_4)_3$ on the organic matter is lower than that of PAC. In the fermented $Fe_2(SO_4)_3$- and $FeCl_3$-combined Ba3 broth groups, the two groups' removal rates were negative for different bacterial fermentation broth dosages. Therefore, the effects of the four coagulants combined with different concentrations of Ba3 broth on the removal of organic matter followed the order of PAC + Ba3 fermented broth >$Al_2(SO_4)_3$ + Ba3 broth > $Fe_2(SO_4)_3$ + Ba3 broth > $FeCl_3$ + Ba3 fermented broth.

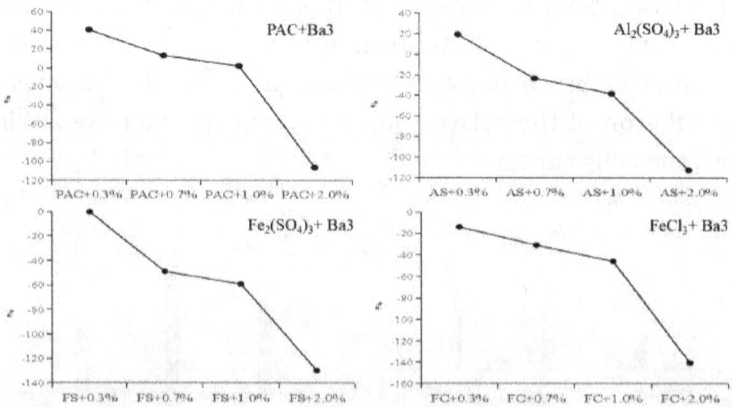

Figure 7. Removal effects on UV254 by the combination of coagulants and fermentation broth.

Figure 8 shows the effect of the combination method of coagulant and Ba3 broth on turbidity. Each group's turbidity removal effect showed a downward trend with the increase in bacterial fermentation broth dosage. Each group had the highest removal rate when the bacterial fermentation broth was added at 0.3%. The removal rates of turbidity in the fermentation broth group of PAC combination Ba3 broth were between 46.3 and 89.1%; the removal rates of turbidity in the fermentation broth group of Al2(SO4)3 combination bacteria were between 55.8 and 83.7%; with Fe2(SO4)3, the turbidity removal rate of the combined Ba3 broth group was between 58.2 and 89.1%; the turbidity removal rates of the FeCl3-combined Ba3 broth group were between 57.8 and 88.4%.

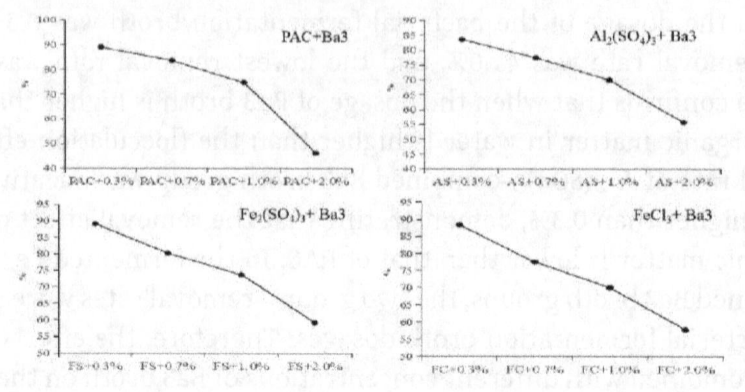

Figure 8. Removal effects on turbidity by the combination of coagulants and fermentation broth.

3.4 The Effect of Combined Coagulant and Ba3 on Cyst Germination Inhibition

The sediment used for algae germination in the simulated sediment was from the Quanzhou section where the red tide of *G. catenatum* had occurred, and the total abundance of phytoplankton in the sediment was 8.04×10^2 cells/g, mainly diatoms and dinoflagellate cysts. The in situ experiment sediment samples were taken at 5, 10, and 15 d to observe the phytoplankton species and abundance in the water phase after germination (Figure 9). The sediment's overlying water was L1 medium with no algae, and the proportion of primary algae diatoms germinated in the three tests was higher. After five days of culture, the abundance of algae in the overlying water of the in situ sediment group (I-1) was approximately 20.3 cells/mL, in which the diatoms reached 17.2 cells/mL. Only an insignificant amount of dinoflagellates germinated. In the group (I-2) with the PAC group, the algae's germination rate was low; only diatoms emerged, and the inhibition rate reached over 70%. However, the phytoplankton abundance of the PAC group (I-2) was 32.5 cells/mL, indicating that mainly diatoms and dinoflagellates still existed. The inhibition rate of this group decreased to 57.2%. After 15 days of culture, the algal abundance of the in situ sediments group was slightly lower than that of the 10-d culture, and the inhibition rate was 61.3%. In the whole experiment period, no algal cell germination was observed in the group of PAC and Ba3 fermented broth (I-3).

In the in situ sediment with added *G. catenatum* cyst (II-1) group culture, diatoms were dominant in the overlying water and we detected only a few dinoflagellates. After 10 d of cultivation, the total abundance of germinated phytoplankton reached 90.5 cells/mL, and the density of *G. catenatum* was significantly higher than that in group I-1, which reached 18.0 cells/mL; the addition of PAC (II-2), compared with the II-1 group, has a particular inhibitory effect on algal cell germination, especially diatoms. After 15 days of cultivation, the algae density in the overlying water was slightly lower than that after 10 days of cultivation, which may be related to the content of nutrients in the water. With the increase in cultivation time, the water's nutrients were gradually consumed, which ultimately decreased the water's nutrients. The PAC-combined Ba3 fermented broth group (II-3) was only detected on the 10th day with an insignificant amount of diatoms appearing, with a removal rate of over 98% (Figure 9). The PAC had a better inhibitory effect on germination in the sediment; there was almost no algae germination. This lack of germination may be because the active substances in the bacterial fermentation broth have a good destruction effect on the algae cells, the experimental water body has a weak exchange flow ability, and the concentration of the bacterial fermentation broth in the water body remains almost unchanged; therefore, the provenance in the sediment will likely be affected by the bacteria to a certain extent; the destruction of active substances will not occur.

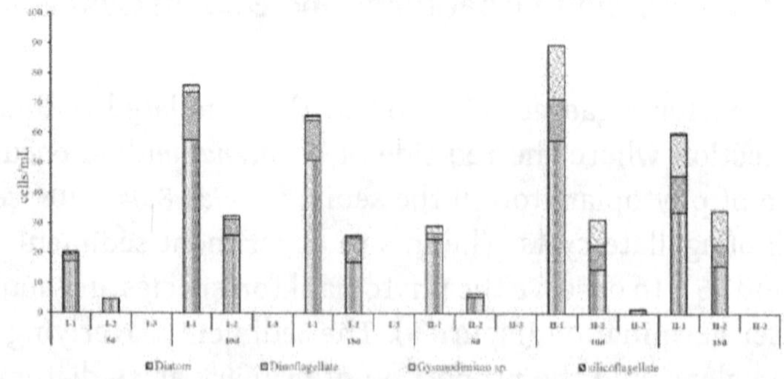

Figure 9. Effect of different algal inhibition method on microcapsule germination in sediment (I) and In situ sediment added with *G. catenatum* cyst (II).

After the L1 medium was added to the sediment, the nitrogen and phosphorus elements in water were essential for the phytoplankton's growth. The initial DIP concentration was 0.13 mg/L, the DIN concentration was 5.333 mg/L, of which the nitrite concentration was 0.042 mg/L, and the ammonium salt concentration was 0.007 mg/L of nitrate. The concentration was 5.283 mg/L. Figure 10 displays the effects of different algae suppression methods on algae cyst germination in the in situ sediments. After five days of PAC administration and PAC-combined Ba3 fermented broth, they dissolved inorganic phosphorus in the water body. The removal effect of the PAC group (I-2) on DIP reached 86.4%. On the 10th day, the removal rate was 87.3%; with the extension of the cultivation time, the removal effect showed a tendency to decrease. On the 15th day, the DIP concentration was 0.011 mg/L and the removal rate dropped to 77.4%. The PAC-combined Ba3 group (I-3) also had a beneficial removal effect on DIP, and the removal effect was slightly higher than that of the PAC group; there was no reduction in removal rate on the 15th day (Figure 10A). Each control group's removal affected the in situ sediment + *G. catenatum* cyst (group II) germination group. Within 5-10 days of the PAC group's control, the removal rate was higher than 85% on the 5th day; on the 15th day, the DIP concentration was 0.009 mg/L, and the removal rate dropped to 81.3%. Compared with group I-2, the DIP content was lower in group II-2, which may consume DIP due to the higher abundance of algae. PAC's removal rate in the combined Ba3 group broth group (II-3) reached more than 87% (Figure 10B).

Figure 10C, D show the effects of different algae inhibitors on dissolved inorganic nitrogen in water and in the in situ sediment germination group. The dissolved inorganic nitrogen in both groups showed a downward trend with time. The cultivation did not transform in each group until the 5th day, but the I-3 group (PAC + Ba3 broth) inorganic nitrogen concentration is higher than that of group I-2 (PAC); when cultured to day 10, group I-3 shows a higher value than those of groups I-2 and I-1. On the 15th day of cultivation, the difference between the groups was more prominent. The dissolved inorganic nitrogen concentration in group I-3 was 2.30 mg/L, while the dissolved inorganic nitrogen concentration in group I-1 was 1.34 mg/L. The lower of the two groups was 0.79 mg/L. Compared with the initial inorganic nitrogen concentration, the decrease rate of inorganic nitrogen was 59.0~85.2% in the entire cultivation cycle under the control of PAC. Figure 10D shows, in the germination group of in situ sediments + *G. catenatum* cysts, that the three groups all have the same tendency of change, and the concentration of dissolved inorganic nitrogen shows a tendency to decrease over time. The concentration of dissolved inorganic nitrogen in the three groups at different incubation times was in the order II-1 > II-3 > II-2 during the entire cultivation cycle. Regarding the decrease rate of group II-2 compared with the initial inorganic nitrogen concentration, the range was between 61.6 and 80.8%, and the removal rate of group II-3 was between 56.6 and 77.8%. Using PAC and PAC combined with Ba3 broth had no noticeable effect on removing dissolved inorganic nitrogen in the water. Compared with group II-2, the addition of bacterial fermentation broth in II-3 increased the content of dissolved inorganic nitrogen in the water body, which was mainly related to the nitrogen-containing compounds in the bacteria's fermentation product. The PAC-combined Ba3 broth group had a beneficial removal effect on the soluble inorganic phosphorus, and the removal rate reached over 87%. The active bacteria substance has nitrogen-containing compounds; therefore, this group's soluble inorganic nitrogen content is higher than that of PAC.

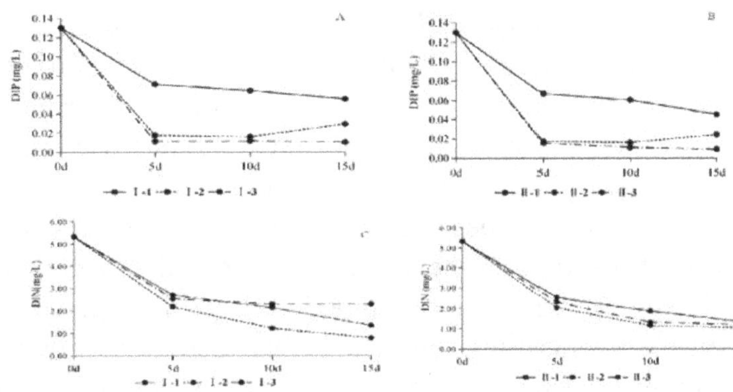

Figure 10. Effect of different algal inhibition methods on (A) dissolved inorganic phosphorus (DIP) and (C) dissolved inorganic nitrogen (DIN) in sediment; and effect of different algal inhibition methods on (B) DIP and (D) DIN in sediment (added *G. catenatum* cyst).

4. DISCUSSION

Our experimental study results revealed that all four coagulant chemicals, PAC, $Al_2(SO_4)_3$, $Fe_2(SO_4)_3$, and $FeCl_3$, exhibited a high removal efficiency against *G. catenatum*, with some differences. However, PAC has the best removal effect, with a low re-solubilization rate, large floc formation, and the most considerable amount of flocculated algal cells. It also has a beneficial removal effect on turbidity and organic matter. Recent studies showed that the high removal efficiencies with five kinds of coagulants are comparable to the results of Liu Lijuan et al. for the control of a lake containing algae bloom [16]. They concluded that PAC has a more impressive algae removal effect by changing the dosage and pH conditions. The algae removal mechanism of polyaluminum salts showed that PAC and algae cells in the water first underwent adsorption and an electrical neutralization reaction, and a bridging network was trapped, so that the flocs were more extensive and, therefore, more prone to settle [19]. Moreover, a pilot test of enhanced coagulation of raw water in the Yangtze River with PAC showed that PAC had a beneficial coagulation effect on the Yangtze River water, forming large flocs and a rapid settling speed [20].

Previous studies have shown that specific bacterial populations can inhibit HAB species' growth through microbial algae suppression into direct algae suppression and indirect algae suppression [21,22]. In one of these studies, Ba3 (*Bacillus* fermentation broth) had a high algae inhibition effect, reaching over 90%. To date, several publications have proven that *Bacillus* sp. can inhibit the growth of harmful algal blooms [17,23]. Compared with other algaecide bacteria isolated from aquatic water, *Bacillus* showed similar or more potent algicidal activity against algae [24]. Zhao et al. [25] isolated four algicidal metabolites from a fermentation broth of *Bacillus* B1 strain against *Phaeocystis globosa*, and all metabolites had a strong alga dissolving effect. In this study, a Ba3 *Bacillus* strain of fermentation broth was used to influence algal growth, and the results showed that the 2-d fermentation broth had the best removal effect on *G catenatum*. Under the microscope, we observed that the cell wall gradually loses its ability to move under the action of the Ba3 broth, the algae cells gradually become transparent, and the contents of the cell overflow due to rupture, though there is no complete algal cell morphology. In terms of their biological safety, anti-compounds are biodegradable. Moreover, when used to control blooms, they appear to be harmless to the environment [26]. These results strongly bolster the claim that Ba3 has potential applications in controlling algae outbreaks.

The coagulant removes algae quickly, and it can coagulate with algae cells to form flocs and settle to the bottom immediately after addition, but the coagulant algae suppression will have a certain back-dissolving phenomenon. The algae killing effect of the fermentation broth is thorough, but the action time is relatively long. However, previous research revealed that PAC combined with algae-lysing bacteria has a high inhibitory effect on *Microcytis aeruginosa* and nutrients in the water [27]. Furthermore, Wang et al. [28] studied the inhibition of the growth of

Scrippsiella trochoidea by combining algae-suppressing bacteria with two modified clays. The results show that the combined algae suppression method can improve the algae suppression effect, increase the algae suppression time, and reduce algae cells' rebound effect. In this study, the combination of coagulants and Ba3 fermented broth was used to suppress algae and remove algae cells in a short time while killing algae cells. The results of the four coagulants combined with *Bacillus* Ba3 fermentation broth, when combined with algae inhibition, showed that the combination of PAC and Ba3 fermented broth had the best effect on inhibiting *G. catenatum* with an inhibition rate of 100% and no rebound. The bacterial fermentation broth is indeed yellow and contains many organic substances. An increased dosage of bacterial fermentation broth leads to increased water turbidity and organic matter content; when the Ba3 broth dosage in the PAC-combined bacterial fermentation broth group was 0.3%, it had the highest removal rate of turbidity and organic matter in water.

Although the red tide community's increase may be significant and alarming, this issue should not be linked to on-site production. Researchers have also proposed other HABs from resting cysts germinating from the bottom sediments, which seriously affect many aspects of the red tide phenomenon [29]. Cysts are also particularly efficacious for community spread. They enable species to survive under adverse conditions, and because their development generally involves sexuality, they promote genetic recombination [30]. In addition, the high abundance of cysts in the sediment may reflect the recent blooming of these species in the area. According to previous studies, many resting cysts will be produced by the end of the blooming cycle. The formation of cysts is one of the predominant factors in bloom termination [31,32]. Several researchers believe that the vegetative cell inoculum size from cyst germination is critical to the beginning of blooming. However, the final bloom size cannot measure the effect of altering cell growth or the aggregation of complex natural conditions because unpredictable variables control these effects. Therefore, it is necessary to control the sediment cyst before germination. Nevertheless, there have only been a few studies on the coagulant effect with combine bacterial fermented broth on cyst formation and germination to control HAB species.

We divided the germination experiment into two groups: in situ sediments and in situ sediments after adding *G. catenatum* cysts. In this study, treatment with the PAC-combined Ba3 fermented broth group significantly removed over 90% of the *G. catenatum* vegetative cells. Recent research has also found that nutrient changes caused by modifying clay can be re-released from the algae matrix into the water, which may contribute to the formation of resting cysts [33]. These results offer a contrast to this study where the PAC-combined Ba3 broth group had a good removal effect on the soluble inorganic phosphorus, and the removal rate reached over 87%. The active bacteria substance has nitrogen-containing compounds; therefore, the content of soluble inorganic nitrogen in this group was higher than that of the PAC group. These treatments not only were effective for removing the vegetative cells but also lowered the nutrients level.

5. CONCLUSIONS

In conclusion, our results show that the feasibility of using appropriate concentrations of PAC combined with Ba3 fermented broth is potentially useful for controlling *G. catenatum* blooms. The effect of bacterial fermentation broth on killing algae is complete, but the time needed for effective action is longer. The active substances in the fermentation broth act on the algae cells to lyse them. The four coagulants combined with the bacterial fermentation broth group had an obvious inhibitory effect on *G. catenatum*. Through the comprehensive analysis of the removal effects of algal cells, turbidity, and organic matter, it was found that after the action of the polyaluminum chloride (50 mg/L) combined with the bacterial fermentation broth (0.3%, *v/v*) for 3 h, the algal inhibition rate was 100%, and the algal cells did not rebound. The inhibition rate of the bacterial fermentation broth and polyaluminum chloride group was more than 98%. Our findings will integrate well into the future studies of controlling toxic dinoflagellate cysts in the benthic environment. Our research also shows that dinoflagellate organisms use nitrogen and phosphorus cues to determine when vegetative growth occurs. Understanding the environmental cues related to algae dormancy is essential for the understanding and management of HABs in aquatic ecosystems.

REFERENCES

1. Zheng, H.; Sun, C.; Hou, X.; Wu, M.; Yao, Y.; Li, F. Pyrolysis of *Arundo donax* L. to produce pyrolytic vinegar and its effect on the growth of dinoflagellate Karenia brevis. *Bioresour. Technol.* **2018**, *247*, 273–281. [CrossRef] [PubMed]
2. Zohdi, E.; Abbaspour, M. Harmful algal blooms (red tide): A review of causes, impacts and approaches to monitoring and prediction. *Int. J. Environ. Sci. Technol.* **2019**, *16*, 1789–1806. [CrossRef]
3. Hallegraeff, G.M.; Blackburn, S.I.; Doblin, M.A.; Bolch, C.J.S. Global toxicology, ecophysiology and population relationships of the chainforming PST dinoflagellate Gymnodinium catenatum. *Harmful Algae* **2012**, *14*, 130–143. [CrossRef]
4. Kudela, R.M.; Gobler, C.J. Harmful dinoflagellate blooms caused by *Cochlodinium* sp.: Global expansion and ecological strategies facilitating bloom formation. *Harmful Algae* **2012**, *14*, 71–86. [CrossRef]

5. Chen, H. Emergency treatment and reflection of red tide event of Gymnodinium catenatum in Fujian sea area in 2017. *J. Fujian Fish.* **2018**, *40*, 308–314.

6. Sun, H.; Zhang, Y.; Chen, H.; Hu, C.; Li, H.; Hu, Z. Isolation and characterization of the marine algicidal bacterium Pseu- doalteromonas S1 against the harmful alga Akashiwo sanguinea. *Mar. Biol.* **2016**, *163*, 66–73. [CrossRef]

7. Kidwell, D. Mitigation of harmful algal blooms: The way forward. *Pices Press* **2015**, *23*, 22–25.

8. Anderson, D.M. Approaches to monitoring, control and management of harmful algal blooms (HABs). *Ocean. Coast. Manag.***2009**, *52*, 342–347. [CrossRef]

9. Kim, Z.H.; Thanh, N.N.; Yang, J.-H.; Park, H.; Yoon, M.-Y.; Park, J.-K.; Lee, C.-G. Improving Microalgae Removal Efficiency Using Chemically-processed Clays. *Biotechnol. Bioprocess. Eng.* **2016**, *21*, 787–793. [CrossRef]

10. Park, J.; Church, J.; Son, Y.; Kim, K.-T.; Lee, W.H. Recent advances in ultrasonic treatment: Challenges and field applications for controlling harmful algal blooms (HABs). *Ultrason. Sonochemistry* **2017**, *38*, 326–334. [CrossRef]

11. Zhuang, L.; Zhao, L.; Yin, P. Combined algicidal effect of urocanic acid, N-acetylhistamine and l-histidine to harmful alga Phaeocystis globosa. *RSC Adv.* **2018**, *8*, 12760–12766. [CrossRef]

12. Shi, X.; Liu, L.; Li, Y.; Xiao, Y.; Ding, G.; Lin, S.; Chen, J. Isolation of an algicidal bacterium and its effects against the harm-ful- algal-bloom dinoflagellate Prorocentrum donghaiense (Dinophyceae). *Harmful Algae* **2018**, *80*, 72–79. [CrossRef]

13. Yu, X.; Cai, G.; Wang, H.; Hu, Z.; Zheng, W.; Lei, X.; Zhu, X.; Chen, Y.; Chen, Q.; Din, H.; et al. Fast-growing algicidal Streptomyces sp U3 and its potential in harmful algal bloom controls. *J. Hazard. Mater.* **2018**, *341*, 138–149.

14. Zhang, F.; Ye, Q.; Chen, Q.; Yang, K.; Zhang, D.; Chen, Z.; Lu, S.; Shao, X.; Fan, Y.; Yao, L.; et al. Algicidal Activity of Novel Marine Bacterium Paracoccus sp Strain Y42 against a Harmful Algal-Bloom-Causing Dinoflagellate, Pro-rocentrum donghaiense. *Appl. Environ. Microbiol.* **2018**, *84*, e01015-18. [CrossRef]

15. Jeong, S. Algicidal Activity of Strain MS-51 against a HAB Causing Alexandrium tamarense. *Int. J. Appl. Environ. Sci.* **2019**, *14*, 655–663.

16. Liu, L.; Wang, L.; Li, M. Research on the enhancement of algae removal and turbidity removal by different coagulants. *J. China Water Supply Drain.* **2010**, *26*, 80–83.

17. Wu, L.; Wu, H.; Chen, L.; Xie, S.; Zang, H.; Borriss, R.; Gao, X. Bacilysin from Bacillus amyloliquefaciens FZB42 Has Specific Bactericidal Activity against Harmful Algal Bloom Species. *Appl. Environ. Microbiol.* **2014**, *80*, 7512–7520. [CrossRef]

18. Figueroa, R.I.; Bravo, I.; Garces, E.; Ramilo, I. Nuclear features and effect of nutrients on Gymnodinium catenatum (Di-nophyceae) sexual stages. *J. Phycol.* **2006**, *42*, 67–77. [CrossRef]

19. Lei, G.; Zhang, X.; Wang, D. The algae removal mechanism by polymerized aluminium salt coagulants and methods to improve algae removal. *J. Water Resour. Prot.* **2007**, *23*, 50–54.

20. Liang, X.; Zhang, M.; Zhao, Y. Pilot test of the enhanced coagulation treatment process of the Yangtze River raw water in Changzhou section. *J. Water Purif. Technol.* **2019**, *38*, 76–81.

21. Teeling, H.; Fuchs, B.M.; Becher, D.; Klockow, C.; Gardebrecht, A.; Bennke, C.M.; Kassabgy, M.; Huang, S.; Mann, A.J.; Wald-mann, J.; et al. Substrate-Controlled Succession of Marine Bacterioplankton Populations Induced by a Phytoplankton Bloom. *Science* **2012**, *336*, 608–611. [CrossRef]

22. Li, D.; Zhang, H.; Chen, X.; Xie, Z.; Zhang, Y.; Zhang, S.; Lin, L.; Chen, F.; Wang, D. Metaproteomics reveals major microbial players and their metabolic activities during the blooming period of a marine dinoflagellate Prorocentrum donghaiense. *Environ. Microbiol.* **2018**, *20*, 632–644. [CrossRef]

23. Sun, P.; Hui, C.; Wang, S.; Khan, R.A.; Zhang, Q.; Zhao, Y.-H. Enhancement of algicidal properties of immobilized Bacillus methylotrophicus ZJU by coating with magnetic Fe3O4 nanoparticles and wheat bran. *J. Hazard. Mater.* **2016**, *301*, 65–73.

24. Yang, F.; Li, X.; Li, Y.; Wei, H.; Yu, G.; Yin, L.; Liang, G.; Pu, Y. Lysing activity of an indigenous algicidal bacterium *Aeromonas* sp against *Microcystis* spp. isolated from Lake Taihu. *Environ. Technol.* **2013**, *34*, 1421–1427. [CrossRef] [PubMed]

25. Zhao, L.; Chen, L.; Yin, P. Algicidal metabolites produced by Bacillus sp strain B1 against Phaeocystis globosa. *J. Ind. Microbiol. Biotechnol.* **2014**, *41*, 593–599. [CrossRef] [PubMed]

26. Perzborn, M.; Syldatk, C.; Rudat, J. Enzymatical and microbial degradation of cyclic dipeptides (diketopiperazines). *AMB Express* **2013**, *3*, 51. [CrossRef]

27. Li, C.; Xu, W.; Ding, G.; Wu, S.; Xie, X. Treatment of Microcystis Aeruginosa by Algicidal Bacteria and Coagulation. *Guangdong Chem. Ind.* **2018**, *45*, 17–20.

28. Wang, Y.; Wu, F.; Zhang, Q. Study on Coupling Method of Modified Clay and Algae Bacterium to Inhibit Sclerotinia striata. *J. Shenzhen Polytech.* **2019**, *18*, 58–63.

29. Garces, E.; Montresor, M.; Lewis, J.; Rengefors, K.; Anderson, D.M.; Barth, H. Phytoplankton Life Cycles and Their Impacts on the Ecology of Harmful Algal Blooms. *Deep-Sea Res. Part II Top. Stud. Oceanogr.* **2010**, *57*, 159–161.

30. Masseret, E.; Grzebyk, D.; Nagai, S.; Genovesi, B.; Lasserre, B.; Laabir, M.; Collos, Y.; Vaquer, A.; Berrebi, P. Unexpected Genetic Diversity among and within Populations of the Toxic Dinoflagellate Alexandrium catenella as Revealed by Nuclear Microsatellite Markers. *Appl. Environ. Microbiol.* **2009**, *75*, 2037–2045. [CrossRef] [PubMed]
31. Wang, Z.; Matsuoka, K.; Qi, Y.; Chen, J. Dinoflagellate cysts in recent sediments from Chinese coastal waters. *Mar. Ecol. Pubbl. Della Stn. Zool. Di Napoli I* **2004**, *25*, 289–311. [CrossRef]
32. Wang, Z.; Qi, Y.; Yang, Y. Cyst formation: An important mechanism for the termination of Scrippsiella trochoidea (Di-nophyceae) bloom. *J. Plankton Res.* **2007**, *29*, 209–218. [CrossRef]
33. Lu, G.; Song, X.; Yu, Z.; Cao, X. Application of PAC-modified kaolin to mitigate Prorocentrum donghaiense: Effects on cell removal and phosphorus cycling in a laboratory setting. *J. Appl. Phycol.* **2017**, *29*, 917–928. [CrossRef]

CHAPTER-10

IDENTIFICATION OF NEW SUB-FOSSIL DIATOMS FLORA IN THE SEDIMENTS OF SUNCHEONMAN BAY, KOREA

1. INTRODUCTION

An estuary can be defined as a semi-enclosed coastal body of water that has a free connection with the open ocean, within which seawater is diluted with freshwater derived from land drainage [1,2]. River mouths, coastal bays, tidal marsh systems, and sounds all fit this definition. Estuaries are transitional zones between freshwater and marine habitats. Due to tides and storms, the water level and salinity vary in estuaries [3]. They are most commonly located in low-relief coastal regions. Estuaries and wetlands are among the most productive aquatic ecosystems, providing a home for both freshwater and marine plants, and a source of nutrients for a variety of animal communities adapted to brackish waters [4,5]. Moreover, they filter out pollutants supplied to the ocean [4,6,7]. Thus, many animals rely on estuaries that have abundant species diversity for food, places to breed, and migration stopovers [8,9] (https://oceanservice.noaa.gov/facts/estuary.html (accessed on 8 February 2021)).

In South Korea, where three sides are surrounded by the sea, there are numerous estuaries such as the Nakdonggang, Keumgang, and Seomjingang. Coastal wetlands cover approximately 2800 km^2, which represents approximately 3% of the total land area [10–12]. Suncheonman Bay, Korea's most representative estuary, is an invasive coastal wetland composed of 21.6 km^2 of tidal flats and 5.4 km^2 of reed fields surrounded by the Yeosu and Goheung Peninsulas [10]. In January 2006, it became the first registered wetland in the Ramsar Convention list in Korea, as it was designated as a "wetland protected area" by the Ministry of Land, Transport and Maritime Affairs in December 2003 [10,13]. In June 2008, Suncheonman Bay was designated as national cultural property, "Myeongseung" number 41, and in 2010, south-western tidal flats in Korea, including Suncheonman Bay (south-western coast tidal flats), were included in the UNESCO World Heritage Tentative List. Such registrations and designations demonstrate the recognition of its ecological and environmental value [14] (https://whc.unesco.org/en/tentativelists/5482 (accessed on 20 April 2021)). In particular, the natural environment and ecosystems within Suncheonman Bay are well preserved, making them a habitat favourable for many species of marine organisms [15]. In these regions, primary producers such as diatoms play a critical role as a food source for large invertebrates and fishes [16].

Diatoms are unicellular algae characterized by a biomineralized (opaline) cell wall that may fossilize and be preserved in the sedimentary record [17,18]. The sub-fossil diatoms herein described consist of Holocene diatom remains not fully involved in the fossilization process. Diatoms thrive in very different environments (e.g., hot springs, polar regions, and fresh, brackish, and marine waters) and are extremely sensitive to physical and chemical changes (e.g., temperature, salinity, and nutrients) in the water [18–23]. Therefore, fossil diatoms represent an excellent source of information about past climate change and its effect on aquatic ecosystems. There have been relatively few studies on sub-fossil diatoms along the southern Korean coast. Marine to brackish sub-fossil diatom assemblages were initially studied in the Pohang and Gampo sediments in 1975 [24], then they were extended to the regions of Bukpyeong [25] and Pohang [26,27] in the East Sea and the regions of the Mankyung-Dongin river estuary [28], Dodaecheon River [29], Ilsan estuary [30], Chollipo [31], Isanpo [32], and Sabsido and Kunsan in the Yellow Sea [33].

The Suncheonman Bay has been used to study various environmental characteristics, including grain size and organic matter in tidal flat sediments [10], seasonal water quality, pollution, environmental safety [13,34], and other geochemical characteristics, as well as local inhabitants such as benthic invertebrates, plants, fishes, birds, bacteria, and fungi [6,14,35–39]. Among the studies conducted to date, the investigation of phytoplankton communities in the Dong Cheon River and Isa Cheon River stream into Suncheonman bay was the most interesting [40]. In this study, we describe a newly recorded sub-fossil diatom assemblage recovered in the sediments of the Suncheonman Bay.

2. MATERIALS AND METHODS

2.1 Coring and Sampling of Sediment

Drilling was carried out using a peat core sampler (52 mm diameter; Peat Sampler, Eijkelkamp Soil & Water, Giesbeek, Netherlands). One sediment core (= SCW03 with a length of 6.0 m) was retrieved from Suncheonman Bay in Korea on 11 June 2018 (Figure 1, Table 1). The core was transported to the laboratory after vacuum packing in a plastic bag to prevent drying and oxidation. Sedimentological description and sub-sampling were performed after the sediment core profile was cut vertically in half [41]. Shell fragments in the SCW03 core were selected for the analysis of chronology and diatoms because the sediment layer was well preserved.

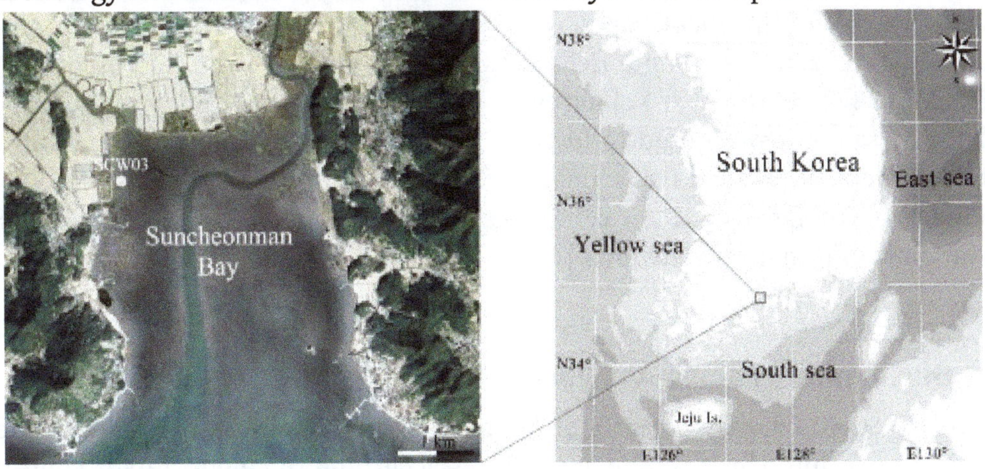

Figure 1. Sampling sites in Suncheonman Bay.

Table 1. Information of sampling sites.

Site	Location	Depth (m)	Latitude (N)	Longitude (E)
SCW03	Haksan-ri Byeollyang-myeon Suncheon-si, Jeollanam-do	6	34°52_10.1266”	127°29_27.5667”

Age dating was performed using an accelerator mass spectrometer (AMS) at the Korean Institute of Geoscience and Mineral Resources (KIGAM), Korea. The estimated ages were calibrated by the OxCal statistical analysis program (http:// c14.arch.ox.ac.uk (accessed on 20 May 2021)).

2.2. Analysis of Chronology

Age dating was performed using an accelerator mass spectrometer (AMS) at the Korean Institute of Geoscience and Mineral Resources (KIGAM), Korea. The estimated ages were calibrated by the OxCal statistical analysis program (http:// c14.arch.ox.ac.uk (accessed on 20 May 2021)).

2. 3 Sample Preparation for Diatom Identification

Thirteen samples of diatoms were collected every 0.5 m along the SCW03 core. Their analysis was conducted according to the following steps: 1 g of sediment was dried at 60°C for 24 h; the siliceous material was boiled with 20 mL of 30% hydrogen peroxide (H2O2) and washed with distilled water to remove organic matter; the treated samples were mounted with Pleurax (Mountmedia, Wako, Japan) and briefly heated using an alcohol lamp for subsequent analysis using a light microscope (LM; Eclipse Ni, Nikon, Tokyo, Japan). Photomicrographs were taken using a digital camera (DS-Ri2, Nikon, Tokyo, Japan). Some remaining peroxide-cleaned samples were filtered using 2.0-µm polycarbonate membrane filters (Nuclepore, Whatman, Maidstone, UK). The membranes were placed on stubs and coated with gold-palladium for analysis using a field emission scanning electron microscope (FE-SEM; MIRA 3, TESCAN, Brno-Kohoutovice, Czech Republic). SEM photomicrographs of all the samples were used to identify the diatoms. Morphological analyses of diatoms were performed using ImageJ v1.32 software (NIS- Elements BR4.50.00, Nikon, Tokyo, Japan) [42]. Taxonomical nomenclature was based on recent taxonomic information guidelines [43].

3. RESULTS

3.1 Sedimentary Facies Analysis

The core SCW03 mostly consists of greenish-grey silty clay and can be divided distinguished into three sedimentary facies according to colour, fossils content, and sedimentary structure (Figure 2A,C). Facies A is characterized by yellowish-brown mottling structures. Shell fragments are not observed in this facies. In Facies B, the yellowish-brown mottling structures are less abundant and shell fragments are observed to occur sporadically. The size of shell fragments are about 2 mm in diameter. Facies C is represented by highly concentrated shells and shell fragments in several-centimeters -thick intervals.

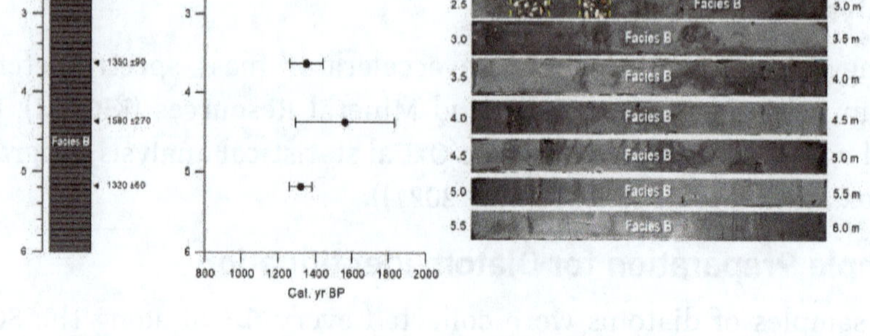

Figure 2. (A) Stratigraphic section, (B) result of age dating, and (C) photographs of core SCW03.

The sediments of core SCW03 are interpreted as deposited in a tidal flat **[44,45]**. It is very similar to the current tidal flat environment. Abundant mottling structures in facies A originate from an oxygen-rich environment, indicating that the sea level has gradually decreased slightly in facies B. In the meantime, highly concentrated shells and shell fragments of facies C are interpreted as sedimentation caused by flooding or storm events [45].

3.2 Age Dating

The results of age dating for five samples in the core SCW03 shows a range from 1170 to 1560 cal. yr BP (Table 2 and Figure 2B).

Table 2. Results of AMS ^{14}C dating and calibrated dates for core SCW03.

Depth (m)	Elevation (m)	14 C yr BP (±1σ)	Cal. yr BP (±2σ)	Laboratory Code	Dating Material
1.59	0.07	1340 ± 20	1240 ± 130	ITg180525	Shell fragments
2.48	-0.82	1240 ± 20	1170 ± 200	ITg180526	Shell fragments
3.64	-1.98	1480 ± 20	1350 ± 90	ITg180527	Shell fragments
4.37	-2.71	1660 ± 20	1560 ± 270	ITg180528	Shell fragments
5.18	-3.52	1410 ± 20	1320 ± 60	ITg180529	Shell fragments

3.3 Diatom Assemblages

A total of 87 diatom species belonging to 52 different genera were identified in the sediments from Suncheonman Bay in Korea (Table 3); of these, six species, namely, *Cyma- tonitzschia marina, Fallacia hodgeana, Navicula mannii, Metascolioneis tumida, Surirella recedens*, and *Thalassionema synedriforme,* were never described before in this area. The diatom flora encountered in this survey was composed of 87 taxa, which were classified into 52 genera. We present information on the valve shape, occurrence depth in sediment, and habitat of 87 species, including the six newly recorded species. Information collected about the newly recorded diatoms identified here included taxonomic information, illustrations, basionyms, synonyms, original description references, depth in the core, distribution, and diagnosis (Table 3).

Table 3. Occurrence (black squares) and habitat of diatom species by depth. A total of 72 diatom species belonging to 52 different genera were identified. Star marks on the specific name are newly recorded species in Korea (6 species: *Cymatonitzschia marina, Fallacia hodgeana, Navicula mannii, Metascolioneis tumida, Surirella recedens,* and *Thalassionema synedriforme*).

	SCW03 (m)	0.1	0.5	1.0	1.5	2.0	2.5	3.0	3.5	4.0	4.5	5.0	5.5	6.0	Habitat	Reference
1	*Achnanthes* sp.						●									-
2	*Actinocyclus octonarius*				●					●					Marine	[46]
3	*Actinoptychus senarius*				●	●	●				●		●	●	Marine	[46]
4	*Amphora pediculus*			●	●	●		●	●	●					Freshwater	[47]
5	*Amphora* sp.	●														-
6	*Auliscus sculptus*	●													Marine	[48]
7	*Bacillaria paxillifera*									●					Marine	[49]
8	*Bacteriastrum* sp.															-
9	*Chaetoceros affinis*	●	●	●	●	●		●			●	●		●	Marine	[50]
10	*Chaetoceros compressus*								●						Marine	[51]
11	*Chaetoceros lorenzianus*											●			Marine	[52]
12	*Chaetoceros* sp.		●			●			●	●	●	●	●	●		-
13	*Chaetoceros resting spores*			●	●	●		●	●	●		●	●	●		-
14	*Climaconeis mabikii*					●									Marine	[53]
15	*Cocconeis placentula*	●						●			●		●		Freshwater	[54]
16	*Cocconeis* sp.										●			●		-
17	*Coscinodiscus asteromphalus*										●			●	Marine	[55]
18	*Coscinodiscus centralis*									●		●			Marine	[56]
19	*Coscinodiscus radiatus*	●	●	●	●		●	●	●	●	●		●	●	Marine	[57]
20	*Coscinodiscus* sp.		●											●		-
21	*Cyclotella litoralis*	●	●	●		●	●	●	●	●	●	●	●	●	Marine/ freshwater	[58]
22	*Cyclotella meneghiniana*													●	Marine/ freshwater	[59]
23	*Cyclotella ocellata*							●	●		●				Freshwater	[60]
24	*Cyclotella* sp.			●												-
25	*Cymatonitzschia marina**					●									Marine	[61-66]
26	*Cymatosira lorenziana*													●	Marine	[67]

27	*Cymatotheca* sp.									•						-
28	*Cymatotheca weissflogii*			•						•		•			Marine	[68]
29	*Delphineis* sp.	•													Marine	-
30	*Diploneis elliptica*	•	•	•	•	•	•	•			•				marine/ freshwater	[69]
31	*Diploneis* sp.	•	•	•						•		•				-
32	*Diploneis weissflogii*			•		•		•		•		•			marine	[70]
33	*Discostella stelligera*											•			freshwater	[71]
34	*Ditylum sol*				•										marine	[72]
35	*Encyonema* sp.				•											-
36	*Epithemia adnata*	•						•							freshwater	[73]
37	*Fallacia hodgeana**													•	Freshwater/ brackish Water	[74]
38	*Fallacia* sp.					•						•		•		-
39	*Fragilaria capucina*	•													marine/ freshwater	[75]
40	*Frustulia vulgaris*							•		•					Freshwater	[76]
41	*Giffenia cocconeiformis*						•			•					Marine	[70]
42	*Giffenia* sp.	•	•	•	•	•	•	•	•	•	•	•	•	•		-
43	*Gomphonema* sp.															-
44	*Gyrosigma accuminatum*	•	•					•							freshwater	[76]
45	*Gyrosigma fasciola*					•						•			Marine	[78]
46	*Gyrosigma* sp.				•	•	•				•	•				-
47	*Gyrosigma turgidum*									•			•		Marine	[79]
48	*Halamphora latecostata*	•											•		Freshwater	[80]
49	*Haslea ostrearia*			•			•			•					Marine	[81]
50	Hyalodiscus subtilis				•										Marine	-
51	*Lyrella* sp.									•						-
52	*Navicula* sp.	•	•			•	•	•	•			•		•		-]
53	*Navicula viridulacalcis*					•		•							Freshwater	[83]
54	*Navicula mannii**						•								brackish water / marine	[70]
55	*Nitzschia sigma*	•		•		•		•				•			Brackish	[84]
56	*Nitzschia* sp.	•	•	•				•		•	•	•		•		-
57	*Paralia sulcata*	•	•	•		•	•	•	•	•	•	•	•	•	Marine	[85]
58	*Parlibellus delognei*					•	•	•		•	•				Marine	[86]

59	*Petrodictyon gemma*					●									marine	[87]
60	*Petroneis marina*	●	●	●		●				●			●		Marine	[88]
61	*Pinnularia* sp.					●										-
62	*Pleurosigma aestuarii*				●	●			●	●	●			●		[89]
63	*Pleurosigma diverse-striatum*											●			Marine	[90]
64	*Pleurosigma normanii*					●		●						●	Marine	[91]
65	*Pleurosigma* sp.													●		-
66	*Pseudonitzschia pungens*	●									●		●		Marine	[92]
67	*Rhaphoneis* sp.												●			-
68	*Rhizosolenia setigera*				●								●		Marine	[81]
69	*Metascolioneis tumida**			●	●	●									Marine	[83]
70	*Sellaphora americana*												●		Freshwater	[94]
71	*Semiorbis* sp.													●		-
72	*Surirella recedens**						●						●	●	Marine	[95-96]
73	*Surirella* sp.			●	●	●		●	●	●		●		●		-
74	*Thalassionema nitzschioides*		●	●		●		●		●		●		●	Marine	[97]
75	*Thalassionema synedriforme**							●							Marine	[98]
76	*Thalassiosira decipiens*			●			●			●					Marine	[99]
77	*Thalassiosira eccentrica*				●			●			●	●	●	●	Marine	[99]
78	*Thalassiosira oestrupii*			●		●				●				●	Marine	[100]
79	*Thalassiosira* sp.		●	●		●	●	●		●	●		●	●		-
80	*Trachyneis aspera*													●	Marine	[101]
81	Triceratium dubium							●							Marine	[102]
82	*Tryblionella acuminata*	●			●				●	●					Marine	[103]
83	*Tryblionella coarctata*					●				●				●	Marine	[104]
84	*Tryblionella granulata*									●		●			Marine	[70]
85	*Tryblionella punctata*							●		●		●		●	marine / freshwater	[105]
86	*Tryblionella* sp.			●		●				●		●				-
87	*Tryblioptychus Cocconeiformis*							●			●	●	●	●	Marine	[106]
	Total	**21**	**14**	**23**	**19**	**31**	**17**	**27**	**14**	**34**	**20**	**27**	**19**	**30**		

Cymatonitzschia marina (F.W.Lewis) Simonsen 1974 (Figure 3A,B)

Basionym: *Cymatopleura marina* F.W.Lewis 1861 [107]

Synonym: *Cymatopleura marina* F.W.Lewis 1861 [107]

Original description: Simonsen 1974: 56, pl. 41: Figures 5–9 [108]\

Description: Valves are observed to be solitary, usually lying in the valve view. Valves are linear lanceolate, with very acute ends. Valves are strictly isopolar, and not constricted in the middle. Overall dimensions include a valve length ranges from 58.42 to 67.94 μm and valve width from 9.14 to 11.28 μm. The valve faces have numerous undulations (9–11), with a distance between two undulations in the ranges from 4.81 to 7.97 μm. Undulations are found to have a nearly trapezoidal shape (Figure 3A). The valve surface has irregular punctate on the undulations. A raphe system is observed running around one side of the valve margin. Striae uniseriate are found to be densely spaced, with approximately 28–29 per 10 μm observed, on one side of the valve margin (Figure 3B).

Depth occurrence in the core: 2.0 m.

Distribution: This species is reported from brackish water to marine environments mainly [61–66]. This taxon is reported from some estuary, e.g., East River, New York and Long Island Sound [109]. *Cymatopleura marina* was first recorded from the Indian Ocean [108].

Differential diagnosis: This genus differs from *Cymatopleura*. The genus *Cymatopleura*, as a member of the Surirellaceae, has a completely different raphe morphology, which runs along the edge of the valve around the entire margin, whereas in *Cymatonitzschia* it is, as in *Nitzschia*, limited to one of the sides [108]

Remarks of raphe: *Cymatonitzschia marina* has an eccentric keeled raphe placed through the edge of the valve, and it appears on one of the sides [108,110].

Fallacia hodgeana (R.M.Patrick and Freese) Y.H.Li and H.Suzuki 2014 (Figure 3C–F) Basionym: *Navicula hodgeana* R.M.Patrick and Freese 1961 [111]

Synonym: *Navicula hodgeana* R.M.Patrick and Freese 1961 [111] Original description: Li et al., 2014 in p. 33 [74]

Description: Valves are observed to be solitary, usually lying in the valve view. Valves are naviculoid and linear-elliptic, with bluntly rounded ends. Overall dimensions include length ranges from 14.73 to 15.42 μm and width ranges from 4.21 to 4.61 μm. The valve face is nearly flat with a slightly curved raphe (Figure 3C,D, arrow). Central raphe endings are proximately hooked (Figure 3E,F, arrowheads). Terminal raphe fissures exhibit a sickle- shaped curve in the same direction. Some parts of the striae are covered with a thin siliceous covering, or conopeum, on the external valve surface. Tow slits opening of the canal, present near the terminal raphe fissures, are also observed (Figure 3F, arrowhead). Areolae are found to be curved upward and were directly connected to the mantle. The finely porous conopeum extends outward from the outer edge of the raphe sterna, running through the surface, and connect to the proximal edge of the mantle. Numerous peg-shaped structures are observed in the nonporous margin of the conopeum along the proximal edge of the mantle. The elongated areolae, with an approximate length and width of 0.28–0.42 μm and 0.15–0.18 μm, respectively, are found on the hyaline area of the valve surface with an undulated junction. The elongated areolae were found in groupings of 12 per 5 μm. Peg-shaped structures, in groups of 12 per 5 μm, are also observed, finely porous on the conopeum 12–13 per 1 μm transversely.

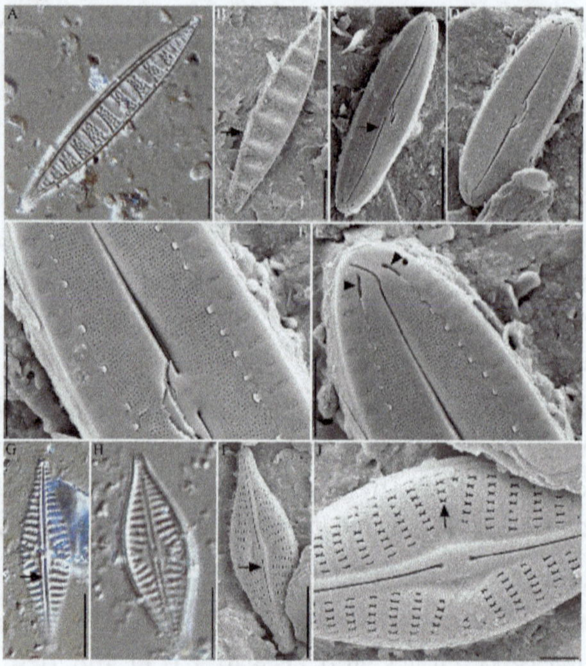

Figure 3. Light microscope (A,G,H) and field emission scanning electron microscopic (B–F,I,J) photomicrographs of diatoms: (A,B) *Cymatonitzschia marina* (the arrow points to raphe), (C–F) *Fallacia hodgeana* (in C the arrow points to raphe; in F the arrow points to the terminal openings), (G–J) *Navicula mannii* with raphe and ribbon-shaped areola (the arrow points to raphe and ribbon-shaped areola). Scale bar = 10 μm.

Depth occurrence in the core: 6.0 m.

Distribution: This species lives in fresh- to brackish-water environments. It was first reported from scrapings of small rock submerged at the edge of a lagoon as *Navicula hodgeana* [111]. Li et al. 2014 collected this species from Edogawa River, Japan. This taxon is known to benthic diatom [74,112,113].

Differential diagnosis: *Fallacia hodgeana* possesses morphological features such as a single H-shaped plastid, depressed lateral sterna interrupting striae that contain round areolae enclosed by hymen; well-developed, finely porous conopeum; and a canal system between the primary silica layer and the conopeum. These features indicate that this species does not belong to the genus *Navicula* or *Pseudofallacia* [74]. This species is related to *Navicula dissipata*. The length-to-breadth ratio is similar, although *N. dissipata* is a larger taxon. The clear central area is narrower in our taxon, and the striae are composed of many fine puncta instead of a few large ones. The median ends of the raphe are close together as in *N. dissipata* [111].

Remarks of raphe: *Fallacia hodgeana* has a slightly curved raphe that terminates at fissures curved in the same way. The distal end fissures were sickle-like and curved in the same direction (Figure 3C arrow). The central endings lie close to each other, and slightly curved slits seem to be promoted from the general valve [74].

Navicula mannii Hagelstein 1939 (Figure 3G–J) Synonym: *Navicula elegantissima* Meister 1935 [114]

Original description: Hagelstein 1939, p. 388, pl. 7, Figures 7 and 8 [115]

Description: Valves are observed to be solitary. Valves are broadly lanceolate, and abruptly constricted toward the ends (Figure 3G–I). Overall dimensions include average length and width ranges from 28.28 to 30.09 μm and from 8.57 to 9.44 μm, respectively. The axial area is narrow, and becomes gradually wider, larger, and rounded toward the central area (Figure 3 I). Raphe is observed to be very slightly curved filiform style with a very thickened and hyaline sternum (Figure 3 I,J, arrow). Striae are very coarse and of low density (9–11 in 10 μm); they are observed to strongly radiate in the middle and then become parallel towards the ends. The central area striae alternate between longer and shorter forms (longer striae 4 areolae, and shorter striae 2 areolae). Areolae are observed to be ribbon-shaped, and approximately 5–6 are found in 2 μm sections (Figure 3 J, arrow).
Depth occurrence in the core: 2.7 cm.

Distribution: *Navicula mannii* was reported in brackish water or marine environments [70,116,117]. Navarro (1983) reported the taxon in tropical temperate waters from the southwestern coast of Puerto Rico [116]. This species was known to neritic, pantropical, and cosmopolitan [116]. Ohtsuka (2005) collected the species from a muddy tidal flat in the Ariake Sea in south-western Japan [117].

Differential diagnosis: Hagelstein (1939) described that the *Navicula mannii* have minutely punctate areolae, but we found the ribbon-shaped areolae on the striae based on SEM observation in this study [115]. Ultrastructural studies of this species are rarely performed using a SEM; this study represents the first example of this approach.

Remarks of raphe: *Navicula mannii* has a straight raphe. Proximal raphe ends have an expanded pore-like shape and bent distal raphe ends (Figure 3G,H,J, arrows). Normally, *Navicula* spp. have a straight raphe system, unlike the raphe shapes of *Navicula cryptocephala* and *Navicula gregaria*, which commonly occur in Korea. *N. mannii* and *N. cryptocephala* both have drop-like internal ends, but *N. mannii* has a more pore-like end than *N. cryp- tocephala* [83,118], with a T-shaped structure. In contrast, *Navicula gregaria* has a different raphe shape than *N. mannii*, which is bent in the same direction as the raphe and exhibits asymmetrical thickening, beside the proximal raphe ends and beside the raphe rib [118].

Metascolioneis tumida (Brébisson ex Kützing) Blanco and Wetzel 2016 (Figure 4A,B)

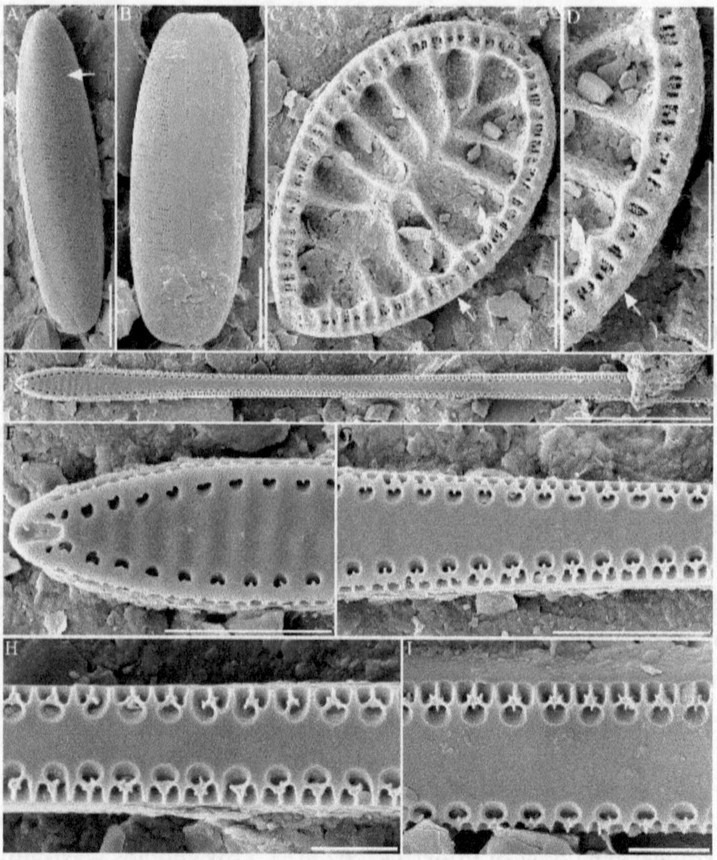

Figure 4. Field emission scanning electron microscopic photomicrographs of diatoms. (A,B) *Metascolioneis tumida* (the arrow points to raphe), (C,D) *Surirella recedens* (the arrow points to raphe), (E–I) *Thalassionema synedriforme*. Scale bar = 10 μm.

Basionym: *Navicula tumida* Brébisson ex Kützing 1849 [119] Synonym: *Navicula tumida* Brébisson ex Kützing 1849 [119] *Scoliopleura tumida* (Brébisson ex Kützing)

Rabenhorst 1864 [120] *Microstigma tumida* (Brébisson) Meister 1919 [121]

Scoliotropis tumida (Brébisson ex Kützing) R.M.Patrick and Freese 1961 [111]

Scolioneis tumida (Brébisson ex Kützing) D.G.Mann 1990 [122]

Navicula jenneri W.Smith 1853 [84]

Scoliopleura jenneri Grunow 1860 [123]

Original description: Blanco, S. and Wetzel, C.E., 2016, pp. 195–205 [124].

Description: Valves are found to be solitary with one-layered valves. Valves are linear lanceolate with bluntly rounded apices (Figure 4A). Overall dimensions include an average length and width ranges from 41.72 to 52.5 µm and from 6.05 to 6.50 µm, respectively. Cells are usually found in a girdle view and twisted about the apical axis (Figure 4B). The valve mantle is relatively deep, and its face curved moderately into mantles. Striae uniseriate (16–17 in 10 µm) with small poroids (12–15 µm in 5 µm) are observed. The raphe system is twisted and sigmoidal in shape (Figure 4A, arrow). The raphe sternum is generally narrow and slightly wider in thcenterre. The raphe is found to be straight with simple raphe endings and straight terminal fissures extending to the valve margin. The girdle consists of several open bands. The band closest to the valve bears one transverse row of poroids 31–32 at 10 µm.

Depth occurrence in the core: 1.0, 1.5, 2.0 m.

Distribution: This species has been found in marine habitats. Stoermer et al. (1999) presented this taxon in a checklist of diatoms as *Navicula tumida* from the marine environment in the Laurentian Great Lakes [125]. Vilicic et al. (2002) reported the species as *Scoliopleura tumida* from the eastern Adriatic Sea [93]. This species was listed to the British marine diatoms as *Scoliopleura tumida* [126,127]. Méléder et al. (2007) reported the taxon as *Scolioneis tumida* in a sediment of mudflat from Bourgneuf Bay, France [128].

Differential diagnosis: Formerly included in *Scolipleura* taxon, but lacking the offset central raphe endings and longitudinal canals of that genus. Distinguishable from *Scoliotropis* by having fewer plastids, which lie against the valves rather than the girdle, by the simple uniseriate striae and raphe structure [122].

Remarks of raphe: *Metascolioneis tumida* (syn. *Scolioneis tumida, Navicula tumida*) has a slightly twisted raphe and raphe sternum that is normally narrow and becomes expanded in the center (Figure 4A, arrow). External central raphe endings have straight fissures along the valve margin. Furthermore, internal central endings are T-shaped and elongated [122].

Surirella recedens A.W.F.Schmidt 1875 (Figure 4C,D)

Homotypic synonym: *Surirella fastuosa* var. *recedens* (A.W. F. Schmidt) Cleve 1878 [129]

Surirella fastuosa var. *typica* f. *recedens* (A. W. F. Schmidt) Deby 1897 [130] Original description: Schmidt and Fricke 1875, pls 17–20. [131]

Description: Valves are found to be solitary, strongly silicified, and lying in the valve or girdle view. Valves are heteropolar with a broadly rounded headpole and cuneate footpole. Overall dimensions include a length and width of 35.08 μm and 22.83 μm, respectively. The valve surface has four costae of 10 μm in length. The valve margin includes fibulae, siliceous braces, 7–8 fibulae, 10 μm. The raphe system runs around the entire valve margin and is located within a canal (Figure 4C,D; arrows). The canal is raised above the valve's surface. One or more potulae are located between the two fibulae. Four or more fibulae are located between the two costae.

Occurring depth in Core: 2.7, 5.5 m.

Distribution: This taxon was known to marine species [95,96]. López-Fuerte and Siqueiros-Beltrones (2016) reported the species as a benthic diatom from coastal waters in Mexico [132] and the Nanaura mudflat in Ariake Sea, Japan [117]. However, *Surirella rece- dens* was found in brackish waters from Cochin Backwater south in the Indian Ocean [133]. Differential diagnosis: *S. recedens* is composed of a heteropolar valve, whereas *Surirella fastuosa* has an isopolar valve [134]. *S. recedens* is smaller overall, in comparison to *S. fastuosa*, and more lanceolate in shape. In addition, *S. fastuosa* has more noticeable apices than *S. recedens* [135]. Goldman (1990) identified the *Surirella* cf. *fastuosa* based on the outline, length, infundibula, and circular pattern of the valve [136].

Remarks of raphe: *Surirella recedens* (Syn. *Surirella fastuosa* var. *recedens, Surella fastuosa* var. *typical* f. *recedens*) *Surirella* sp. has a raphe that is located along the margin of the valve (Figure 4D, arrow). A raphe positioned within a canal might be elevated above the valve surface in several species [137]. *S. recedens* shows a representative *Surirellaceae* raphe system, positioned along the margin of the valve (Figure 4D, arrow).

Thalassionema synedriforme (Greville) G.R.Hasle 1999 (Figure 4E–I) Basionym: *Asterionella synedriformis* Greville 1865 [138]

Synonym: *Asterionella synedriformis* Greville 1865 [138]

Thalassionema javanicum Grunow Hasle in Hasle and Syvertsen 1996 [139] Depth occurrence in the core: Hasle, G.R. 1999, pp. 54–59, 23 figures [140].

Description: The valves are heteropolar spatulate, linear, long, slightly wider in the middle, and constricted towards the head pole, rather than towards the foot pole (Figure 4E). Valve width ranges from 2.66 to 2.87 μm in the middle part of the valve and from 4.03 to 4.40 μm in the middle part of the footpole. An apical spine is located in the head pole, not the foot-pole part of the valve. The valve face is flat, with a wide sternum and slight undulation (5–6 in 5 μm) at the foot pole (Figure 4F). Areolae are placed within the valve face and valve mantle. Areolae are heart-shaped (5–6 in 5 μm) near the foot pole part, but similar to Y- or flower-shaped occlusions (5–6 in 5 μm) toward the middle part of the valve (Figure 4F–I). Labiate processes are placed at each pole of the valve. An opening of the labiate process places the external apex in the foot pole [141].

Depth occurrence in the core: 3.0 m.

Distribution: *Thalassionema synedriforme* was known to marine species. Hasle (2001) mentioned the species is restricted in tropical and subtropical waters [98]. This species was recorded for the first time from Argentinean coastal waters [142].

Differential diagnosis: *Asterionella synedriformis* Greville is the basionym of *Thalas- sionema synedriforme* [98]. Frenguelli (1941) mentioned that he found valves with 9–10 areo- lae within 10 μm and illustrated a linear, fragmentary specimen that, according to the valve outline and areola density, might also be attributed to *Thalassionema frauenfeldii*, but not to *T. synedriforme* [143]. Additionally, in the case of *Thalassiothrix javanica* (Grunow), Hustedt and Frenguelli illustrated a specimen that was slightly heteropolar with 6–7 areolae within 10 μm, which differs in areolae density and valve outline from *Thalassionema synedriforme* (12–16 areolae in 10 μm, according to Hasle 2001) [98].

4. DISCUSSION

4.1 Diatoms in SCW03

In this study, the analysis of sub-fossil diatoms among core samples obtained from Suncheon Bay, Korea, was carried out. Within the SCW03 core sample, a total of 52 genera and 87 species of sub-fossil diatoms were identified, with locations ranging from the surface to within 6 m of the basement, and among them, six species of newly recorded sub-fossil diatom never recorded before were found. At a depth of 4.0 m, the maximum variety of diatom samples was observed, with 23 genera 34 species, while depths of 0.5 m and 3.5 m revealed the lowest variety of diatoms, with 12 genera 14 species and 10 genera 14 species, respectively (Table 3). The highest and lowest taxonomic richness occur at depths of 4.0 m (23 genera, 34 species) and 3.5 m (10 genera, 14 species), respectively (Table 3).

Among the diatoms observed, *Amphora* sp., *Auliscus sculptus,* and *Fragilaria capucina* were found only at 0.1 m depths; therefore, they are hypothesized to have only recently entered the Suncheonman Bay (Table 3). *Cymatosira lorenziana*, *Fallacia hodgeana*, *Pleurosigma* sp., *Semiorbis* sp., and *Trachyneis aspera* are no longer observed since they appeared to only occur at 6.0 m depth. It is hypothesized that this is due to climate, environmental, or topographical changes in the Suncheonman Bay habitat. Intriguingly, *C. lorenziana* is mainly found in warm waters, whereas *T. aspera* is mainly distributed in the Antarctic area, with almost opposite habitat characteristics; however, the causative environmental changes in Suncheonman Bay could not be identified in this study [144–146]. Nevertheless, identifying past environmental changes in Suncheon Bay is an important source of information for the prediction of future environmental changes, and should be investigated further.

Most of the marine and brackish species occur at a depth of 5.0 to 6.0 m of the SCW03 core sample, and the emergence of freshwater species gradually increased within depth ranges of 3.0 to 4.5 m. Interestingly, only marine and brackish species appeared at the 2.5 m section (Table 3). Marine and some freshwater species appeared within occur the range of 1.0 m to 2.0 m depths, while only marine and brackish species appeared at 0.5 m depth; finally, freshwater species increased again at 0.1 m depth. These changes in the flora of diatoms assemblage composition observed suggest that the sediments recovered from SCW03 were deposited in a marine area that was scarcely influence by freshwater inflow in the past (3.0–4.5 m depth; about 1260–1830 yr BP) and experienced a renewed influx of fresh water in recent times (0.1 m depth; about 1340 ± 20 yr BP) (Tables 2 and 3, Figure 2) [147,148]. More accurate results and solid interpretations of the triggers responsible for such environmental changes require further studies of comprehensive environmental change, including a quantitative analysis of diatoms and a chronological analysis of core samples.

4.2 New Recorded Taxa from Korea

In this study, we identified six species newly recorded in Suncheonman Bay area: Cymatonitzschia marina, *Fallacia Hodgeana, Navicula mannii, Metascolioneis tumida, Surirella recedens*, and *Thalassionema synedriforme*. *N. mannii* was first identified by light microscopy in 2005; however, in this study, we observed the ultra structure of *N. mannii* and discovered a ribbon-shaped areolae by Fe-SEM [117]. The ribbon-shaped areolae of *N. mannii* are the first to ever be recorded.

Unrecorded sub-fossil diatoms in Suncheonman Bay were discovered at 1.0 m depths. This study is therefore meaningful because if we study the modern composition of diatoms in Suncheonman Bay or Korea, we would estimate six previously unrecorded diatoms were extinct in this area.

REFERENCES

1. Pritchard, D.W. *What Is an Estuary: Physical Viewpoint*; American Association for the Advancement of Science: Washington, DC, USA, 1967.
2. Smol, J.P.; Stoermer, E.F. *The Diatoms: Applications for the Environmental and Earth Sciences*; Cambridge University Press: Cambridge, UK, 2010.
3. Ralston, D.K.; Geyer, W.R. Response to channel deepening of the salinity intrusion, estuarine circulation, and stratification in an urbanized estuary. *J. Geophys. Res. Ocean.* **2019**, *124*, 4784–4802. [CrossRef]
4. Selleslagh, J.; Amara, R. Environmental factors structuring fish composition and assemblages in a small macrotidal estuary (eastern English Channel). *Estuar. Coast. Shelf Sci.* **2008**, *79*, 507–517. [CrossRef]
5. Elliott, M.; O'reilly, M.G.; Taylor, C.J.L. The forth estuary: A nursery and overwintering area for North Sea fishes. *Hydrobiologia* **1990**, *195*, 89–103. [CrossRef]
6. Ye, S.J.; Jeong, J.M.; Kim, H.J.; Park, J.M.; Huh, S.H.; Baeck, G.W. Fish assemblage in the tidal creek of Sangnae-ri Suncheon, Korea. Korean J. Ichthyol. **2014**, 26, 74–80.
7. Kecinski, M.; Messer, K.D.; Peo, A.J. When cleaning too much pollution can be a bad thigns: A field experiment of consumer demand for oysters. *Ecol. Econ.* **2018**, *146*, 686–695. [CrossRef]
8. Canham, R.; Flemming, S.A.; Hope, D.D.; Drever, M.C. Sandpipers go with the flow: Correlations between estuarine conditions and shorebird abundance at an important stopover on the Pacific Flyway. *Ecol. Evol.* **2021**, *11*, 2828–2841. [CrossRef] [PubMed]
9. Laprise, R.; Dodson, U.J. Environmental variability as a factor controlling spatial patterns in distribution and species diversity of zooplankton in the St. Lawrence Estuary. *Mar. Ecol. Prog. Ser.* **1994**, *107*, 67–81. [CrossRef]
10. Jang, S.G.; Cheong, C.J. Characteristics of grain size and organic matters in the tidal flat sediments of the Suncheon Bay (Korean). J. Korean Soc. Mar. Environ. Energy. **2010**, 13, 198–205.
11. Du, G.; Son, M.; Yun, M.; An, S.; Chung, I.K. Microphytobenthic biomass and species composition in intertidal flats of the Nakdong River estuary, Korea. *Estuar. Coast. Shelf Sci.* **2009**, *82*, 663–672. [CrossRef]
12. Kim, T.I.; Choi, B.H.; Lee, S.W. Hydrodynamics and sedimentation induced by large-scale coastal developments in the Keum River Estuary, Korea. *Estuar. Coast. Shelf Sci.* **2006**, *68*, 515–528. [CrossRef]

13. Jang, S.G.; Cheong, C.J. Seasonal characteristics of seawater quality in the Suncheon Bay. *Ecotoxicol. Environ. Saf.* **2010**, *12*, 47–57.

14. Kim, K.; Lee, K.J.; Han, B.H. Environmental ecological status of Suncheon bay and its application to the criteria of UNESCO world nature heritage. *Korean J. Environ. Eco.* **2013**, *27*, 625–641. [CrossRef]

15. Park, H.J.; Kwak, J.H.; Kang, C.K. Trophic consistency of benthic invertebrates among diversified vegetational habitats in a temperate coastal wetland of Korea as determined by stable isotopes. *Estuar. Coast.* **2015**, *38*, 599–611. [CrossRef]

16. Kim, H.K.; Kwon, Y.S.; Kim, Y.J.; Kim, B.H. Distribution of epilithic diatoms in estuaries of the Korean Peninsula in relation to environmental variables. *Water* **2015**, *7*, 6702–6718. [CrossRef]

17. Leterme, S.C.; Prime, E.; Mitchel, J.; Brown, M.H.; Ellis, A.V. Diatom adaptability to environmental change: A case study of two Cocconeis species from high-salinity areas. *Diatom Res.* **2013**, *28*, 29–35. [CrossRef]

18. Cho, A.; Cheong, D.; Kim, J.C.; Shin, S.; Park, Y.H.; Katsuki, K. Delta formation in the Nakdong River, Korea, during the Holocene as inferred from the diatom assemblage. *J. Coast. Res.* **2017**, *33*, 67–77. [CrossRef]

19. Laugaste, R.; Pork, M. Diatoms of lake peipsi-pihkva: A floristic and ecological review. *Hydrobiologia* **1996**, *338*, 63–76. [CrossRef]

20. Ryu, E.; Lee, S.J.; Yang, D.Y.; Kim, J.Y. Paleo-environmental studies of the Korean peninsula inferred from diatom assemblages. *Quat. Int.* **2008**, *176*, 36–45. [CrossRef]

21. Lee, S.D.; Lee, H.; Park, J.; Yun, S.M.; Lee, J.-Y.; Lim, J.; Park, M.; Kwon, D. Late Holocene diatoms in sediment cores from the Gonggeomji wetland in Korea. *Diatom Res.* **2020**, *35*, 195–229. [CrossRef]

22. Lee, H.; Lee, S.D.; Lee, J.-Y.; Lim, J.; Kwon, D.; Park, M. Holocene paleo-environmental changes and characteristic of diatom distribution in Upo Wetland of Korea. *Korean J. Environ. Ecol.* **2020**, *53*, 109–137. [CrossRef]

23. Oh, S.H.; Koh, C.H. Distribution of diatoms in the surficial sediments of the Mangyung-Dongjin tidal flat, west coast of Korea (Eastern Yellow Sea). *Mar. Bio* **1995**, *122*, 487–496. [CrossRef]

24. Lee, Y.G. Neogene diatoms of Pohang and Gampo areas, Kyongsangbug-do, Korea. *J. Geological. Soc. Korea* **1975**, *11*, 99–113.

25. Lee, Y.G. On the fossil diatoms in the Bukpyeong Formation, Bukpyeong area, Gangweon-do, Korea. *J. Geological. Soc. Korea* **1977**, *13*, 23–40.

26. LEE, Y.G. Micropaleontological study of Neogene strata of south-eastern Korea and adjacent sea floor. *J. Paleontol. Soc. Korea* **1986**, *2*, 83–113.
27. Lee, Y.G. Neogene paleotemperature oscillations in the Pohang Basin, Korea. *JKESS* **1988**, *9*, 203–216.
28. Lee, Y.G.; Park, Y.A.; Choi, J.Y. Sedimentary facies and micro-paleontological study of tidal sediments off the Mankyung-Dongjin River estuary, west coast of Korea. *J.Korean Soc. Oceanogr.* **1995**, *30*, 77–90.
29. Hwang, S.I.; Yoon, S.O.; Jo, W.R. The change of the depositional environment on Dodaecheon River basin during the Middle Holocene. *J. Geol. Soc. Korea* **1997**, *32*, 403–420.
30. Hwang, S. The Holocene depositional environment and sea-level change at Ilsan area. *J. Geol. Soc. Korea* **1998**, *33*, 143–163.
31. Ryu, E.; Nahm, W.H.; Yang, D.Y.; Kim, J.Y. Diatom floras of a western coastal wetland in Korea: Implication for late Quaternary paleo-environment. *J. Geol. Soc. Korea* **2005**, *41*, 227–239.
32. Yi, S.; Ryu, E.; Kim, J.Y.; Nahm, W.H.; Yang, D.Y.; Shin, S.C. Late Holocene paleo-environmental changes inferred from palynological and diatom assemblages in Isanpo area, Ilsan, Gyeonggi-do, Korea. *J. Geol. Soc. Korea* **2005**, *41*, 295–322.
33. Bak, Y.S. Mid-Holocene sea-level fluctuation inferred from diatom analysis from sediments on the west coast of Korea. *Quat. Int.* **2015**, *384*, 139–144. [CrossRef]
34. Park, T.M.; Kweon, D.H. Study of the heavy metal pollution of the Suncheon Bay tidal flat. *J. Korea Soc. Environ. Adm.* **2001**, *7*, 467–472.
35. Kamimura, S.; Itoh, H.; Ozeki, S.; Kojima, S. Molecular diversity of Cerithidea gastropods inhabiting Suncheon Bay, and the Japanese and Ryukyu Islands. *Plankton Benthos Res.* **2010**, *5*, 250–254. [CrossRef]
36. Park, Y.K.; Yoo, M.L.; Heo, H.S.; Lee, H.W.; Park, S.H.; Jung, S.C.; Park, S.S.; Seo, S.G. Wild reed of Suncheon Bay: Potential bio-energy source. *Renew Energy* **2012**, *42*, 168–172. [CrossRef]
37. Choi, S.C.; Choi, D.G.; Hwang, J.S.; Kim, J.G.; Choo, Y.S. Solute patterns of four halophytic plant species at Suncheon Bay in Korea. *J. Ecol. Environ.* **2014**, *37*, 131–137. [CrossRef]
38. You, Y.H.; Park, J.M.; Lee, M.-C.; Kim, J.G. Phylogenetic analysis and diversity of marine bacteria isolated from rhizosphere soils of halophyte in Suncheon Bay. *Microbiol. Biotechnol. Lett.* **2015**, *43*, 65–78. [CrossRef]

39. You, Y.H.; Yoon, H.; Kang, S.M.; Shin, J.H.; Choo, Y.S.; Lee, I.J.; Lee, J.M.; Kim, J.G. Fungal diversity and plant growth promotion of endophytic fungi from six halophytes in Suncheon Bay. *J. Microbiol. Biotechnol.* **2012**, *22*, 1549–1556. [CrossRef]

40. Noh, K.H.; Kim, J.H.; Chung, Y.C. Species composition and dynamics of phytoplankton community in Dong Cheon and Isa Cheon flowed into Suncheon Bay. *Korean J. Limnol.* **1991**, *24*, 153–163.

41. Lee, H.; Yun, S.M.; Lee, J.-Y.; Lee, S.D.; Lim, J.; Cho, P.Y. Late Holocene climate changes from diatom records in the historical Reservoir Gonggeomji, Korea. *J. Appl. Phycol.* **2018**, *30*, 3205–3219. [CrossRef]

42. Schneider, C.A.; Rasband, W.S.; Eliceiri, K.W. NIH image to imageJ: 25 years of image analysis. *Nat. Methods* **2012**, *9*, 671–675. [CrossRef] [PubMed]

43. Guiry, M.D.; Guiry, G.M. AlgaeBase [online]. World-Wide Electronic Publication, National University of Ireland: Galway, Ireland, 2020. Available online: https://www. algaebase.org (accessed on 8 December 2020).

44. Lim, J.; Lee, J.-Y.; Hong, S.; Park, S.; Lee, E.; Yi, S. Holocene coastal environmental change and ENSO-driven hydroclimatic variability in East Asia. *Quat. Sci. Rev.* **2019**, *220*, 75–86. [CrossRef]

45. Lee, H.; Lee, J.-Y.; Shin, S. Middle Holocene Coastal Environmental and Climate Change on the Southern Coast of Korea. *Appl. Sci.* **2021**, *11*, 230. [CrossRef]

46. Hoppenrath, M. A revised checklist of planktonic diatoms and dinoflagellates from Helgoland (North Sea, German Bight). *Helgol. Mar. Res.* **2004**, *58*, 243–251. [CrossRef]

47. Vilbaste, S. Benthic diatom communities in Estonian rivers. *Boreal. Environ. Res.* **2001**, *6*, 191–203.

48. Tsoy, I.; Prushkovskaya, I.; Aksentov, K.; Astakhov, A. Environmental changes in the Amur Bay (Japan/East Sea) during the last 150 years revealed by examination of diatoms and silicoflagellates. *Ocean. Sci. J.* **2015**, *50*, 433–444. [CrossRef]

49. Jahn, R.; Schmid, A.M.M. Revision of the brackish-freshwater diatom genus *Bacillaria Gmelin* (Bacillariophyta) with the description of a new variety and two new species. *Eur. J. Phycol.* **2007**, *42*, 295–312. [CrossRef]

50. Myklestad, S. Production of carbohydrates by marine planktonic diatoms. I. Comparison of nine different species in culture. *J. Exp. Mar. Biol. Ecol.* **1974**, *15*, 261–274. [CrossRef]

51. Tabassum, A.; Saifullah, S. The planktonic diatom of the genus Chaetoceros Ehrenberg from northwestern Arabian Sea bordering Pakistan. *Pak. J. Bot.* **2010**, *42*, 1137–1151.

52. Lewis, J.; Harris, A.; Jones, K.; Edmonds, R. Long-term survival of marine planktonic diatoms and dinoflagellates in stored sediment samples. *J. Plankton Res.* **1999**, *21*, 343–354. [CrossRef]

53. Park, J.; Lee, J.H.; Khim, J.S. The identity of 'Berkeleya scopulorum'from Northeast Asia: Report on Climaconeis mabikii sp. nov. from temperate marine waters with notes on biogeography of the genus. *Ocean. Sci. J.* **2016**, *51*, 591–598. [CrossRef]

54. Tang, T.; Qu, X.; Li, D.; Liu, R.; Xie, Z.; Cai, Q. Benthic algae of the Xiangxi river, China. *J. Freshw. Ecol.* **2004**, *19*, 597–604. [CrossRef]

55. Sumper, M. A phase separation model for the nanopatterning of diatom biosilica. *Science* **2002**, *295*, 2430–2433. [CrossRef]

56. Choudhury, A.K.; Pal, R. Phytoplankton and nutrient dynamics of shallow coastal stations at Bay of Bengal, Eastern Indian coast. *Aquat. Ecol.* **2010**, *44*, 55–71. [CrossRef]

57. Al-Harbi, S.M. Phytoplankton composition of *ROPME Sea Area (Arabian Gulf). *Mar. Scienes* **2005**, *16*, 105–114. [CrossRef]

58. Romero, O.E.; Thunell, R.C.; Astor, Y.; Varela, R. Seasonal and interannual dynamics in diatom production in the Cariaco Basin, Venezuela. *Deep Sea Res. Part. I Oceanogr. Res. Pap.* **2008**, *56*, 571–581. [CrossRef]

59. Lacerda, S.; Koening, M.; Neumann-Leitão, S.; Flores-Montes, M. Phytoplankton nyctemeral variation at a tropical river estuary (Itamaracá-Pernambuco-Brazil). *Braz. J. Biol.* **2004**, *64*, 81–94. [CrossRef]

60. Schlegel, I.; Scheffler, W. Seasonal development and morphological variability of Cyclotella ocellata (Bacillariophyceae) in the eutrophic Lake Dagow (Germany). *Int. Rev. Hydrobiol.* **1999**, *84*, 469–478.

61. Felício-Fernandes, G.; de Souza-Mosimann, R.M. Diatomáceas no sedimento do manguezal de Itacorubi-Florianópolis, Santa Catarina, Brasil. *Insul. Rev. Botânica* **1994**, *23*, 149–215.

62. Lam, N.N.; Hai, D.; Ho, T. Species composition of diatoms (Bacillariophyceae) in Van Phong-Ben Goi Bay, Centrol Viet Nam. Collect. Mar. Res. Work. **1999**, 9, 179–195.

63. Lewis, F.W. Original communications: Notes on new and rarer species of diatomaceæ of the United States sea board. *J. Cell Sci.* **1862**, *2*, 155–161. [CrossRef]

64. Licea, S.; Moreno Ruiz, J.; Luna, R. Checklist of diatoms (Bacillariophyceae) from the Southern Gulf of Mexico: Data-Base (1979–2010) and new records. *J. Biodivers. Endanger. Species* **2016**, *4*, 1–7.

65. Liu, R.; Liu, J.Y. *Checklist of Biota of Chinese Seas*; Institute of Oceanology, Chinese Academy of Sciences: Beijing, China, 2008; pp. 1–1267.

66. Natilj, L.; Damsiri, Z.; Chaouti, A.; Loudiki, M.; Khalil, K.; Elkalay, K. Spatio-temporal patterns of the micro-phytoplankton community structure and distribution in a North African lagoon. *J. Mater. Environ. Sci.* **2016**, *7*, 4419–4434.

67. McGee, D.; Laws, R.A.; Cahoon, L.B. Live benthic diatoms from the upper continental slope: Extending the limits of marine primary production. *Mar. Ecol. Prog. Ser.* **2008**, *356*, 103–112. [CrossRef]

68. Gómez, F.; Wang, L.; Hernández-Becerril, D.U.; Lisunova, Y.O.; Lopes, R.M.; Lin, S. Molecular phylogeny suggests transfer of Hemidiscus into Actinocyclus (Coscinodiscales, Coscinodiscophyceae). *Diatom Res.* **2017**, *32*, 21–28. [CrossRef]

69. Akar, B.; S¸ahin, B. Diversity and ecology of benthic diatoms in Karagöl lake in Karagöl-Sahara National Park (S¸avs¸at, Artvin, Turkey). *Turk. J. Fish. Aquat. Sci.* **2017**, *17*, 15–24. [CrossRef]

70. Park, J.; Khim, J.S.; Ohtsuka, T.; Araki, H.; Witkowski, A.; Koh, C.H. Diatom assemblages on Nanaura mudflat, Ariake Sea, Japan: With reference to the biogeography of marine benthic diatoms in Northeast Asia. *Bot. Stud.* **2012**, *53*, 105–124.

71. Saros, J.; Anderson, N. The ecology of the planktonic diatom Cyclotella and its implications for global environmental change studies. *Biol. Rev.* **2015**, *90*, 522–541. [CrossRef]

72. Gómez, F.; Wang, L.; Lin, S. Molecular phylogeny suggests the affinity of the planktonic diatoms Climacodium and Bellerochea (Lithodesmiales, Mediophyceae). *Diatom Res.* **2018**, *33*, 349–354. [CrossRef]

73. Battegazzore, M.; Gallo, L.; Lucadamo, L.; Morisi, A. Quality of the main watercourses in the Pollino National Park (Apennine Mts, S Italy) on the basis of the diatom benthic communities. *Studi Trent. Sci. Nat. Acta Biol.* **2003**, *80*, 89–93.

74. Li, Y.; Suzuki, H.; Nagumo, T.; Tanaka, J. Morphology and Ultrastructure of Fallacia hodgeana (Bacillariophyceae). *J. JAP. Bot.* **2014**, *89*, 27–34.

75. Lewis, R.J.; Jensen, S.I.; DeNicola, D.M.; Miller, V.I.; Hoagland, K.D.; Ernst, S.G. Genetic variation in the diatom Fragilaria capucina (Fragilariaceae) along a latitudinal gradient across North America. *Plant Syst. Evol.* **1997**, *204*, 99–108. [CrossRef]

76. Dickman, M.D.; Peart, M.R.; Wai-Shu Yim, W. Benthic diatoms as indicators of stream sediment concentrationin Hong Kong. *Int. Rev. Hydrobiol.* **2005**, *90*, 412–421. [CrossRef]

77. Round, F.E. The ecology of benthic algae. In *Algae and Man*; Springer: Boston, MA, USA, 1964; pp. 138–184.

78. Underwood, G.; Phillips, J.; Saunders, K. Distribution of estuarine benthic diatom species along salinity and nutrient gradients. *Eur. J. Phycol.* **1998**, *33*, 173–183. [CrossRef]

79. Reid, G.; Williams, D.M. Systematics of the Gyrosigma balticum complex (Bacillariophyta), including three new species. *Physiol. Res.* **2003**, *51*, 126–142. [CrossRef]

80. Stepanek, J.G.; Kociolek, J.P. Several new species of Amphora and Halamphora from the western USA. *Diatom Res.* **2013**, *28*, 61–76. [CrossRef]

81. Rowland, S.; Allard, W.; Belt, S.; Massé, G.; Robert, J.M.; Blackburn, S.; Frampton, D.; Revill, A.; Volkman, J. Factors influencing the distributions of polyunsaturated terpenoids in the diatom, Rhizosolenia setigera. *Phytochemistry* **2001**, *58*, 717–728. [CrossRef]

82. Dayala, V.; Salas, P.; Sujatha, C. Spatial and seasonal variations of phytoplankton species and their relationship to physicochemical variables in the Cochin estuarine waters, Southwest coast of India. *Indian J. Mar. Sci.* **2014**, *43*, 943–953.

83. de Vijver, B.V.; Jarlman, A.; Lange-Bertalot, H. Four new Navicula (Bacillariophyta) species from Swedish rivers. *Cryptogam. Algol.* **2010**, *31*, 355–367.

84. Smith, W.; West, T. A Synopsis of the British Diatomaceæ: With Remarks on Their Structure, Functions and Distribution; and Instructions for Collecting and Preserving Specimens; Smith and Beck, Pub.: London, UK, 1853; Volume 1, p. 1853.

85. McQuoid, M.R.; Hobson, L.A. Assessment of palaeoenvironmental conditions on southern Vancouver Island, British Columbia, Canada, using the marine tychoplankter Paralia sulcata. *Diatom Res.* **1998**, *13*, 311–321. [CrossRef]

86. Kawamura, T.; Hirano, R. Seasonal changes in benthic diatom communities colonizing glass slides in Aburatsubo Bay, Japan. *Diatom Res.* **1992**, *7*, 227–239. [CrossRef]

87. Teanpisut, K.; Patarajinda, S. Species diversity of marine planktonic diatoms around Chang Islands, Trat Province. *Kasetsart J.* **2007**, *41*, 114–124.

88. Al-Handal, A.Y.; Thomas, E.W.; Pennesi, C. Marine benthic diatoms in the newly discovered coral reefs, off Basra coast, Southern Iraq. *Phytotaxa* **2018**, *372*, 111–152. [CrossRef]

89. Hagerthey, S.E.; Defew, E.C.; Paterson, D.M. Influence of Corophium volutator and Hydrobia ulvae on intertidal benthic diatom assemblages under different

nutrient and temperature regimes. *Mar. Ecol. Prog. Ser.* **2002**, *245*, 47–59. [CrossRef]

90. Siqueiros Beltrones, D.A.; Martínez, Y.J. Prospective floristics of epiphytic diatoms on Rhodophyta from the Southern Gulf of Mexico. *CICIMAR Ocean.* **2017**, *32*, 35–49. [CrossRef]
91. Chen, Y.C. Immobilization of twelve benthic diatom species for long-term storage and as feed for post-larval abalone Haliotis diversicolor. *Aquaculture* **2007**, *263*, 97–106. [CrossRef]
92. Hillebrand, H.; Sommer, U. Nitrogenous nutrition of the potentially toxic diatom Pseudonitzschia pungens f. multiseries Hasle. *J. Plankton Res.* **1996**, *18*, 295–301. [CrossRef]
93. Vilic˘ic′, D.; Marasovic′, I.; Miokovic′, D.J.A.B.C. Checklist of phytoplankton in the eastern Adriatic Sea. *Acta Bot. Croat.* **2002**, *61*, 57–91.
94. Szczepocka, E.; Rakowska, B. Diatoms in the biological assessment of the ecological state of waters using the Czarna Staszowska River as an example. *Oceanol. Hydrobiol. St.* **2015**, *44*, 254. [CrossRef]
95. Villac, M.C.; Kaczmarska, I.; Ehrman, J.M. Diatoms from ship ballast sediments (with consideration of a few additional species of special interest). In *Diatom Monographs*; Koeltz Botanical Books: Glashütten, Germany, 2016; Volume 18, p. 557.
96. Foged, N. Diatoms in Alaska. *Bibl. Phycol.* **1981**, *53*, 1–318.
97. Burns, D. Distribution of planktonic diatoms in Pelorus Sound, South Island, New Zealand. *N. Z. J. Mar. Freshwater Res.* **1977**, *11*, 275–295. [CrossRef]
98. Hasle, G.R. The marine, planktonic diatom family Thalassionemataceae: Morphology, taxonomy and distribution. *Diatom Res.* **2001**, *16*, 1–82. [CrossRef]
99. Hasle, G.R. The biogeography of some marine planktonic diatoms. *Deep Sea Res. Oceanogr. Abstr.* **1976**, *23*, 319-IN6. [CrossRef]
100. Barron, J.A. Planktonic marine diatom record of the past 18 my: Appearances and extinctions in the Pacific and Southern Oceans. *Diatom Res.* **2003**, *18*, 203–224. [CrossRef]
101. Palmisano, A.C.; SooHoo, J.B.; White, D.C.; Smith, G.A.; Stanton, G.R.; Burckle, L.H. Shade Adapted Benthic Diatoms Beneath Antarctic Sea Ice. *J. Phycol.* **1985**, *21*, 664–667. [CrossRef]
102. Fernandes, L.; de Souza-Mosimann, R. Triceratium moreirae sp. nov. and *Triceratium dubium* (Triceratiaceae-Bacillariophyta) from estuarine environments of Southern Brazil, with comments on the genus Triceratium CG Ehrenberg. *Rev. Bras. Biol.* **2001**, *61*, 159–170. [CrossRef]

103. Suphan, S.; Peerapornpisal, Y. Fifty three new record species of benthic diatoms from Mekong River and its tributaries in Thailand. Chiang. Mai. J. Sci. **2010**, 37, 326–343.

104. Álvarez-Blanco, I.; Blanco, S. *Benthic Diatoms from Mediterranean Coasts*; Schweizerbart'sche Verlagsbuchhandlung: Stuttgart, Germany, 2014; Volume 60, pp. 3–4.

105. Petrov, A.; Nevrova, E. Database on Black Sea benthic diatoms (Bacillariophyta): Its use for a comparative study of diversity pecularities under technogenic pollution impacts. *Ocean. Biodivers. Inform.* **2007**, *202*, 153–165.

106. Prasad, A.; Nienow, J.; Livingston, R. The marine diatom genus Tryblioptychus Hendey (Thalassiosiraceae, Coscinodiscophyceae): Fine structure, taxonomy, systematics and distribution. *Diatom Res.* **2002**, *17*, 291–308. [CrossRef]

107. Lewis, F.W. Notes on New and Rarer Species of Diatomaceae of the United States Seaboard; Merrihew, Thompson: Philadelphia, PA, USA, 1861.

108. Simonsen, R. The diatom plankton of the Indian Ocean Expedition of R/V "Meteor" 1964–1965. *Gebruì der Borntraeger* **1974**, *19*, 1–107.

109. Boyer, C.S. *Diatomaceae of Philadelphia and Vicinity*; Press of J. B. Lippincott Company: Philadelphia, PA, USA, 1916.

110. Al-Yamani, F.Y.; Saburova, M.A. Marine phytoplankton of Kuwait's waters volume 2 Diatoms. *Kuwait Inst. Sci. Res.* **2019**, *351*, 1–336.

111. Patrick, R.M.; Freese, L.R. Diatoms (Bacillariophyceae) from Northern Alaska. *Proc. Acad. Nat. Sci. USA* **1961**, *112*, 129–293.

112. Li, Y.; Suzuki, H.; Nagumo, T.; Tanaka, J.; Sun, Z.; Xu, K. Three new species of Fallacia from intertidal sediments in Japan. *Diatom Res.* **2019**, *34*, 75–83. [CrossRef]

113. Witkowski, A.; Lange-Bertalot, H.; Metzeltin, D. Diatom flora of marine coasts. In *Iconographia Diatomologica: Annotated Diatom Micrographs*; Lange-Bertalot, H., Ed.; Gantner Verlag: Oxford, UK, 2000; Volume 7.

114. Meister, Berichte der Schweizerischen Botanischen Gesellschaft. *Und Neue Kieselalgen*; Seltene, Kommissionsverlag von Rascher & Co.: Zurich, Switzerland, 1935; Volume 44, pp. 87–108.

115. Hagelstein, R. The Diatomaceae of Porto Rico and the Virgin Islands. *Sci. Surv. Porto Rico. Virgin. Isl.* **1939**, *8*, 313–450.

116. Navarro, J. A survey of the marine diatoms of Puerto Rico VI. suborder Raphidineae: Family Naviculaceae (Genera Haslea, Mastogloia and Navicula). *Bot. Mar.* **1983**, *26*, 119–136. [CrossRef]

117. Ohtsuka, T. Epipelic diatoms blooming in Isahaya Tidal Flat in the Ariake Sea, Japan, before the drainage following the Isahaya-Bay Reclamation Project. *Phycol. Res.* **2005**, *53*, 138–148. [CrossRef]

118. Cox, E.J. Studies on the diatom genus Navicula Bory. VII. The identity and typification of Navicula gregaria DonKin, N. Cryptocephala Kutz. and related taxa. *Diatom Res.* **1995**, *10*, 91–111. [CrossRef]

119. Kützing, F.T. *Species Alagrum*; F. A. Brockhaus: Lipsiae, Germany, 1849.

120. Rabenhorst, L. *Flora Europaea Algarum Aquae Dulcis et Submarinae*; Apud E. Kummerum: Lipsiae, Germany, 1864; Volume 1, pp. 1864–1868.

121. Meister, F. Zur Pflanzengeographie der Schweizerischen Bacillariacee. In *Botanische Jahrbücher für Syster Matik, Pflanzengeschichte und Pflanzengeographie*; Switzerland, 1919; Volume 55, pp. 125–159. Available online: https://www.biodiversitylibrary.org/item/ 136876#page/137/mode/1up (accessed on 28 May 2021).

122. Round, F.E.; Crawford, R.; Mann, D. *The Diatoms: Biology and Morphology of the Genera*; Cambridge University Press: Cambridge, UK, 1990; pp. 1–747.

123. Grunow, A. *Über Neue Oder Ungenügend Bekannte Algen, Erste Folge, Diatomeen, Familie Naviculaceen*; Verhandlungen der Kaiserlich- Königlichen Zoologisch-Botanischen Gesellschaft in Wein: Austria, 2021; Volume 10, pp. 503–582. Available online: https: //www.zobodat.at/pdf/VZBG_10_0503-0582.pdf (accessed on 27 May 2021).

124. Blanco, S.; Wetzel, C.E. Replacement names for botanical taxa involving algal genera. *Phytotaxa* **2016**, *266*, 195–205. [CrossRef]

125. Stoermer, E.F.; Russell, G.; Kreis, J.; Andresen, N.A. Checklist of Diatoms from the Laurentian Great Lakes. II. *J. Great Lakes Res.* **1999**, *25*, 515–566. [CrossRef]

126. Hendey, N.I. A preliminary check-list of British marine diatoms. *J. Mar. Biol. Ass. UK* **1954**, *33*, 537–560. [CrossRef]

127. Hendey, N.I. A revised check-list of British marine diatoms. *J. Mar. Biol. Ass. UK* **1974**, *54*, 277–300. [CrossRef]

128. Méléder, V.; Rincé, Y.; Barillé, L.; Gaudin, P.; Rosa, P. Spatiotemporal changes in microphytobenthos assemblages in a macrotidal flat (Bourgneuf Bay, France) 1. *J. Phycol.* **2007**, *43*, 1177–1190. [CrossRef]

129. Cleve, P.T. *Diatoms from the West. Indian Archipelago*; Norstedt: Stockholm, Sweden, 1878; Volume 5.

130. Deby, J. Le genre Surirella. Travail posthume, traduit, mis en ordre et publié par M. le docteur Henri Van Heurck. In *Annales de la Société Belge de Microscopie*; A. Manceaux, Libraire-editeur: Belgium, 1897; Volume 9, pp. 147–177.

131. Schmidt, A.; Fricke, F. *Atlas der Diatomaceen-Kunde*; CH Kain: O. R. Reisland, Germany, 1875; Volume 1.

132. López-Fuerte, F.O.; Siqueiros-Beltrones, D.A. A checklist of marine benthic diatoms (Bacillariophyta) from Mexico. *Phytotaxa* **2016**, *283*, 201–258. [CrossRef]

133. Gopinathan, C. On new distributional records of plankton diatoms from the Indian Seas. *J. Mar. Biol. Ass. India* **1975**, *17*, 223–240.

134. Watanabe, T.; Mayama, S.; Idei, M. Overlooked Heteropolarity in Surella cf. fastuosa (Bacillariophyta) and Relationships between Valve Morphogenesis and Auxospore Development. *J. Phycol.* **2012**, *48*, 1265–1277. [CrossRef] [PubMed]

135. Al-Handal, A.Y.; Compere, P.; Riaux-Gobin, C. Marine benthic diatoms in the coral reefs of Reunion and Rodrigues Islands, West Indian Ocean. *Micronesica* **2016**, *2016*, 1–77.

136. Goldman, N.; Paddock, T.B.B.; Shaw, K.M. Quantitative Analysis of Shape Variation in Populations of Surirella fastuosa. *Diatom Res.* **1990**, *5*, 25–42. [CrossRef]

137. Spaulding, S.; Edlund, M. Surirella. In Diatoms of North America. Available online: https://diatoms.org/genera/surirella (accessed on 2 March 2021).

138. Greville, R.K. Transactions of The Microscopical Society. In *Descriptions of New and Rare Diatoms. Series XIV*; Wiley: Hoboken, NJ, USA, 1865; Volume 13, pp. 1–10.

139. Hasle, G.R.; Syvertsen, E.E. Marine diatoms. In *Identifying Marine Diatoms and Dinoflagellates*; Academic Press: San Diego, CA, USA, 1996.

140. Hasle, G. Thalassionema synedriforme comb. nov. and Thalassiothrix spathulata sp. nov., two marine, planktonic diatoms from warm waters. *Phycologia* **1999**, *38*, 54–59. [CrossRef]

141. Sugie, K.; Suzuki, K. A new marine araphid diatom, Thalassionema kuroshioensis sp. nov., from temperate Japanese coastal waters. *Diatom Res.* **2015**, *30*, 237–245. [CrossRef]

142. Sar, E.A.; Sunesen, I.; Fernández, P.V. Marine diatoms from Buenos Aires coastal waters (Argentina). II. Thalassionemataceae and Rhaphoneidaceae. *Rev. Chil. Hist. Nat.* **2007**, *80*, 63–79. [CrossRef]

143. Frenguelli, J. XVI Contribución al conocimiento de las diatomeas argentinas. In *Diatomeas del Río de La Plata*; Uniccersida Nacional de la Plata/ Instituto de Meseo: La Plata, Argentina, 1941; Volume 3, pp. 213–334.

144. Kaleli, M.A.; Kociolek, J.P.; Solak, C.N. Taxonomy and distribution of diatoms on the Turkish Mediterranean Coast, Dalyan (Mug˘ la). *Mediterr. Mar. Sci.* **2020**, *21*, 201–215. [CrossRef]

145. Sutherland, D.L. Surface-associated diatoms from marine habitats at Cape Evans, Antarctica, including the first record of living Eunotogramma marginopunctatum. *Polar Bio* **2008**, *31*, 879–888. [CrossRef]

146. Rivkin, R.B.; Putt, M. Photosynthesis and cell division by Antarctic microalgae: Comparison of benthic, planktonic and ice algae. *J. Phycol.* **1987**, *23*, 223–229. [CrossRef]

147. Lee, Y.G.; Kim, S.; Jeong, D.U.; Kim, J.K.; Woo, H.J. Effects of Heavy Rainfall on Sedimentation in the Tidal Salt Marsh of Suncheon Bay, South Korea. *J. Coast. Res.* **2013**, *29*, 566–578. [CrossRef]

148. Muylaert, K.; Sabbe, K.; Vyverman, W. Spatial and Temporal Dynamics of Phytoplankton Communities in a Freshwater Tidal Estuary (Schelde, Belgium). *Estuar. Coast. Shelf Sci.* **2000**, *50*, 673–682. [CrossRef]

CHAPTER-11

FIRST REPORT OF THE MARINE BENTHIC DINOFLAGELLATE BYSMATRUM SUBSALSUM FROM KOREAN TIDAL POOLS

1. INTRODUCTION

The genus *Bysmatrum* M.A. Faust and K.A. Steidinger was erected to separate three benthic *Scrippsiella* species, *Scrippsiella arenicola* T. Horiguchi and R.N. Pienaar, *S. Subsalsa* (Ostenfeld) K.A. Steidinger and Balech and *S. caponii* T. Horiguchi and R.N. Pienaar [1]. These species share a number of morphological characterizations of thecal plates: a lack of contact between 2a and 3a, the presence of six cingular plates, and the posterior sulcal plate that does not touch the cingulum [1]. Based on such morphological characteristics, six *Bysmatrum* species (including *Bysmatrum subsalsum* (Ostenfeld) M.A. Faust and K.A. Steidinger as the type species have been reported so far [1–6] and, within the genus, the *Bysmatrum* species) can be distinguished from each other by the differences in cell size and shape, plate ornamentation, cingulum displacement, the morphology of the apical pore complex (APC), nucleus position, habitat and habitus [1,4,5,7,8]. However, molecular data for *Bysmatrum* species are as yet available for only five species; *B. arenicola* T. Horiguchi and R. N. Pienaar, *B. austrafrum* Dawut, Sym, Suda and Horiguchi, *B. granulosum* L. Ten–Hage, Since the morphological descriptions of *Bysmatrum* species by Faust and Steidinger [1], Anglès et al. [7] documented the morphological details and molecular phylogeny of *B. sub- salsum* collected from different locations of the Mediterranean Basin, and concluded that the strains of *B. subsalsum* are morphologically indistinguishable but genetically distinct, probably due to the differences in habitat, physiology or life-history traits. Since then, Luo et al. [8] documented that *B. subsalsum* strains from Malaysia and the French Atlantic formed a subclade (ribotype B) that can be distinguished from a clade (ribotype A) consisting of stains from the Mediterranean

Sea and that there is cryptic diversity within *B. subsalsum,* based on the genetic distance shown in ITS sequences among *Bysmatrum* species [7]. This indicates that additional molecular data (including morphological descriptions) are needed to clarify the cryptic diversity of *B. subsalsum,* with strains established from water bodies of various geographical regions.

Bysmatrum subsalsum is a cosmopolitan species [10], and its occurrences have been recorded in the Aral Sea, the Mediterranean Sea, the Caribbean Sea, the Gulf of Mexico, the French and Portuguese Atlantic coast and the East and South China Sea [7,8]. However, in the Korean coastal area, this species has not been found, and only *B. gregarium* (as *B. caponii*) isolated from plankton samples has been recorded, with its morphology and molecular information [9]. In the tidal pools of the southern area of Korea, we observed dense patches caused by un-identified organisms. These were isolated, and strains were established and the morphological features were examined using light and scanning electron microscopy (SEM). The observations revealed that the species was identical to *B. subsalsum.* In this study, we describe the morphological details of Korean strains of *B. subsalsum* and report on their molecular characterization, based on concatenated small subunit (SSU) and large subunit (LSU) rRNA gene sequences.

2. MATERIALS AND METHODS

2.1 Sampling and Culture

In June 2017, dense patches of phytoplankton were observed in the tidal pools of Geoje Island (34°59′34.3" N, 128°41′44.1" E) and Jeju Island (33°29′26.5" N, 126°25′12.8" E), Korea.

Water temperature and salinity were 23.0°C and 34 psu in the tidal pool of Geoje Island, and 10.5°C and 24.5 psu in the tidal pool of Jeju Island, respectively. The water samples from the pools were collected using the 50 mL conical tubes and were then transported to the laboratory for observation of the causative organism. In the laboratory, the single mass colonies from the samples were isolated using a glass micropipette on an inverted microscope (Eclipse Ts2R, Nikon, Tokyo, Japan), and transferred into a 24-well tissue plate containing f/2 medium adjusted to a salinity of 32. The isolated colonies were maintained at 24°C in a 12:12 LD cycle under a photon irradiance of 100 μmol photons·m¯2·s¯1. After sufficient growth, the cells were transferred to a culture flask. A monoclonal culture was successfully established from the cells collected in the tidal pool of Geoje Island and deposited in the Library of Marine Samples, Korea Institute of Ocean Science and Technology, as strain number LIMS-PS-2685 (=MABIK PD00002006). The isolate from the tidal pool of Jeju Island was also established as a strain; however, it is currently unavailable because of unexpected cell death.

2.2 Morphological Observation

Live cells were isolated and photographed at 1000X magnification using an AxioCam 512c digital camera (Carl Zeiss, Göttingen, Germany) on an upright microscope (Zeiss Axio Imager2, Carl Zeiss, Göttingen, Germany). For fluorescence microscopy, approximately 1 mL of cell culture was transferred to a 1.5-mL micro-centrifuge tube, and 4′,6-diamidino- 2-phenylindole (DAPI) stain (Sigma-Aldrich, St. Louis, MO, USA) was added at a final concentration of 10 μg mL^{-1}. Cells were then incubated in the dark at room temperature for 1 h, and viewed and photographed through a Zeiss Filterset 49 (emission: BP 365–445; beamsplitter: FT 395).

For scanning electron microscopy, a 20 mL aliquot of a dense culture was fixed in glutaraldehyde and paraformaldehyde with a final concentration of 2% (*w/v*). The aliquot containing fixed cells was filtered through a polycarbonate membrane filter (5 μm pore size), without applying additional pressure and rinsed three times with distilled water to remove the salt. After rinsing, the sample was dehydrated in an ethanol series (10, 30, 50, 70, 90 and 100% ethanol, followed by two 100% ethanol steps) and dried using a critical point dryer (BAL-TEC, CPD 300, Balzers, Germany). Finally, the sample was coated with gold-palladium and examined using a field emission-scanning electron microscope (JSM-7610F, Jeol, Japan).

2.3 DNA Extraction and Sequencing

The mass colonies were harvested via centrifugation for 10 min at room temperature, and the supernatant was discarded. Genomic DNA was extracted from cell pellets using a DNeasy Plant Mini Kit (Qiagen, Hilden, Germany), according to the manufacturer's protocol. The SSU rRNA gene was amplified using primers (ATF01: 5′-YAC CTG GTT GAT CCT GCC AGT AG-3′ and ATR01: 5′-RMW TGA TCC TTC YGC AGG TTC ACC-3′) [11], and the D1–D3 regions of the LSU rRNA gene were amplified using previously described primers (D1R: 5′-ACC CGC TGA ATT TAA GCA TA-3′ [12] and 28r691: 5′-CTT GGT CCG TGT TTC AAG AC-3′) [11]. PCR reactions were conducted in a volume of 50 μL; 2.0 μL gDNA; 5 μL 10x buffer; 0.2 mM dNTP; 0.1 μM each primer; 0.25 U Taq polymerase (Takara Ex Taq, Takara, Seoul, Korea); and ddH_2O to a final volume of 50 μL. The PCR condition was as follows: 94°C for 4 min, 30 cycles of 94 °C for 1 min, 54 °C for 1 min, and 72°C for 1 min and then 1 extension cycle at 72°C for 10 min. PCR products were purified with QIAqucik PCR purification kit (Qiagen, Hilden, Germany). All rRNA gene sequencing was performed using an ABI PRISM 3700 DNA Analyzer (Applied Biosystems, Foster City, CA, USA). The sequences were trimmed and assembled into contigs using Geneious Prime (https://www.geneious.com (accessed on 2 January 2020)).

2.1 Phylogenetic Analysis

The sequences were aligned using MAFFT in Geneious plugin with default settings, and some regions manually adjusted using the Geneious Prime. The phylogenetic analysis includes all the published sequences of *Bysmatrum* species. A total of 50 sequences of the order Peridinales were selected, while the family Peridiniaceae and Protoperidiniaceae were used as outgroups. The sequence alignments of SSU and LSU rRNA were concatenated with introduced gaps. The phylogenetic tree for the concatenated sequence alignment was inferred using maximum likelihood (ML) analyses via RAxML version 8 [13], and using Bayesian inference (BI) through MrBayes version 3.2 [14]. The general time reversible (GTR) model with parameters accounting for γ-distributed rate variation across sites (G) was used in all analyses, taking into account 6-class gamma. The GTR+G substitution model was selected using the Akaike information criterion (AIC) as implemented in jModelTest version 2.1.4 [15]. Bootstrap analyses for ML were carried out with 1000 replicates to evaluate statistical reliability. The Markov chain Monte Carlo (MCMC) method for BI was used with four runs for 10 million generations, sampling every 100 generations. The first 10% of trees were deleted as burn-in, and a majority rule consensus tree was constructed to examine the posterior probabilities of each clade. The final trees were visualized with FigTree v1.4.4 (http://tree.bio.ed.ac.uk / software/figtree/ (accessed on 5 January 2020)).

3. RESULTS

3.1 Morphology of Vegetative Cell and Resting Cyst of Bysmatrum subsalsum

Cells had a conical epitheca and a round hypotheca, were slightly obliquely dorsoventrally flattened and the longitudinal flagella were visible (Figure 1a–f). They were 37–52 μm in length and 31–44 μm in width (n = 50). An apical stalk consisting of a transparent gelatinous matrix was observed on the apex of the cells (Figure 1b,c). A yellowish eyespot was located on the right side of the sulcus (Figure 1d,e). Rod-shaped chloroplasts were observed in the periphery of the cell (Figure 1g,h). The DAPI-stained nucleus was large, commonly elongated (Figure 1h) and sometimes horseshoe-shaped by the different observation angle (Figure 1i). The position of a nucleus was posteriorly in the hypotheca (Figure 1g–i).

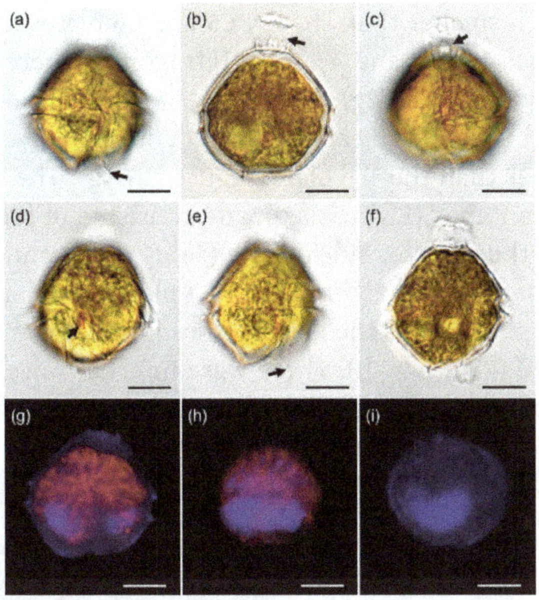

Figure 1. Light micrographs of *Bysmatrum subsalsum* (Strain: LIMS-2685). (a) Surface focus of ventral view showing the longitudinal flagella (arrow), sulcus and cingulum. (b) Deeper focus of ventral view showing the outline of the cell and apical stalk (arrow). (c) Surface focus of ventral-apical view showing the apical stalk (arrow). (d) Deeper focus of ventral-lateral view showing a red eyespot (arrow). (e) Surface focus of ventral-right lateral view showing the longitudinal flagella (arrow). (f) Deeper focus of dorsal view. (g) Epifluorescence image of ventral view of DAPI-stained cell showing the position of the nucleus. (h) Epifluorescence image of dorsal view of DAPI-stained cell showing the position of the nucleus. (i) Epifluorescence image of antapical view of DAPI-stained cell showing the shape of the nucleus. Scale bars = 5 μm.

SEM observation revealed that the cells have a plate formula of Po, Cp, X, $4'$, 3a, 7”, 6C, 4S, $5'''$, $2''''$ (Figure 2). The thecal plates were covered with strong reticulations (Figure 2a–e). Apical pore complex (APC) was tear-shaped and included a Po plate, a round cover plate (Cp) and a canal plate (X) with thick margins formed by the raised borders of the apical plates (Figure 2f). The first apical plate ($1'$) was pentagonal and asymmetrical with shorter anterior sutures than the posterior ones, and surrounded five plates: $2'$, $4'$, $1''$, $7''$ and Sa (Figure 2a,d,e). Three intercalary plates (1a, 2a and 3a) were similar in size, and 1a and 2a contacted each other; however, 3a was separated from 1a and 2a. (Figure 2b,c,e). Plate 1a was elongate and rectangular, whereas plates 2a and 3a were hexagonal and pentagonal, respectively (Figure 2b,c,e). The precingular plates were symmetrically distributed (Figure 2e). The first precingular plate ($1''$) was pentagonal and smaller than the others (Figure 2a,e). The cingulum was deeply excavated and descended by about one cingulum width (Figure 2a–d). Six cingulum plates were observed; plates C1,

C2 and C3 were much smaller than the C4, C5 and C6 plates (Figure 2a–d). In post cingular series, plates $1^{!!!}$, $2^{!!!}$ and $4^{!!!}$ were tetragonal, whereas plates $3^{!!!}$ and $5^{!!!}$ are pentagonal in shape (Figure 2g). The plate $1^{!!!}$ was much smaller than the other plate (Figure 2a–c,g). Two antapical plates ($1^{!!!!}$ and $2^{!!!!}$) have pentagonal shapes, and plate $2^{!!!!}$ is larger than the $1^{!!!!}$ plate (Figure 2a–c,g). The sulcus was wide and did not contact the antapex (Figure 2a,g), and consisted of four major plates with inconspicuous lists: the anterior sulcal plate (Sa) is narrow and elongated; the left sulcal plate (Sl) in narrow and right sulcal (Sr) plate are triangular (Figure 2a,h); the posterior sulcal plate (Sp) is the largest sulcal plate, wider than long, and does not contact the plate (Figure 2g,h). There were internal sulcal lists (Isl) emerging from left side of the Sr plate (Figure 2h). The left sulcal list (Lsl) that had emerged from the lower side of plate $1^{!!!}$ was visible (Figure 2h).

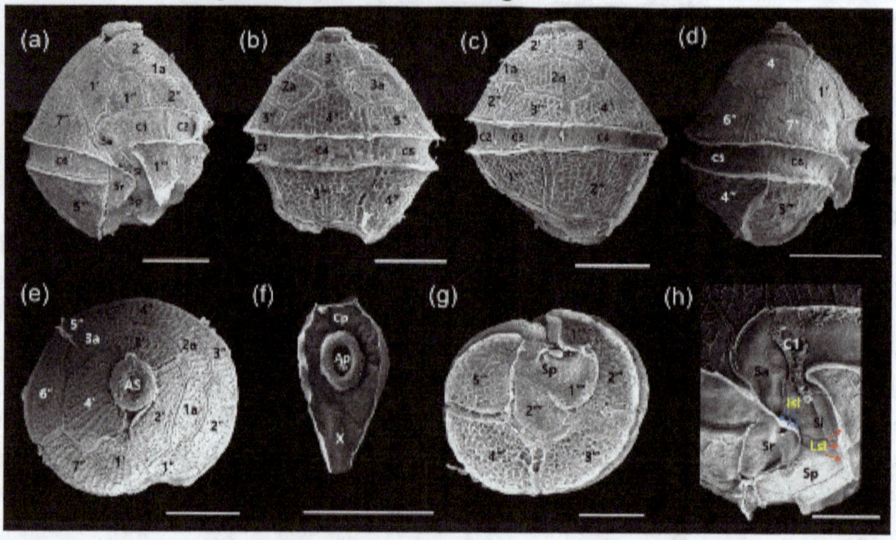

Figure 2. Scanning electron micrographs of vegetative cells of *Bysmatrum subsalsum* strain LIMS-2685 from Korea. (a) Ventral view. (b) Dorsal view. (c) Dorsal-left lateral view. (d) Ventral-right lateral view. (e) Apical view, showing a centrally located raised dome (apical stalk: AS) and epithecal plate pattern. (f) Detail of apical pore complex showing the apical pore (AP), cover plate (Cp) and canal plate (X). (g) Antapical view, showing hypothecal plate pattern. (h) Detail of the sulcal plates showing anterior sulcal plate (Sa), right sulcal (Sr) and left sulcal (Sl) plates, posterior sulcal plate (Sp), left (Lsl; red arrows) and internal (Isl; blue arrows) sulcal lists. Scale bars = 10 μm (a–e,g); 5 μm (f,h).

Spherical cysts were observed under culture conditions (Figure 3). The cysts were 29.5–34.6 μm in diameter (n = 25). The cyst was grayish in colour, and an orange accumulation body was visible (Figure 3a–c). The cyst wall was smooth, without any distinguishing features on the surface (Figure 3d).

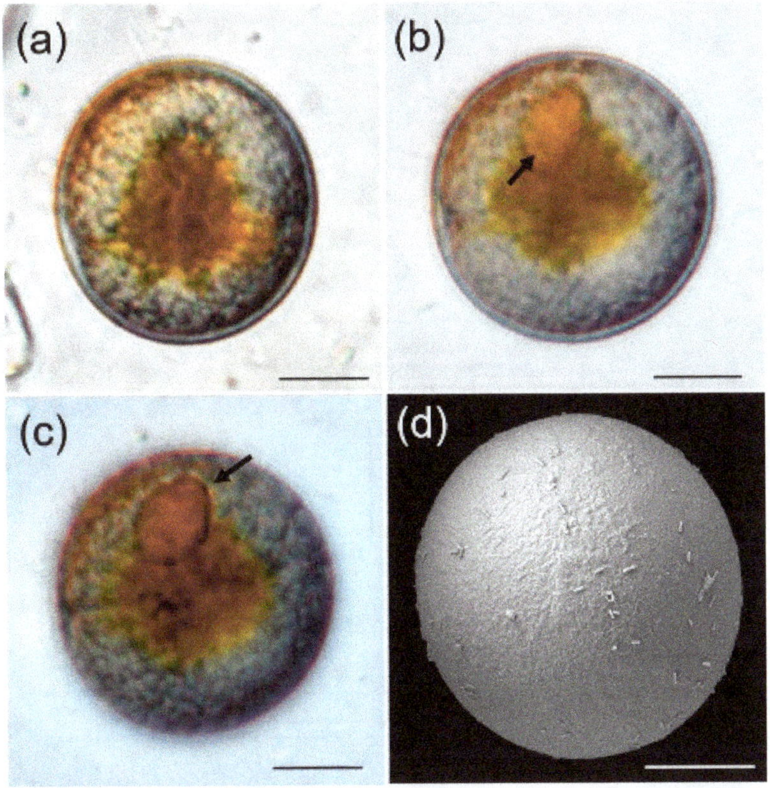

Figure 3. Light and scanning electron micrographs of cysts of *Bysmatrum subsalsum* strain LIMS-2685. (a) Light micrographs of spherical cyst. (b,c) Light micrographs of cysts showing large, orange, accumulation body (arrows). (d) Scanning electron micrograph showing a smooth organic wall without any ornamentation. Scale bars = 5 µm.

3.2 Molecular Phylogeny

Bayesian inference (BI) and maximum likelihood (ML) based on the concatenated SSU and LSU rRNA gene sequences yielded similar phylogenetic trees. The genus *Bysmatrum* was a monophyletic group, with maximum support (Figure 4). *Bysmatrum arenicola* was the base of *Bysmatrum* species, and *B. subsalsum* was branched from *B. gregarium,* with moderate supports (BT/PP = 68/0.92). Two ribotypes (ribotype A and B) of *B. subsalsum* were identified; ribotype A was comprised only of strains from the Mediterranean Sea with maximum support (BT/PP = 100/1.00), whereas ribotype B included the strains from China, Malaysia, the French Atlantic and Korea. *Bysmatrum austrafrum* (HG321) was nested between the two ribotypes. In ribotype B, the Korean strains of *B. subsalsum* were closely related to the Malaysian strain (TBBYS03) (BT/PP = 80/1.00).

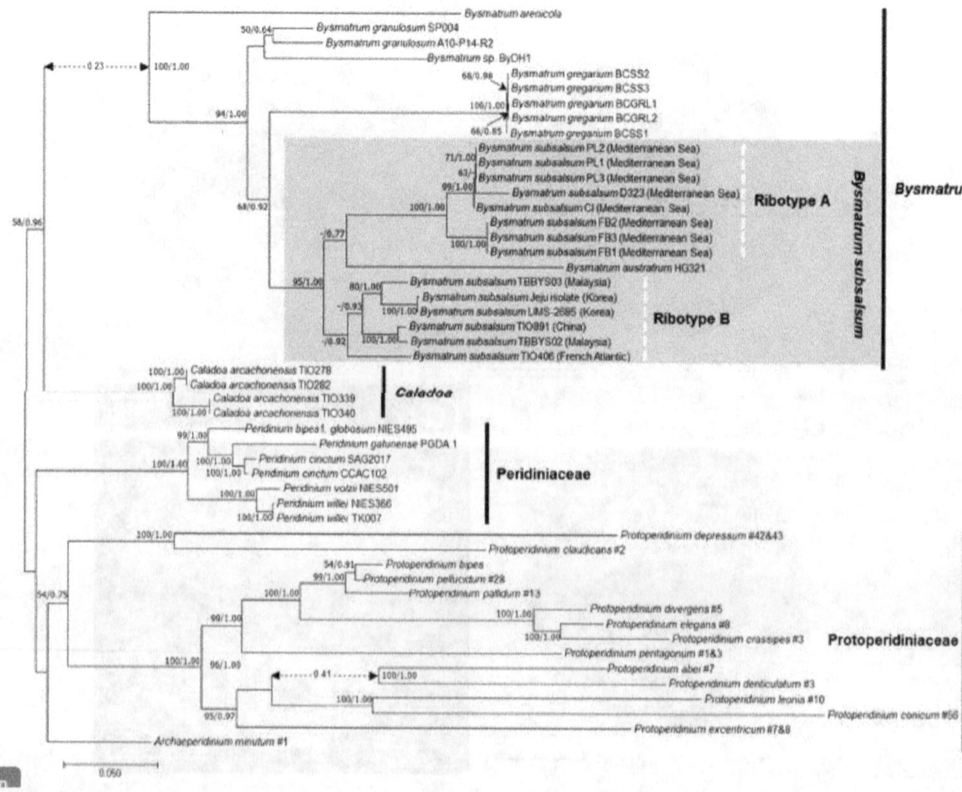

Figure 4. Phylogeny of *Bysmatrum subsalsum* inferred from concatenated SSU and partial LSU rRNA gene sequences using maximum-likelihood (ML). Ribotypes are labelled according to designations by Luo et al. [8]. Numbers on branches are statistical support values to clusters on the right of them (left: ML bootstrap support (BT) values; right: Bayesian posterior probabilities (PP)). Bootstrap support values > 50% and Bayesian posterior probabilities > 0.7 are shown. Branch lengths are drawn to scale, with the scale bar indicating the number of nucleotide substitutions per site.

4. DISCUSSION

4.1 Morphological Comparisons of Korean Isolates of Bysmatrum Subsalsum with Other Isolates of B. Subsalsum, and B. Austrafrum

According to Anglès et al. [7] and Luo et al. [8], the genetic sequences of *B. subsalsum* show large intraspecific differences, clustering two well-differentiated clades (ribotype A and B). The two clades of *B. subsalsum* based on SSU and LSU sequences were also shown in this study, and the Korean isolates of *B. subsalsum* were nested in the ribotype B and clustered with the isolates from China, Malaysia and the French Atlantic (Figure 4). The Korean isolates of *B. subsalsum* were morphologically characterized by the separa- tion of plates 2a and 3a, the tear-shaped APC, an elongated rectangular 1a plate and the general morphology, such as cell size, plate ornamentation and nucleus position, which coincides with that of *B. subsalsum* in previous studies (Luo et al. [8] and reference therein). Luo et al. [8] reported the differences in the number of sulcal lists among strains of *B. sub- salsum* in two clades; the French strain has the right sulcal list (Three sucal lists), whereas the Malaysian strains are characterized by the absence of the right sulcal list (Two sulcal lists). In Korean strains, the right sulcal list was not present. In addition, the number of sulcal lists varied among specimens collected from other geographical regions [16–19]. Anglès et al. [7] concluded that despite a certain degree of morphological variation (such as cell size, APC morphology and size, and cingulum displacement), cells from the two clades of *B. subsalsum* exhibit similar morphological characteristics. Luo et al. [8] also con- firmed the morphological similarities in the Malaysian and French strains of *B. subsalsum*. Consequently, the prominent morphological characteristics for clarifying the two clades of *B. subsalsum* still remain unclear.

Recently, Luo et al. [20] recorded *B. austrafrum* between two clades of *B. subsalsum*, with strong support in the phylogeny, which is in agreement with our result. *Bysmatrum austrafrum* was first described by Dawut et al. [6]. This species has a typical plate of Po, X, $4^{!}$, 3a, 7”, 6C, 4S, $5^{!!!}$, $2^{!!!!}$ for the genus *Bysmatrum* and is morphologically characterized by dimensions of 25–45 μm long and 20–42.5 μm wide, arranged reticulation in the thecal plates, the possession of an equatorially positioned cingulum and a cingulum displaced by a distance exceeding its own width. Based on these morphological features, Dawut et al. [6] reported that *B. austrafrum* is similar to *B. subsalsum*, and concluded that *B. austrafrum* can be distinguished from *B. subsalsum* by differences in cell size, the shape of the APC and apical plate $1^{!}$. However, cell sizes recorded in *B. austrafrum* were recorded in the other specimens of *B. subsalsum* (see Table 2 in Luo et al. [8]), and the shapes of the APC and apical plate $1^{!}$ of *B. austrafrum* quite resembled those of Korean isolates and the specimens of *B. subsalsum* recorded by Luo et al. [8]. In addition, although Dawut et al. [6] did not consider the absence or presence of the right sulcal list, *B. austrafrum* does not have the right sulcal list in their description. Consequently, it is quite difficult to distinguish *B. Subsalsum* from *B. austrafrum*, based on their morphological features. Nevertheless, different types of eyespots have been reported in *B. subsalsum* and *B. austrafrum* [6,8]. Based on the types of eyespots suggested by Moestrup and Daugbjerg [21], *B. subsalsum* has the Type B eyespot, whereas *B. austrafrum* presents the Type A eyespot. This may be a useful characteristic for distinguishing *B. subsalsum* from *B. austrafrum*. However, as the Type A eyespot has not been reported in other *Bysmatrum* species, such as *B. granulosum* and *B. gregarium*, more isolates of *B. Austrafrum* need to be examined for clarifying the type of eyespot within *Bysmatrum* species.

4.2 Morphology of Bysmatrum subsalsum Cyst

Cyst morphology can be helpful to understand the diversity within the genus (e.g., Li et al. [22]). In *Bysmatrum* species, a cyst–theca stage relationship has only been established through germination experiments for Mediterranean *B. subsalsum* [7]. Two types of cysts of *B. subsalsum* have been described in culture and sediments; in the culture spherical to ovoidal cysts without any ornamentations were observed [7,19], whereas cysts with the typical plate pattern, which are morphologically similar to vegetative cells of *B. subsalsum*, were identified in sediments [8,23]. In our study, the spherical cysts were also observed from the culture. *Bysmatrum subsalsum* from the Mediterranean Sea (ribotype A) could produce both cyst types in culture [7,19], and from the French strain of *B. subsalsum* (ribotype B), only cysts with the thecal plate were described [8]. In previous studies, differences in cyst types have been recorded between cultures and natural sediments. For example, unarmored dinoflagellate *Margalefidinium polykrikoides* (formerly *Cochlodinium polykrikoides*) produced two different types of cyst (a spherical cyst without ornamentation in culture and an ornamented cyst in sediments) [24,25]. Tang and Gobler [24] suggested that the ornaments and spines of *M. polykrikoides* might be caused by biotic or chemical processes in sediments. However, this does not seem to be in accordance with *B. subsalsum*, because two different types of cysts were reported in sediments.

Similar morphological features between cyst and vegetative cell of unarmored dinoflagellate *M. polykrikoides* have been reported [26,27], and the cyst was identified as a temporary cyst that can be the short-term stage. The temporary cyst of *M. polykrikoides* is surrounded by a transparent and thin hyaline membrane, indicating in sediments the unarmored temporary cysts may be destroyed because of geochemical processes related to organic matter degradation or the attack of viruses or bacteria into the cell. However, temporary cysts with thecal plates (armored cyst) (such as cyst of *B. subsalsum*) may be protected from the environmental conditions in sediments. If so, the cyst with typical thecal plates of *B. subsalsum* from sediments or culture may be temporary, because the resting cysts of dinoflagellate usually have distinct morphological features that can be distinguished from their vegetative cell (e.g., Matsuoka and Fukuyo [28]).

4.3 Phylogenetic Position of Bysmatrum subsalsum

In the phylogenetic tree, the Korean isolates within *B. subsalsum* were nested in the ribotype B consisting of the isolates from China, Malaysia and the French Atlantic, whereas the ribotype A includes only the isolates from the Mediterranean Sea. Iwataki et al. [29] documented that in the phylogenetic tree for *M. polykrikoides* isolates, the ribotypes can be useful for characterizing the geographical distribution pattern of *M. polykrikoides*. In our study, the ribotype A also represents the isolates of *B. subsalsum* from the Mediterranean Sea, which is in agreement with the conclusion by Iwataki et al. [29]. However, the ribotype B is a mixture of isolates originated from different geographic regions, although it mostly includes the Asian isolates. This indicates that the cells originating from France might be transferred from the Asian areas. Benthic dinoflagellates such as *Bysmatrum* species usually have a restricted distribution, possibly because of benthic, epiphytic behaviour. Nevertheless, *B. susalsum* has been reported from various samples, including plankton samples, and their resting cysts are also present in the sediments (Luo et al. [8] and reference therein). This suggests that the vegetative cells and resting cysts of *B. subsalsum* can artificially be introduced into other coastal areas, possibly caused by ballast ship waters (e.g., Hallegraeff [30]).

4.4 Since the discovery of two ribotypes of *B. subsalsum*, Luo et al. [20] identified

B. austrafrum between two clades of *B. subsalsum* in a phylogenetic tree based on concatenated SSU and LSU rDNA sequences, which is in agreement with our result. In the phylogenetic tree, *B. subsalsum* and *B. austrafrum* were grouped, with strong support (BT/PP = 95/1.00) (Figure 4), indicating that the two species may be conspecific. This result can be supported by morphological similarity between the two species. However, as there are differences in the types of eyespots between the two species, *B. subsalsum* can also be a species that distinguished from *B. austrafrum*. If so, it is possible that *B. subsalsum* is not monophyletic. This finding supports the idea that there is cryptic diversity within B. subsalsum [7].

4.4 Environmental Conditions in Relation to the Growth of Bysmtrum subsalsum

Although *Bysmatrum* species are considered benthic dinoflagellates, the occurrence of *B. subsalsum* has been reported in the plankton, with floating detritus and on sand and macroalgae [18]. Luo et al. [8] recorded *B. subsalsum* in plankton and sediment samples, and Anglès et al. [7] found *B. subsalsum* in mangrove detritus and salt marshes. In Korean isolates of *B. subsalsum*, the species is considered to inhabit tidal pools, and *B. austrafrum* was also discovered in tidal pools in South Africa [6]. In comparison with *B. subsalsum*, other *Bysmatrum* species, such as *B. teres*, *B. granulosum*, *B. arenicola* and *B. gregarium* have been reported in the restricted habitats (Luo et al. [8] and reference therein). This indicates that *B. subsalsum* in contrast to other *Bysmatrum*, may have broad environmental tolerance and a worldwide distribution.

Environmental conditions in relation to the growth of *B. subsalsum* have been rarely reported. Anglès et al. [7] documented that, based on the occurrence of *B. subsalsum* re- ported in previous studies, the species has a wide salinity tolerance and is able to grow under a wide range of sea water temperatures, with a preference for salinities > 30 and temperatures > 20 C. López-Flores et al. [31] recorded that the cell abundances of *B. subsalsum* (as *Scrippsiella subsalsa*) decreased sharply in October, when salinity reached values > 30 and water temperature was low (15.6 °C). In this study, however, the dense patch of *B. subsalsum* was observed at 10.5 °C, indicating that it has a wider temperature tolerance than previously known. Further studies are also needed to clarifying the morphology (vegetative cell and cyst) of the genus *Bysmatrum* and, in particular, of *B. subsalsum*.

REFERENCES

1. Faust, M.A.; Steidinger, K.A. *Bysmatrum* gen. nov. (Dinophyceae) and three new combinations for benthic scrippsielloid species. *Phycologia* **1998**, *37*, 47–52. [CrossRef]
2. Horiguchi, T.; Pienaar, R.N. Validation of *Bysmatrum arenicola* Horiguchi et Pienaar, sp. nov. (Dinophyceae). *J. Phycol.* **2000**, *36*, 237. [CrossRef]
3. Ten-Hage, L.; Quod, J.-P.; Turquet, J.; Coute, A. *Bysmatrum granulosum* sp. nov., a new benthic dinoflagellate from the southwestern Indian Ocean. *Eur. J. Phycol.* **2001**, *36*, 129–135. [CrossRef]
4. Murray, S.; Hoppenrath, M.; Larsen, J.; Patterson, D.J. *Bysmatrum teres* sp. nov., a new sand-dwelling dinoflagellate from north-western Australia. *Phycologia* **2006**, *45*, 161. [CrossRef]
5. Hoppenrath, M.; Murray, S.A.; Chomérat, N.; Horiguchi, T. *Marine Benthic Dinoflagellates-Unveiling Their Worldwide Biodiversity*; E. Schweizerbartsche Verlagsbuchhandlung: Stuttgart, Germany, 2014; 276p.
6. Dawut, M.; Sym, S.D.; Suda, S.; Horiguchi, T. *Bysmatrum austrafrum* sp. nov. (Dinophyceae), a novel tidal pool dinoflagellate from South Africa. *Phycologia* **2018**, *57*, 169–178. [CrossRef]
7. Anglès, S.; Reñé, A.; Garcés, E.; Lugliè, A.; Sechi, N.; Camp, J.; Satta, C.T. Morphological and molecular characterization of *Bysmatrum subsalsum* (Dinophyceae) from the western Mediterranean Sea reveals the existence of cryptic species. *J. Phycol.* **2017**, *53*, 833–847. [CrossRef]
8. Luo, Z.; Lim, Z.F.; Mertens, K.N.; Gurdebeke, P.; Bogus, K.; Carbonell-Moore, M.C.; Vrielinck, H.; Leaw, C.P.; Lim, P.T.; Chomérat, N.; et al. Morpho-molecular diversity and phylogeny of *Bysmatrum* (Dinophyceae) from the South China Sea and France. *Eur. J. Phycol.* **2018**, *53*, 318–335. [CrossRef]
9. Jeong, H.J.; Jang, S.H.; Kang, N.S.; Yoo, Y.D.; Kim, M.J.; Lee, K.H.; Yoon, E.Y.; Potvin, E.; Hwang, Y.J.; Kim, J.I.; et al. Molecular characterization and morphology of the photosynthetic dinoflagellate *Bysmatrum caponii* from two solar saltons in western Korea. *Ocean Sci. J.* **2012**, *47*, 1–18. [CrossRef]

10. Karen, A.S.; Tangen, K. Dinoflagellates. In *Identifying Marine Diatoms and Dinoflagellates*; Tomas, C.R., Ed.; Academic Press: San Diego, CA, USA, 1996; pp. 387–584.

11. Ki, J.-S.; Han, M.-S. Molecular analysis of complete SSU to LSU rDNA sequence in the harmful dinoflagellate *Alexandrium tamarense* (Korean isolate, HY970328M). *Ocean Sci. J.* **2005**, *40*, 43–54. [CrossRef]

12. Orsini, L.; Sarno, D.; Procaccini, G.; Poletti, R.; Dahlmann, J.; Montresor, M. Toxic *Pseudo-nitzschia multistriata* (Bacillariophyceae) from the Gulf of Naples: Morphology, toxin analysis and phylogenetic relationships with other *Pseudo-nitzschia* species. *Eur. J. Phycol.* **2002**, *37*, 247–257. [CrossRef]

13. Stamatakis, A. RAxML version 8: A tool for phylogenetic analysis and post-analysis of large phylogenies. *Bioinformatics* **2014**, *30*, 1312–1313. [CrossRef] [PubMed]

14. Ronquist, F.; Teslenko, M.; van der Mark, P.; Ayres, D.L.; Darling, A.; Höhna, S.; Larget, B.; Liu, L.; Suchard, M.A.; Huelsenbeck, J.P. MrBayes 3.2: Efficient bayesian phylogenetic inference and model choice across a large model space. *Syst. Biol.* **2012**, *61*, 539–542. [CrossRef] [PubMed]

15. Darriba, D.; Taboada, G.L.; Doallo, R.; Posada, D. jModelTest 2: More models, new heuristics and parallel computing. *Nat. Methods* **2012**, *9*, 772. [CrossRef]

16. Balech, E. Tercera contribución al conocimiento del género Peridinium Museo Argentino de ciencias naturales 'Bernadino Rivadavia'e Instituto nacional de investigacion de las ciencias naturales. *Rev. Hydrobiol.* **1964**, *4*, 179–195.

17. Steidinger, K.A.; Balech, E. *Scrippsiella subsalsa* (Ostenfeld) comb. nov. (Dinophyceae) with a discussion on *Scrippsiella*. *Phycologia* **1977**, *16*, 69–73. [CrossRef]

18. Faust, M.A. Morphology and ecology of the marine benthic dinoflagellate *Scrippsiella subsalsa* (Dinophyceae). *J. Phycol.* **1996**, *32*, 669–675. [CrossRef]

19. Gottschling, M.; Soehner, S.; Zinssmeister, C.; John, U.; Plötner, J.; Schweikert, M.; Aligizaki, K.; Elbrächter, M. Delimitation of the Thoracosphaeraceae (Dinophyceae), including the calcareous dinoflagellates, based on large amounts of ribosomal RNA sequence data. *Protist* **2012**, *163*, 15–24. [CrossRef]

20. Luo, Z.; Mertens, K.N.; Nézan, E.; Gu, L.; Pospelova, V.; Thoha, H.; Gu, H. Morphology, ultrastructure and molecular phylogeny of cyst-producing *Caladoa arcachonensis* gen. et sp. nov. (Peridiniales, Dinophyceae) from France and Indonesia. *Eur. J. Phycol.* **2019**, *54*, 235–248. [CrossRef]

21. Moestrup, Ø.; Daugbjerg, N. On dinoflagellate phylogeny and classification. In *Unravelling the Algae: The past, Present, and Future of Algal Systematics*; Brodie, J., Lewis, J., Eds.; CRC press: Boca Raton, FL, USA, 2007; pp. 215–230.

22. Li, Z.; Mertens, K.N.; Gottschling, M.; Gu, H.; Söhner, S.; Price, A.M.; Marret, F.; Pospelova, V.; Smith, K.F.; Carbonell-Moore, C.; et al. Taxonomy and Molecular Phylogenetics of Ensiculiferaceae, fam. nov. (Peridiniales, Dinophyceae), with Consideration of their Life-history. *Protist* **2020**, *171*, 125759. [CrossRef]

23. Limoges, A.; Mertens, K.N.; Ruíz-Fernández, A.-C.; de Vernal, A. First report of fossilized cysts produced by the benthic *Bysmatrum subsalsum* (Dinophyceae) from a shallow Mexican lagoon in the Gulf of Mexico. *J. Phycol.* **2015**, *51*, 211–215. [CrossRef]

24. Tang, Y.Z.; Gobler, C.J. The toxic dinoflagellate *Cochlodinium polykrikoides* (Dinophyceae) produces resting cysts. *Harmful Algae* **2012**, *20*, 71–80. [CrossRef]

25. Li, Z.; Han, M.-S.; Matsuoka, K.; Kim, S.-Y.; Shin, H.H. Identification of the resting cyst of *Cochlodinium polykrikoides* Margalef (Dinophyceae, Gymnodiniales) in Korean coastal sediments. *J. Phycol.* **2015**, *51*, 204–210. [CrossRef]

26. Shin, H.H.; Li, Z.; Yoon, Y.H.; Oh, S.J.; Lim, W.-A. Formation and germination of temporary cysts of *Cochlodinium polykrikoides* Margalef (Dinophyceae) and their ecological role in dense blooms. *Harmful Algae* **2017**, *66*, 57–64. [CrossRef]

27. Li, Z.; Matsuoka, K.; Shin, H.H. Revision of the life cycle of the harmful dinoflagellate *Margalefidinium polykrikoides* (Gymnodiniales, Dinophyceae) based on isolates from Korean coastal waters. *J. Appl. Phycol.* **2020**, *32*, 1863–1873. [CrossRef]

28. Matsuoka, K.; Fukuyo, Y. *Technical Guide for Modern Dinoflagellate Cyst Study*; Japan Society for the Promotion of Science: Tokyo, Japan, 2000; pp. 1–29.

29. Iwataki, M.; Kawami, H.; Mizushima, K.; Mikulski, C.M.; Doucette, G.J.; Relox, J.R.; Anton, A.; Fukuyo, Y.; Matsuoka, K. Phylogenetic relationships in the harmful dinoflagellate *Cochlodinium polykrikoides* (Gymnodiniales, Dinophyceae) inferred from LSU rDNA sequences. *Harmful Algae* **2008**, *7*, 271–277. [CrossRef]

30. Hallegraeff, G.M. Transport of toxic dinoflagellates via ships' ballast water: Bioeconomic risk assessment and efficacy of possible ballast water management strategies. *Mar. Ecol. Prog. Ser.* **1998**, *168*, 297–309. [CrossRef]

31. López-Flores, R.; Garcés, E.; Boix, D.; Badosa, A.; Brucet, S.; Masó, M.; Quintana, X.D. Comparative composition and dynamics of harmful dinoflagellates in Mediterranean salt marshes and nearby external marine waters. *Harmful Algae* **2006**, *5*, 637–648. [CrossRef]

CHAPTER-12

NEW RECORDS OF THE DIATOMS (*BACILLARIOPHYCEAE*) FROM THE COASTAL LAGOONS IN KOREA

1. INTRODUCTION

Lagoons are a form of riparian terrain where the river and sea water meet, but they have unique ecological characteristics that distinguish them from estuaries because, in lagoons, the entrance to the coast is blocked by sand dunes [1,2]. The lagoons distributed along the Korean coast are known to have been formed by a combination of rising sea levels during the postglacial age and the development of sandbars or sand dunes [3,4].

The East Sea coast boasts the cleanest marine environments in Korea and is blessed with natural resources to sustain the area. There are 18 lagoons distributed along a 112 km stretch of the East Sea coast [5]. These lagoons provide beautiful natural scenery and have unique ecological value with their brackish water lakes, important migratory bird habitats, and are also economically valuable as tourist attractions [4,6].

Hwajinpo Lake is the largest lake in Korea with a circumference of 16 km and is a typical high-salinity brackish lake with a high ecological value [4,7]. Hyangho Lagoon hosts a community of aquatic plants covering approximately 20% of its area. It belongs to a heavily landlocked lake, and, unlike other lagoons, freshwater lakes are distributed behind it, serving as a sheltered habitat for organisms living in the lake [8]. Cheonjinho Lake's characteristic brackish water has disappeared due to the blocked inflow of seawater; it has a wide variety of free-floating and floating-leaved plants, and a high proportion of aquatic and wetting plants [9,10]. Gyeongpoho is the representative lagoon of the East Sea coast for the endangered prickly waterlily, *Euryale ferox*, which appeared 40 years ago and is known as a favoured habitat for various migratory birds [11,12]. The Maeho lake has a high ecological conservation value due to its diverse distribution of biological resources and ecosystem services, such as being designated as a cultural heritage reserve. However, the lake area is decreasing due to the nearby expansion of farmland and subsequent soil erosion, causing sedimentation [4]. Gapyeongri wetland is one of the smallest lagoons, well-hidden behind a widely developed coastal sand dune [13]. Yeomgaeho Lake is also small, but is a typical natural lagoon, suitable as a migratory bird habitat; however, land reclamation is occurring rapidly in the area [13]. Regardless, the lagoons are worthy of research because they have ecologically diverse characteristics and high biodiversity [14,15].

Diatoms are recognized worldwide as one of the most suitable biological components for water quality assessments because they continuously occur in aquatic ecosystems and respond quickly to environmental changes [16,17]. Diatoms can occur wherever there is water, can attach to all substrates, such as gravel and plants, and can live in various water environments [18,19]. However, research on diatoms in lagoons is poor. Even though data from low-salinity environments are readily available, those from high-salinity environments are limited [20].

There are many reports of new and unrecorded diatoms in Korea [21–23]. Species collected in environments such as lagoons, estuaries, and sedimentary soils with unique habitats other than freshwater environments are being reported [23–25].

In this study, we describe taxonomic information on 11 previously unrecorded diatom species found in six lagoons in Korea, and provide details on the ecological characteristics, reference specimen, basionym, synonym, and distribution of each.

2. MATERIALS AND METHODS

2.1 Collecting and Fixation of Samples

Samples were collected from six lagoons: Hwajinpo, Hyangho, Maeho, Gapyeongri wetland, Cheonjinho, and Gyeongpoho, adjacent to the East Sea in Korea from 2018–2020. The epilithic diatoms were scraped off the upper surface of the gravel using a small chisel or toothbrush. The collected wild cells were immediately fixed with Lugol's solution (iodine-iodide solution), 4% neutralized formalin, or 2% glutaraldehyde [26]. Environ- mental parameters such as water temperature (°C), pH, dissolved oxygen (mg/L), electric conductivity (μs/cm^3), and salinity (PSU) were measured using a portable multi-parameter water quality meter (Pro DSS, YSI, Yellow Springs, OH, USA) in the field.

2.2 Pretreatment and Observation of Diatoms

To remove organic matter, the fixed samples were boiled in equal amounts of mixed HCl and KMnO4 at 70 °C until the samples were slightly coloured, and were then rinsed in distilled water to remove residual acid, following a modified method of Hasle & Fryxell [27]. The samples were then rinsed in distilled water to remove any residual acid. The samples, some fixed and some cleaned, were mounted in Pleurax (Mountmedia, Wako, Japan) for observation using a light microscope (LM) (Eclipse Ni; Nikon, Tokyo, Japan), equipped with Nomarski differential interference contrast optics (DIC) and a digital camera (DS-Ri2; Nikon, Tokyo, Japan). In addition, the morphological characteristics of various foci on the valve face on *Mastogloia elliptica* were observed with a microscope.

The terminology was according to that of Anonymous [28], Ross et al. [29], and Round et al. [30].

3. RESULTS

3.1 Identification of Diatom Species in Korean Lagoons

A total of 11 previously unrecorded diatom species were identified from the six lagoon sites surveyed in this study (Figure 1). Two species each belonged to the genera *Gomphonema* and *Pinnularia*, with one species represented by the genera *Diploneis*, *Mastogloia*, *Cosmioneis*, *Haslea*, *Gyrosigma*, *Navicula*, and *Tryblionella*, respectively. According to site, four of the un- recorded species (*Mastogloia elliptica*, *Cosmioneis citriformis*, *Haslea crucigera*, and *Pinnularia bertrandii*) were detected at Hyangho, the two *Gomphonema* species (*G. guaraniarum* and *G. italicum*) were detected in the Gapyeongri wetland, and two species (*Navicula freesei* and *Tryblionella littoralis* var. *tergestina*) appeared in Yeomgaeho. One unrecorded species was, respectively, detected at Gyeongpoho (*Diploneis didyma*), Cheonjinho (*Pinnularia nodosa* var. *percapitata*), and Maeho (*Gyrosigma sinense*) (Table 1).

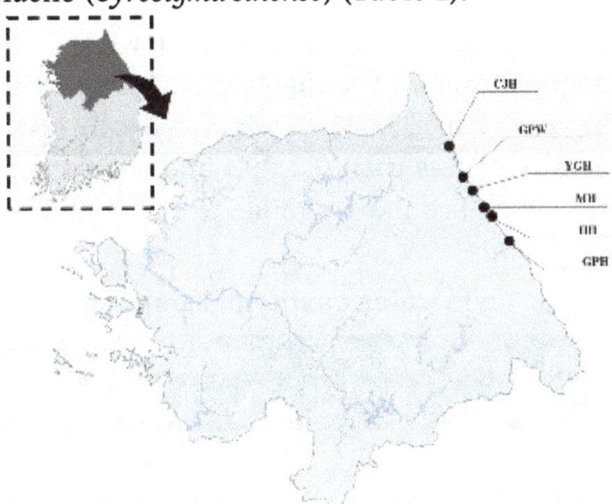

Figure 1. Sampling sites along the lagoons (CJH: Chunjinho, GPW: Gapyeongri wetland, YGH: Yeomgaeho, MH: Maeho, HH: Hyangho, GPH: Gyeongpoho).

Table 1. Information of sampling sites.

Site	Location	Latitude (N)	Longitude (E)	Collected Species
GPH	Jeo-dong, Gangneung-si, Gangwon-do	34°47_32.43"	128°53_48.89"	Diploneis didyma
HH	Hyangho-ri, Jumunjin-eup, Gangneung-si, Gangwon-do	37°54_35.2"	128°48_22.2"	Mastogloia elliptica; Cosmioneis citriformis; Haslea crucigera; Pinnularia bertrandii
CJH	Bongpo-ri, Toseong-myeon, Goseong-gun, Gangwon-do	38°15_14.97"	128°33_20.47"	Pinnularia nodosa var. percapitata
MH	Maeho-gil, Hyeonnam-myeon, Yangyang-gun, Gangwon-do	37°57_8.17"	128°46_8.88"	Gyrosigma sinense
GPW	Gapyeong-ri, Sonyang-myeon, Yangyang-gun, Gangwon-do	38°6_8.46"	38°6_8.46"	Gomphonema guaraniarum; Gomphonema italicum
YGH	Yeounpo-ri, Sonyang-myeon, Yangyang-gun, Gangwon-do	38°2_14.12"	128°42_1.73"	Navicula freesei; Tryblionella littoralis var. tergestina

3.2. Environmental Characteristics at the Sampling Sites

The pH of the six sites ranged from 7.6 to 8.5, showing a weakly basic range. The salinity of Gyeongpoho was in the range of 22.1–25.4 PSU and the electrical conductivity was higher than 35,000 μS/cm³. By contrast, Yeomgaeho and Cheonjinho had a salinity of less than 0.2 PSU and an electrical conductivity of less than 360 μS/cm³, which are more similar to the conditions of a freshwater environment (Table 2).

Table 2. Field water quality of sampling sites.

Site	Year	Temperature (°C)	DO (mg/L)	Conductivity (μS/cm3)	Salinity (PSU)	pH
GPH	2018	20.8	8.5	35,261	22.1	8.4
	2019	16.0	10.4	39,820	25.4	8.5
HH	2018	17.8	11.0	8926	5.0	8.1
	2020	28.6	-	60	0.03	8.2

Site	Year	Temperature (°C)	DO (mg/L)	Conductivity (μS/cm3)	Salinity (PSU)	pH
CJH	2019	16.0	9.5	360	0.2	8.7
	2020	14.0	9.2	152	0.1	8.1
MH	2018	20.2	10.9	15,969	9.4	8.1
	2020	15.8	10.8	5767	3.9	7.9
GPW	2020	13.3	16.2	1528	0.9	8.0
YGH	2020	8.3	11.5	288	0.1	7.6

3.3. Species Descriptions

3.3.1. Cosmioneis citriformis R.L. Lowe & A.R. Sherwood 2010

Original description: Lowe and Sherwood 2010, p. 24, Figures 10–18 [31]. Dimensions: Length Range 29–33 μm, width range 12–15 μm, 17–20 striae in 10 μm. Description: The valves are citriform with broadly rounded margins. The ends are short, narrow, and rostrate. The axial area is narrow and expands to a rounded central area. The proximal raphe ends are slightly expanded. The distal raphe ends are curved in the same direction. Internally, the raphe ends are anchor-shaped. Striae are punctuated and radiate, curving from the margin towards the center of the valve. There are 17 striae in 10 μm in the centre that become finer at the ends (20 in 10 μm). There are 16 areolae in 10 μm, which are round towards the valve ends, becoming transversely elongated near the centre (Figure 2A).

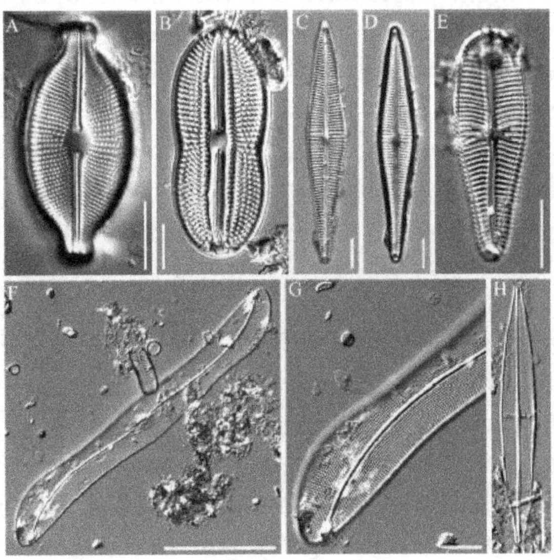

Figure 2. Light micrographs of diatoms. (A) *Cosmioneis citriformis* (B) *Diploneis didyma* (C,D) *Gomphonema guaraniarum* (E) *Gomphonema italicum* (F,G) *Gyrosigma sinense* (H) *Haslea crucigera*. Scale bars. 2–6, 8, 9:10 μm; 7:50 μm.

Ecology and distribution: Distributed in freshwater habitats, benthic taxon. Hyangho lake is the focus in this study (Table 1).

3.3.2. *Diploneis didyma* (Ehrenberg) Ehrenberg 1839

Original description: Ehrenberg 1839, pls. 1–4 [32].

Basionym: *Pinnularia didyma* Ehrenberg 1844 [33].

Dimensions: Length Range 60–74 μm, width range 25–28 μm, 8–10 striae in 10 μm. Description: The cells are solitary and free, the valves are elliptical or linear-elliptical, with or without a median constriction, and the apices are rounded or broadly cuneate. The central nodule is prominent, often large, quadrate, and strongly formed; the central area is small, reduced, and produces longitudinal extensions that are usually described as "horns" which lie on either side of the raphe and enclose it as solid ribs. Beyond the horns, there are thinner, usually narrow, and depressed areas, typically referred to as "furrows." These may be hyaline and structureless, may contain a row of large puncta, or may be crossed by faintly transverse costae. Beyond the furrows, on each side of the raphe, some specimens have a lunate area in each segment, which is usually referred to as the "lunula." This may be crossed by costae or alveoli, which may or may not bear a single or double row of puncta; these are frequently more developed and closer to the valve margin. In some specimens, transverse costae may be absent. Chromatophores are usually found as two deeply crenulated bodies that lie along the girdle. The valves are panduriform and slightly constricted in the middle, dividing the valve surface into two tongue-shaped segments. The central nodule is subquadrate or almost circular in some cases and protrudes to form two horns on either side of the raphe or median line, respectively. The valve surface is costate, transverse in the middle, but slightly curving. Radiating lines are found towards the apices that are crossed by numerous undulating longitudinal lines (Figure 2B).

Ecology and distribution: Distributed in marine-brackish habitats, benthic taxon. Gyeongpoho lake is the focus in this study (Table 1).

3.3.3. *Gomphonema guaraniarum* Metzeltin & Lange-Bertalot 2007

Description: The valves are rhombic-lanceolate, with less rounding of the apical and basal ends. The raphe-sternum is narrow and linear. The central area is unilaterally expanded, which is limited by a shortened median striae. The raphe is slightly sinuous with proximal ends that are punctuated and are slightly curved towards the stigma. The transapical striae slightly radiate parallel to the central region. Areolae are inconspicuous. There is a stigma at the end of the central stria. Under scanning electron microscopy (SEM), the stigma appears delicate and rounded, and striae are uniseriate with the areolae rounded to elongate longitudinally. The ends of the raphe dilate into pores, and the distal ends are curved, extending to the valve mantle. The pore field is formed by rounded poroids disposed on both sides of the terminal raphe fissure (Figure 2C,D).

Ecology and distribution: Distributed in freshwater habitats, epiphytic taxon. Gapyeon- gri Wetland is the focus in this study (Table 1).

3.3.4 Gomphonema italicum Kützing 1844

Original description: Metzeltin & Lange-Bertalot 2007,p. 147, pl. 212, Figures 9–14 [34]. Dimensions: Length Range 58–77 μm, width range 10–12 μm, 10–12 striae in 10 μm.Description: The valves are rhombic-lanceolate, with less rounding of the apical and basal ends. The raphe-sternum is narrow and linear. The central area is unilaterally expanded, which is limited by a shortened median striae. The raphe is slightly sinuous with proximal ends that are punctuated and are slightly curved towards the stigma. The transapical striae slightly radiate parallel to the central region. Areolae are inconspicuous. There is a stigma at the end of the central stria. Under scanning electron microscopy (SEM), the stigma appears delicate and rounded, and striae are uniseriate with the areolae rounded to elongate longitudinally. The ends of the raphe dilate into pores, and the distal ends are curved, extending to the valve mantle. The pore field is formed by rounded poroids disposed on both sides of the terminal raphe fissure (Figure 2C,D).

Ecology and distribution: Distributed in freshwater habitats, epiphytic taxon. Gapyeo-ngri Wetland is the focus in this study (Table 1).

3.3.5. *Gyrosigma sinense* (Ehrenberg) Desikachary 1988

Original description: Desikachary 1988, pp. 1–13, pls. 401–621 [36].

Basionym: *Navicula sinensis Ehrenberg* 1847 [37].

Description: The valves are linear-sigmoid, with inflated central and distal portions. The colour in resin and standardized dark-field microscopy is bright blue. The raphe sternum has a double curvature and is rotated towards the internal central raphe node, strongly eccentric at the ends, where it is markedly displaced owing to its concavity. The central area is rhombic and rotated. The terminal areas are triangular and strongly displaced away from the apices so that they are in a completely lateral position. The central external raphe fissures have isomorphic deflection patterns crossing the striae. The apical structure shows one very long and one short apical microforamina segment on opposite sides of the raphe sternum (Figure 2F,G).

Ecology and distribution: Distributed in marine habitats, benthic taxon. Maeho lake is the focus in this study (Table 1).

3.3.6. *Haslea crucigera* (W. Smith) Simonsen 1974

Original description: Simonsen 1974, pp. 47 [38].

Basionym: *Schizonema crucigerum* W. Smith 1856 [39].

Dimensions: Length Range 95–97 μm, width range 11–12 μm, 17–20 striae in 10 μm. Description: Living cells have slightly curved, narrowly rectangular frustules in girdle view. The valves are lanceolate to linear–lanceolate. Two band-like plastids lie against the girdle on each side of the cell. The internal margin of the plastids usually appears to be slightly undulating because of the presence of small, obliquely inserted, and rod-shaped pyrenoids. In the cleaned cells, the raphe appears straight and central. Transapical striae are visible under light microscopy and are crossed by more delicate longitudinal striae. The central two or three transapical virgae are thickened, forming a pseudostauros. Under SEM, the external valve surface is covered with closely spaced longitudinal strips of silica separated by narrow slits, which merge with a continuous peripheral slit near the apices. The external raphe fissures are slightly expanded and turn to one side centrally before sharply deflecting to the same side at the poles. Internally, the raphe slits open laterally in the raphe sternum, except at the center where the endings are straight and approximate, and at the apices, where they are slightly expanded in a slightly raised helictoglossa. An accessory rib on the primary side of the valve flanges over the raphe sternum, obscuring it for much of its length. The internal areola arrangement is similar to that of the other taxa, but with fewer longitudinal striae. On both sides of the raphe, the three central virgae thicken, forming a pseudostauros. The thickened virgae are fused with an accessory rib on the primary side of the valve and with a shorter thinner rib on the secondary side of the valve. The thickening of the virgae extends further across the valve and is more even compared with that in *H. salstonica* (Figure 2H).

Ecology and distribution: Distributed in brackish inland water habitats, benthic taxon.

Hyangho lake is the focus in this study (Table 1).

3.3.7 Mastogloia elliptica (C. Agardh) Cleve 1983

Original description: (C. Agardh) Cleve 1983, pl. 185, Figures 24–27 [40]. Basionym: *Frustulia elliptica* C. Agardh 1824 [41].

Dimensions: Length Range 17–57 μm, width range 8–12 μm, 15–18 striae in 10 μm.

Description: The valves are elliptical to linear-elliptical with convex to nearly parallel sides, and blunt, obtusely rounded apices. The axial area is narrow and barely wider than the raphe, and the central area is circular. The raphe branches are lateral and sinuous, with weakly expanded proximal ends. There are numerous tecta of approximately the same size. The striae consist of single rows of areolae that radiate throughout (Figure 3A–I). The raphe is a straight line or an undulate shape. The striae of *Mastogloia elliptica* are parallel in shape, and the raphe is straight (Figure 3A). In addition, due to the complex valvocopula structure, pertaca and striae are visible at the same time (Figure 3B–D), with seven symmetrical pertaca on each side (Figure 3E–I).

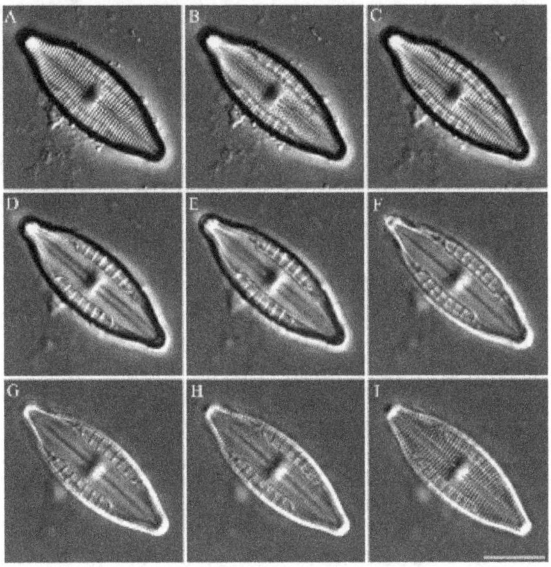

Figure 3. Light micrographs of *Mastogloia elliptica* in various depth of field ranges in microscope. (A–C,I) surface focus of valve face with uniseriate striae (D–H) Deeper focus with similar sized numerous partecta (arrows). Scale bar:10 μm.

Ecology and distribution: Distributed in marine and freshwater habitats, benthic taxon. Hyangho lake is the focus in this study (Table 1).

3.3.8 *Navicula freesei* R.M. Patrick & Freese 1961

Original description: Patrick, R.M. & Freese 1961, p. 206, pl. 2, Figure 14 [42]. Dimensions: Length Range 61–77 μm, width range 13–15 μm, 8–9 striae in 10 μm.

Description: The valves are lanceolate with rounded apices. The striae are radiated at the center, and are parallel and slightly convergent at the apices. The central area is asymmetrically rounded, whereas the axial area is narrow and straight. Striae on one or both sides of the central area are irregularly spaced. The raphe is lateral and straight. The distal raphe fissures form curved hooks onto the mantle. The proximal raphe ends form an elongated central poroid. A thickened central nodule is evident. Girdle bands have not yet been defined (Figure 4A).

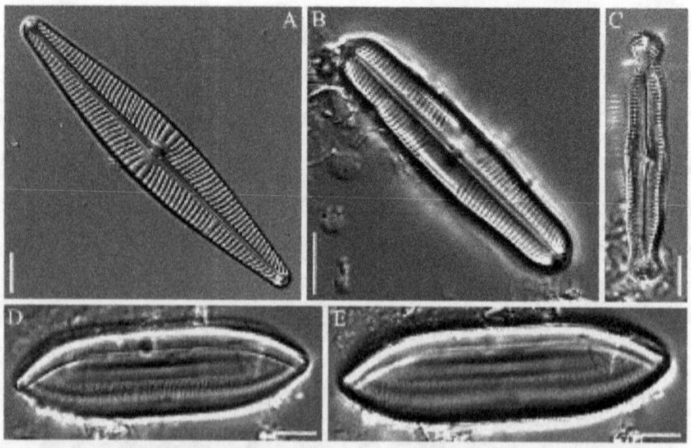

Figure 4. Light micrographs of diatoms. (A) Navicula freesei (B) Pinnularia bertrandii (C) Pinnularia nodosa var. percapitata (D,E) Tryblionella littoralis var. tergestina. Scale bars:10 μm.

Ecology and distribution: Distributed in freshwater habitats, benthic taxon. Yeomgaeho is the focus in this study (Table 1).

3.3.9 *Pinnularia bertrandii* Krammer 2000

Original description: Krammer 2000, pp. 122, 226, pl. 91, Figures 22–30 [43]. Dimensions: Length Range 14–27 μm, width range 4–6 μm, 17–18 striae in 10 μm.

Description: The valves are linear-elliptical to linear-lanceolate with weakly convex sides. The ends are subcapitate nearly covering the breadth of the valve body, which are broadly rounded. The raphe is filiform to slightly lateral, the central pores are small and slightly laterally bent, and the terminal fissures are distinct. The axial area is narrowed and the lanceolate is widened from the ends to the fascia. The striae in the middle weakly radiate and moderately converge at the ends and longitudinal bands are absent (Figure 4B).

Ecology and distribution: Distributed in freshwater habitats, benthic taxon. Hyangho lake is the focus in this study (Table 1).

3.3.10 *Pinnularia nodosa* var. percapitata Krammer 2000

Original description: Krammer 2000, pp. 57, Figures 26:9-12; 27:9, 10 [43]. Dimensions: Length Range 47–61 μm, width range 9–10 μm, 8–10 striae in 10 μm. Description: The valves are linear with triundulate margins. In the largest specimens, the central undulation is wider than the distal undulations. Apices are distinctly capitated. Axial areas are about one-third of the valve's width and widen from the apices towards the valve center. The central area is a bilateral fascia. The surface of the valve is mottled along either side of the raphe and continues into the central area. The raphe is straight. The proximal raphe ends are bent to one side and terminate in small, tear-shaped pores. Distal raphe fissures are shaped like question marks. The striae are weakly radiated at the valve center and become strongly convergent at the apices (Figure 4C).

Ecology and distribution: Distributed in freshwater habitats, benthic taxon. Chunjinho lake is the focus in this study (Table 1).

3.3.11 *Tryblionella littoralis* var. *tergestina* (Grunow) Snoeijs 1998

Original description: Snoeijs & Balashova 1998, pp. 1–144, Figure 1, pls. 101 [44]. Basionym: *Nitzschia littoralis* var. *tergestina* Grunow 1880 [45].

Dimensions: Length Range 30–100 μm, width range 12–30 μm, 30–38 striae in 10 μm. Description: The valves are broadly elliptical-lanceolate to linear-elliptical, with a very slight central constriction of the keel. Apices are cuneate, narrowed, and rounded. Striae are difficult to resolve under light microscopy (Figure 4D,E).

Ecology and distribution: Distributed in freshwater habitats, benthic taxon. Yeomgaeho is the focus in this study (Table 1).

4. DISCUSSION

Reservoir salt concentrations fluctuate over time; thus, lagoons create a unique ecosystem by mixing inland freshwater with intrusions of seawater. Accordingly, freshwater life with resistance to salt mixes with marine life in the lagoon ecosystem [3]. In this study, 11 previously unrecorded diatom species were discovered in six lagoons in Korea. *Diploneis didyma* was detected in Gyeongpoho. This species is mainly found in harbors but has also been reported in freshwater. At the time of the survey, Gyeongpoho had a high electrical conductivity above 30,000 μS/cm^3 and the salinity was 22 PSU, demonstrating environmental conditions similar to those of seawater (Tables 2 and 3). *Cosmioneis* predominantly inhabits brackish water zones [46–48] but has also been found in alkaline freshwater environments [30,49]. We detected *Cosmioneis citriformis* in Hyangho, which has very low salinity (0.03 PSU) and an electrical conductivity of 60 μS/cm^3, demonstrating similarity to the conditions of a freshwater environment. However, the measurements taken in the survey of 2018 showed a salinity of 5.0 PSU and an electrical conductivity of 8925 μS/cm^3 at this site, and *C. citriformis* was not found at that time (Tables 2 and 3). *Haslea crucigera* is a benthic species and is mainly found in high-salinity water [50]; however, during the 2018 survey, the salinity at Hyangho was 5.0 PSU and the electrical conductivity was 8926 μS/cm^3. Therefore, *H. crucigera* appears to prefer a water environment with some salt, although the salinity at this site remained lower than that of seawater throughout the survey period (Tables 2 and 3). *Pinnularia bertrandii* is a benthic species that is mainly detected in freshwater environments [51]. During the survey in 2020, the environmental conditions of the water at Hyangho were similar to those of a freshwater environment with a salinity of 0.03 PSU and an electrical conductivity of 60 μS/cm^3 (Tables 2 and 3). *Pinnularia nodosa* var. *percapitata* is also a benthic species that is mainly found in freshwater environments, mostly in streams, reservoirs, and small lakes, with a pH ranging from 6.1 to 7.8 (slightly acidic to neutral), and an electric conductivity from 22 to 169 μS/cm^3 [43]. At the time of sampling, the water of Chunjinho had very low salinity of 0.1–0.2 PSU, with a pH of 8.1 to 8.7 and a relatively low electrical conductivity of 152–360 μS/cm^3 (Tables 2 and 3). *Gyrosimga sinense* is a benthic species that has a seawater-based and widespread distribution under high water temperatures [52]. At the time of sampling, the water temperature of Maeho was relatively high, ranging from 15.8°C to 20.2°C, with salinity ranging from 3.9 to 9.4 PSU, and electrical conductivity ranging from 5767 to 15,969 μS/cm^3 (Tables 2 and 3). The genus *Gomphonema* is mostly attachable and is mainly found in freshwater environments, representing the largest genus in freshwater environments among diatoms of the world's most broadly distributed species [53,54]. In particular, *Gomphonema* can grow by attaching to aquatic plants because of the mucous stem secreted from the pore field at the end of the valve [30,55,56]. *G. guaraniarum* is a freshwater species and its ecological properties are rarely described in the literature. *G. italicum* is also a freshwater species with a relatively high electrical conductivity of 13,250 μS/cm^3 and is known to emerge in environments with a weakly basic pH [57]. In this study, *G. guaraniarum* and *G. italicum* were detected at the Gapyeong-ri wetland, with a water temperature of 13.3°C, a DO of 16.2 mg/L, a pH of 8.0, a salinity of 0.9 PSU, and an electrical conductivity of 1528 μS/cm^3, indicating that they inhabit an environment with freshwater-like conditions (Tables 2 and 3). *Navicula freesei* was found in freshwater and brackish environments, mainly with neutral to weakly basic conditions [42]. *Tryblionella littoralis* var. *tergestina* is an epipelic diatom and marine species [44]. In this study, these two species were identified at Yeomgaeho, with the water environment being similar to a freshwater environment at the time of sampling (a salinity 0.1 PSU, a pH of 7.6, and an electrical conductivity of 288 μS/cm^3) (Tables 2 and 3). This suggests that *Tryblionella littoralis* var. *tergestina* adapted to the desalinated lagoon. *Mastogloia* is a benthic and epiphytic diatom, which was identified in both freshwater and marine environments; however, most species of this genus appear in marine environments and prefer weakly basic water bodies [58,59]. Patrick and Reimer [60] report *M. elliptica* as a halophilic to mesohalobic taxon characteristic of coastal areas, but is also found in inland lakes with some salinity. In Europe, Krammer and Lange-Bertalot [61] report *M. elliptica* from brackish waters in coastal areas and from saline inland waters. At the time of sampling, the pH of Hyangho was 8.2 and the salinity was 0.03 PSU, exhibiting a water en- vironment close to that of the freshwater environment (Tables 2 and 3). *Mastogloia* has oval to linear oval convex valves, with complex silica chambers on both sides, called pertaca, that secrete mucus, and living cells that have two plastids [39,62]. This genus appears to be closely related to the genus *Aneumastus* but with a more complex valvocopula [63]. Patrick and Reimer [60] reported that *M. elliptica* is a halophilic to mesohalobic taxon characteristic of coastal areas but is also found in inland lakes with some salinity. In Europe, Krammer and Lange-Bertalot [61] reported *M. elliptica* from brackish waters in coastal areas and from saline inland waters.

Table 3. Information of New record diatoms in Korea.

No.	Species	Habitat	Distribution	This Study		Reference
				Salinity	Habitat	
1	*Diploneis didyma*	Benthic	Marine	22.1–25.4	Epilithic	[64,65]
2	*Mastogloia elliptica*	Benthic	Freshwater	0.03–5.0	Epilithic	[59]
3	*Cosmioneis citriformis*	Epiphytic	Marine	0.03–5.0	Epilithic	[31]
4	*Haslea crucigera*	Benthic	Freshwater	0.03–5.0	Epilithic	[50]
5	*Pinnularia bertrandii*	Benthic	Freshwater	0.03–5.0	Epilithic	[51]
6	*Pinnularia nodosa* var. *percapitata*	Benthic	Marine	0.1–0.2	Epilithic	[66]
7	*Gyrosigma sinense*	Benthic	Freshwater	3.9–9.4	Epilithic	[52]
8	*Gomphonema guaraniarum*	Benthic	Freshwater	0.9	Epilithic	[67]
9	*Gomphonema italicum*	Benthic	Terrestrial	0.9	Epilithic	[67]
10	*Navicula freesei*	Benthic	Freshwater	0.1	Epilithic	[42]
11	*Tryblionella littoralis* var. *tergestina*	Benthic	Marine	0.1	Epilithic	[44]

REFERENCES

1. Dominguez, J.M.; Martin, L.; Bittencourt, A.C. Sea-Level History and Quaternary Evolution of River Mouth–Associated Beachridge Plains Along the East–Southeast Brazilian Coast: A Summary. In *Sea-Level Fluctuation and Coastal Evolution*; Nummedal, D., Pilkey, O.H., Howard, J.D., Price, W.A., Eds.; SEPM Special Publication: Broken Arrow, OK, USA, 1987; Volume 41, pp. 115–127.
2. Müller, A.; Mathesius, U. The palaeoenvironments of coastal lagoons in the southern Baltic Sea, I. The application of sedimentary Corg/N ratios as source indicators of organic matter. *Palaeogeogr. Palaeoclimatol. Palaeoecol.* **1999**, *145*, 1–16. [CrossRef]
3. Hwang, S.I.; Yoon, S.O. Geomorphic characteristics of coastal lagoons and river basins, and sedimentary environment at river mouths along the middle east coast in the Korean peninsular. *Korean J. Geomorphol. Assoc.* **2008**, *15*, 17–33.

4. Moon, B.R.; Jeon, H.J.; Jeon, S.L.; Lee, J.S.; Shin, J.E.; Ahn, J.H.; Yang, Y.W.; Hyun, M.S.; Kim, M. Seasonal Variations of Water Quality and Phytoplankton of 4 Lagoons in the East Coast of Korea. *Int. J. Environ. Sci. Technol.* **2015**, *24*, 1101–1121.
5. Yum, J.G.; Takemura, K.; Tokuoka, T.; Yu, K.M. Holocene environmental changes of the Hwajinpo Lagoon on the eastern coast of Korea. *J. Paleolimnol.* **2003**, *29*, 155–166. [CrossRef]
6. Kwon, S.Y.; Lee, J.I.; Kim, D.J.; Kim, B.C.; Heo, W.M. The limnological survey of a coastal lagoon in Korea (3): Lake Hwajinpo. Korean J. Ecol. Environ. **2004**, 37, 12–25.
7. Heo, W.M.; Kim, B.C.; Jun, M.S. Evaluation of eutrophication of lagoons in the eastern coast of Korea. *Korean J. Limnol.* **1999**, *32*, 141–151.
8. Kwon, S.Y.; Lee, J.I.; Kim, D.J.; Kim, B.C.; Heo, W.M. The Limnological survey of a coastal lagoon in Korea (2): Lake Hyangho. Korean J. Ecol. Environ. **2004**, 37, 1–11.
9. Kim, H.Y. Change of the vegetation structure according to hydrosere processes and the evaluation of successional state of the 18 lagoon located in the east coastal region, Korea. Master's Thesis, Gangneung University, Gangneung, Korea, 2009; pp. 1–137.
10. Jeong, Y.I.; Hong, B.R.; Kim, Y.C.; Lee, K.S. Distribution, life history and growth characteristics of the Utricularia japonica Makino in the east coastal lagoon, Korea. *Korean J. Ecol. Environ.* **2016**, *49*, 110–123. [CrossRef]
11. Kim, T.K.; Kim, C.S. An Analysis of Factors Influencing the Landscape of Gyeong Po Lake and the Establishment of Criteria for Height Control. *J. Korean Inst. Landsc. Archit.* **2009**, *37*, 104–113.
12. Park, S.D.; Lee, S.; Shin, S.S.; Yoon, B.M. Effects of Operation of the Kyeongpo Retarding Basin on Flood Water Levelin Kyeongpo Lake. *J. Wet. Res.* **2016**, *18*, 413–423. [CrossRef]
13. Yoon, S.; Hwang, S.; Park, C.; Jin, M. Landscape changes during the 20th century of Ssangho, Gapy-eongri wetland, Gunggaeho and Yeongaeho, Yang yang-gun, Gangwon province. *J. Korean Geomorph. Assoc.* **2010**, *17*, 41–52.
14. Falcão, M.; Vale, C. Nutrient dynamics in a coastal lagoon (Ria Formosa, Portugal): The importance of lagoon–sea water exchanges on the biological productivity. *Cienc. Mar.* **2003**, *29*, 425–433. [CrossRef]
15. Bizzarro, J.J. A review of the physical and biological characteristics of the Bahia Magdalena lagoon complex (Baja California Sur, Mexico). *Bull. South. Calif. Acad. Sci.* **2008**, *107*, 1–24. [CrossRef]

16. Salomoni, S.E. Diatomáceas epilíticas indicadoras da qualidade de água na bacia do Rio Gravataí, Rio Grande do Sul, Brasil. Ph.D. Thesis, Ppgern Ufscar, São Carlos, São Paulo, Brasil, 2004; pp. 1–230.

17. Cho, I.H.; Kim, H.K.; Lee, M.H.; Kim, Y.J.; Lee, H.; Kim, B.H. The Effect of Monsoon Rainfall Patterns on Epilithic Diatom Communities in the Hantangang River, Korea. *Water* **2020**, *12*, 1471. [CrossRef]

18. Wunsam, S.; Cattaneo, A.; Bourassa, N. Comparing diatom species, genera and size in biomonitoring: A case study from streams in the Laurentians (Quebec, Canada). *Freshwater Biol.* **2002**, *47*, 325–340. [CrossRef]

19. Joung, S.H.; Park, H.K.; Lee, H.J.; Lee, S.H. Effect of climate change for diatom bloom at winter and spring season in Mulgeum Station of the Nakdong River, South Korea. *J. Korean Soc. Water Environ.* **2013**, *29*, 155–164.

20. Sylvestre, F.; Beck-Eichler, B.; Duleba, W.; Debenay, J.P. Modern benthic diatom distribution in a hypersaline coastal lagoon: The Lagoa de Araruama (RJ), Brazil. *Hydrobiologia* **2001**, *443*, 213–231. [CrossRef]

21. Park, J.S.; Yun, S.M.; Lee, S.D.; Lee, J.B.; Lee, J.H. New Records of the Diatoms (Bacillariophyta) in the Brackish and Coastal Waters of Korea. *Korean J. Environ. Biol.* **2017**, *35*, 215–226. [CrossRef]

22. Lee, H.; Yun, S.M.; Lee, J.Y.; Lee, S.D.; Lim, J.; Cho, P.Y. Late Holocene climate changes from diatom records in the historical Reservoir Gonggeomji, Korea. *J. Appl. Phycol.* **2018**, *30*, 3205–3219. [CrossRef]

23. Lee, S.D.; Lee, H.; Park, J.; Yun, S.M.; Lee, J.Y.; Lim, J.S.; Park, M.R.; Kwon, D.Y. Late Holocene diatoms in sediment cores from the Gonggeomji Wetland in Korea. *Diatom Res.* **2020**, *35*, 195–229. [CrossRef]

24. Park, J.S.; Lee, S.D.; Kang, S.E.; Lee, J.H. New records of the marine pennate diatoms in Korea. *J. Ecol. Environ.* **2014**, *37*, 231–244. [CrossRef]

25. Lee, J.H.; Park, J.S. Newly recorded diatom species in marine and fresh water of Korea. *J. Ecol. Environ.* **2015**, *38*, 545–562. [CrossRef]

26. Sournia, A. Quelques nouvelles données sur le phytoplancton marin et la production primaire à Tuléar (Madagascar). *Hydrobiologia* **1968**, *31*, 545–560. [CrossRef]

27. Hasle, G.R.; Fryxell, G.A. Diatoms: Cleaning and mounting for light and electron microscopy. *Trans. Am. Microsc. Soc.* **1970**, *89*, 469–474. [CrossRef]

28. Anonymous. Proposals for a standardization of diatom terminology and diagnoses. Nova Hedwigia, Beiheft. **1975**, 53, 323–354.

29. Ross, R.; Cox, E.J.; Karayeva, N.I.; Mann, D.G.; Paddock, T.B.B.; Simonsen, R.; Sims, P.A. An amended terminologyfor the siliceous components of the diatom cell. *Nova Hedwig. Beiheft.* **1979**, *64*, 513–533.

30. Round, E.E.; Crawford, R.M.; Mann, D.G. *The Diatoms. Biology and Morphology of the Genera*; Cambridge University Press: Cambridge, UK, 1990; pp. 1–747.
31. Lowe, R.L.; Sherwood, A.R. Three new species of *Cosmioneis* (Bacillariophyceae) from Hawaii. *Proc. Acad. Nat. Sci. Phila.* **2010**, *160*, 21–28. [CrossRef]
32. Ehrenberg, C.G. Über die Bildung der Kreidefelsen und des Kreidemergels durch unsichtbare Organismen. In *Abhandlungen der Königlichen Akademie der Wissenschaften zu Berlin*; Physikalische Klasse: Göttingen, Germany, 1839; Volume 1838, pp. 59–147, pls.
33. Ehrenberg, C.G. Mittheilung über 2 neue Lager von Gebirgsmassen aus Infusorien als Meeres-Absatz in Nord-Amerika und eine Vergleichung derselben mit den organischen Kreide-Gebilden in Europa und Afrika. In *Bericht über die zur Bekanntmachung geeigneten Verhandlungen der Königlich-Preussischen Akademie der Wissenschaften zu Berlin*; Physikalische Klasse: Göttingen, Germany, 1844; Volume 1844, pp. 57–97.
34. Metzeltin, D.; Lange-Bertalot, H. Tropical Diatoms of South America II. Special remarks on biogeography disjunction. *Iconogr. Diatomol.* **2007**, *18*, 1–877.
35. Kützing, F.T. *Die Kieselschaligen Bacillarien oder Diatomeen*, Nordhausen: Zu finden bei W.: Köhne, Germany, 1844; pp. [i-vii], [1]-152, pls 1–30.
36. Desikachary, T.V. Marine diatoms of the Indian Ocean region. In *Atlas of Diatoms*; Madras: Madras Science Foundation: Tamil Nadu, India, 1988; Volume V, pp. 1–13.
37. Ehrenberg, C.G. Mittheilung uber vor Kurzem von dem Preuß. Seehandlungs Schiffe, der Adler, aus Canton Mitgebrachte Verkaufliche Chinesische Blumen-Cultur-Erde; Bericht über die zur Bekanntmachtung geeigneten Verhandlungen der Koniglich Preussischen Akademie der Wissenschaften zu Berlin: Berlin, Germany, 1847; pp. 476–485.
38. Simonsen, R. The diatom plankton of the Indian Ocean Expedition of R/V Meteor 1964-5. *Meteor Forschungsergebnisse, Reihe D Biologie* **1974**, *19*, 1–107.
39. Smith, W. A Synopsis of the British Diatomaceae; with Remarks on Their Structure, Functions and Distribution; and Instructions for Collecting and Preserving Specimens. A–E; John van Voorst: London, UK, 1856; Volume 2, pp. [i-vi]–xxix, pls 32–60, 61–62.
40. Schmidt, A. *Atlas der Diatomaceen-kunde*; ser. IV: Heft 47; O.R. Reisland: Leipzig, Germany, 1983; pp. 185–188.
41. Agardh, C.A. *Systema Algarum*; Literis Berlingianis [Berling]: Lundae, Sweden, 1824; pp. [i]-xxxvii, [1]-312.
42. Patrick, R.M.; Freese, L.R. Diatoms (Bacillariophyceae) from Northern Alaska. *Proc. Acad. Nat. Sci. Phila.* **1961**, *112*, 129–293.

43. Krammer, K. The genus Pinnularia. In *Diatoms of the European Inland Waters and Comparable Habitats*; Lange-Bertalot, H., Ed.; Ruggell: A.R.G. Gantner Verlag K.G.: Berlin/Heidelberg, Germany, 2000; Volume 1, pp. 1–703.

44. Snoeijs, P.; Balashova, N. Intercalibration and distribution of diatom species in the Baltic Sea. In *The Baltic Marine Biologists Publication No. 16e*; Opulus Press: Uppsala, Sweden, 1998; Volume 5, pp. 1–144.

45. Cleve, P.T.; Grunow, A. Beiträge zur Kenntniss der arctischen Diatomeen. *Kongliga Svenska Vetenskaps-Akademiens Handlingar* **1880**, *17*, 1–121.

46. Hällfors, G. *Checklist of Baltic Sea Phytoplankton Species (Including Some Heterotrophic Protistan Groups)*; Helsinki Commission, Baltic Marine Environment Protection Commission: Helsinki, Finland, 2004; pp. 1–210.

47. Sylvestre, F.; Guiral, D.; Debenay, J.P. Modern diatom distribution in mangrove swamps from the Kaw Estuary (French Guiana). *Mar. Geol.* **2004**, *208*, 281–293. [CrossRef]

48. Marshall, H.G.; Burchardt, L.; Lacouture, R. A review of phytoplankton composition within Chesapeake Bay and its tidal estuaries. *J. Plank. Res* **2005**, *27*, 1083–1102. [CrossRef]

49. Kociolek, J.P.; Spaulding, S.A. Symmetrical naviculoid diatoms. In *Freshwater Algae of the United States*; Sheath, B., Wher, J., Eds.; Academic Press: Cambridge, MA, USA, 2003; pp. 637–653.

50. Massé, G.; Rincé, Y.; Cox, E.J.; Allard, G.; Belt, S.T.; Rowland, S.J. *Haslea salstonica* sp. nov. and *Haslea pseudostrearia* sp. nov.(Bacillariophyta), two new epibenthic diatoms from the Kingsbridge estuary, United Kingdom. *Comptes Rendus de l'Académie des Sciences-Series III-Sciences de la Vie* **2001**, *324*, 617–626. [CrossRef]

51. Kheiri, S.; Solak, C.N.; Edlund, M.B.; Spaulding, S.; Nejadsattari, T.; Asri, Y.; Hamdi, S.M.M. Biodiversity of diatoms in the Karaj River in the Central Alborz, Iran. *Diatom Res.* **2018**, *33*, 355–380. [CrossRef]

52. Jahn, R.; Sterrenburg, F.A. *Gyrosigma sinense* (Ehrenberg) desikachary: Typification and emended species description. *Diatom Res.* **2003**, *18*, 61–67. [CrossRef]

53. Zhang, W.; Xu, X.Y.; Kociolek, J.P.; Wang, L.Q. *Gomphonema shanghaiensis* sp. nov., a new diatom species (Bacillariophyta) from a river in Shanghai, China. *Phytotaxa* **2016**, *278*, 29–38. [CrossRef]

54. Abarca, N.; Zimmermann, J.; Kusber, W.H.; Mora, D.; Van, A.T.; Skibbe, O.; Jahn, R. Defining the core group of the genus *Gomphonema* Ehrenberg with molecular and morphological methods. *Bot. Lett.* **2020**, *167*, 114–159. [CrossRef]

55. Wojtal, A. Diatoms of the genus Gomphonema Ehr. [Bacillariophyceae] from a karstic stream in the Krakowsko-Czestochowska Upland. *Acta Soc. Bot. Pol.* **2003**, *72*. [CrossRef]

56. Cox, E.J. Diatoms, Diatomeae (Bacillariophyceae sl, Bacillariophyta). *Syllabus Plant Fam.* **2015**, *2*, 64–103.

57. Rybak, M.; Noga, T.; Poradowska, A. Diversity in anthropogenic environment–permanent puddle as a place for development of diatoms. *J. Ecol. Eng.* **2019**, *20*, 165–174. [CrossRef]

58. Krammer, K. Bacillariophyceae. I. Teil. Naviculaceae. In Susswasserflora von Mitteleuropa. Springer: Berlin/Heidelberg, Germany, 1986; Volume 2, 876p.

59. Vidakovic´, D.; C´iric´, M.; Krizmanic´, J. First finding of a genus *Haslea Simonsen* in Serbia and new diatom taxa for the country's flora in extreme and unique habitats in the Vojvodina Province. *Bot. Serbica* **2020**, *44*, 3–9. [CrossRef]

60. Patrick, R.; Reimer, C.W. The diatoms of the United States. Exclusive of Alaska and Hawaii. Volume 1: Fragilariaceae, Eunotiaceae, Achnanthaceae, Naviculaceae. *Monogr. Acad. Nat. Sci. Phila.* **1966**, *13*, 1–688.

61. Krammer, K.; Lange-Bertalot, H. Bacillariophyceae. 1. Teil: Naviculaceae. In *Süsswasserflora von Mitteleuropa, Band 2/1*; Ettl, H., Gerloff, J., Heynig, H., Mollenhauer, D., Eds.; Gustav Fisher Verlag: Jena, Germany, 1986; pp. 1–876.

62. Pennesi, C.; Poulin, M.; De Stefano, M.; Romagnoli, T.; Totti, C. New insights to the ultrastructure of some marine *Mastogloia* species section Sulcatae (Bacillariophyceae), including *M. neoborneensis* sp. nov. *Phycologia* **2011**, *50*, 548–562. [CrossRef]

63. Maltsev, Y.; Andreeva, S.; Podunay, J.; Kulikovskiy, M. Description of *Aneumastus mongolotusculus* sp. nov.(Bacillariophyceae, Mastogloiales) from Lake Hovsgol (Mongolia) on the basis of molecular and morphological investigations. *Nova Hedwig.* **2019**, *148*, 21–33. [CrossRef]

64. Donadel, L.; Torgan, L. Subtropical Coastal Lagoon from Southern Brazil: Environmental Conditions and Phytobenthic Commu- nity Structure. In *Lagoon Environments Around the World-A Scientific Perspective*; IntechOpen: London, UK, 2019; pp. 189–205.

65. Cartaxana, P.; Ruivo, M.; Hubas, C.; Davidson, I.; Serôdio, J.; Jesus, B. Physiological versus behavioral photoprotection in intertidal epipelic and epipsammic benthic diatom communities. *J. Exp. Mar. Biol. Ecol.* **2011**, *405*, 120–127. [CrossRef]

66. Genkal, S.I.; Yarushina, M.I. Materials on the flora of Bacillariophyta in aquatic ecosystems of the Yarayakha River basin (Yamal Peninsula). *Contemp. Probl. Ecol.* **2016**, *9*, 306–317. [CrossRef]

67. Medeiros, G.; Amaral, M.W.W.; Ferreira, P.C.; Ludwig, T.V.; Bueno, N.C. *Gomphonema* Ehrenberg (Bacillariophyceae, Gomphone- mataceae) do rio São Francisco Falso, Paraná, Brasil. *Biota Neotrop.* **2018**, *18*, e20180568. [CrossRef]

CHAPTER-13

MOLECULAR AND MORPHOLOGICAL CHARACTERIZATION OF *COLACONEMA FORMOSANUM* SP. NOV. (COLACONEMATACEAE, RHODOPHYTA)

1. INTRODUCTION

Florideophyceae is the most speciose class in Rhodophyta (red algae), with about 6900 species distributed worldwide [1]. Nemaliophycidae is one of the most diverse subclasses among red algae [2], including small and filamentous algae, growing as epiphytes or endophytes attached to rocks, macroalgae, and some marine animals. Their reproduction occurs mainly by monosporangia, which exist widely in both marine and freshwater systems [3,4].

At present, 11 orders are included in Nemaliophycidae: Acrochaetiales (Feldmann, 1953), Balbianiales (Sheath and Müller, 1999), Balliales (Choi et al., 2000), Batrachospermales (Pueschel and Cole, 1982), Colaconematales (Harper and Saunders, 2002), Corynodactylales (Saunders et al., 2017), Entwisleiales (Scott et al., 2013), Nemaliales (Schmitz, 1892), Palmariales (Guiry, 1978), Rhodachlyales (West et al., 2008), and Thoreales (Müller et al., 2002) [1]. Interestingly, due to the ambiguous taxonomy in previous morphological evidence, the term "Acrochaetioid algae" is currently used to indicate some genera within Nemaliophycidae, such as *Acrochaetium* Nägeli, *Audouinella* Bory, *Balbiania* Sirodot, *Colaconema* Batters, *Grania* (Rosenvinge) Kylin, *Rhodochorton* Nägeli, and *Liagorophila* Yamada, although it is only used to describe the morphological affinity rather than phylogenetic affinity [5,6].

Even though morphology alone provides limited information for taxonomy due to the simple vegetative and reproduction branches of the Acrochaetiales [7], some phenotypic and biological attributes remain reliable, such as: life cycle pattern, germination pattern (Type I, Type II), reproduction development, vegetative features (cell size, plastid morphology, number of pyrenoids), attachment type (epiphyte or endophyte), biochemical characteristics (phycoerythrin type), and even ultrastructural observations (pit plug) [5,6,8]. The distinguishable characteristics of genus *Colaconema* were proposed to be prostrate, irregular filaments; monosporangiate; cup-like cells at terminals and intercalares; an unfixed plastid in each cell; and a single or more pyrenoids per cell. Most of these characteristics are unchallenged but inaccurate, and this has been the case for decades [7,9].

When the molecular phylogeny is considered, many species within Nemaliophycidae have obtained better taxonomic support. For instance, Lam et al. (2016) redefined the phylogeny of Nemaliophycidae using the *psa*A, *psb*A, *psa*B, EF2, *cox1*, *cob*, 18S rDNA, and 28S rDNA genes [2]; Saunders et al. (2018) resolved the family level assignments of the Acrochaetiales–Palmariales complex using the mitochondrial cytochrome c oxidase 1 gene (COI-5P) [10]; Soares et al. (2020) established the new species *Rhodachlya westii* using the *rbc*L gene [3]; Skriptsovar et al. (2020) assessed the phylogenetic relationships within Palmariaceae using the internal transcribed spacer (ITS) and *rbc*L genes [11]; Araújo et al. (2014) and Montoya et al. (2020) identified the endophytic algae *C. infestans* and *C. daviesii* through morphology and *rbc*L and COI-5P genes, respectively [12,13].

Members of the order Colaconematales are usually epiphytes or endophytes and are only present in marine regions [7]. There are currently 51 taxonomically accepted species in the genus *Colaconema*, which has a circumtropical distribution [1]. Records have shown that the host selection mechanism of *Colaconema* species appears to be random, e.g., on *Kappaphycus alvarezii* (Doty) L.M. Liao, *Chondracanthus chamissoi* (C. Agardh) Kützing, and *Nizamuddinia zanardinii* (Schiffner) P.C. Silva [12–14]. The endophyte phenomenon negatively affects the host and is seen seasonally, however the invasion mechanism is not clear [12,13].

Our observations in the field found an endophyte filamentous alga attached to *Sarcodia suae* S.-M.Lin & C.Rodríguez-Prieto (Plocamiales), an economically important red alga in Taiwan, mainly as a foodstuff with abundant dietary fiber, nutrition, and antioxi- dants [15,16].

However, there has been no information available about the endophyte algae from *S. suae* in Taiwan before. This study presents the third record of a *Colaconema* species in Taiwan (the other two species being *C. secundatum* (Lyngbye) Woelkerling and *C. tenuissimum* (Collins) Woelkerling) discovered attached to a local red alga of economic importance, *S. suae*. To fully understand the features of this endophytic alga, we conducted morphological and molecular analyses to clarify its taxonomic position.

2. MATERIALS AND METHODS

Sample collection: Infected thalli of *Sarcodia suae* were collected at an intertidal zone in Wan-li-tong, Pingtung, Taiwan, in July 2014 (21°59′41″ N, 120°42′22″ E).

Morphological studies: Samples of the endophytic filamentous algae were fixed in 4% formalin/seawater for morphological observations. Photomicrographs were obtained using an Olympus BX53 microscope (Olympus, Tokyo, Japan) with a DP72 digital camera and cellSens software (Olympus). The voucher specimens were deposited in the National Taiwan Ocean University (NTOU) Algal Cultivation and Biotech Laboratory, Taiwan.

Sample cultivation: Samples of endophytic isolates were cultured in the laboratory for 21 days to increase the algal biomass for molecular studies. The alga was cultured in a beaker with 800 mL Provasoli-enriched seawater (PES) culture medium [17] (sterilized) to which GeO_2 (1 mL L^{-1} PES) was added to inhibit the growth of diatoms. The culture conditions were as follows: temperature of 25 ± 2 °C, salinity of 32 ± 1 psu, and a 12:12 h light/dark cycle with an irradiance of 50 μmol photons m^{-2} s^{-1} provided by 9W, 900 Lumen 6500 K white LED bulbs (Everlight, New Taipei City, Taiwan).

DNA extraction: Fragments of silica-dried thalli were ground in a sterile mortar with liquid nitrogen. Total genomic DNA was extracted from 40 mg of algal tissue using the Wizard Genomic DNA Purification Kit (Promega, Madison, WI, USA) following the manufacturer's protocol, with minor modifications in centrifugation at 16,000× *g* for 10 min at 94 °C (Allegra X-12R; Beckman Coulter, La Brea, CA, USA), and in rehydration with 40 μL Tris-EDTA buffer solution. The partial *rbc*L gene was amplified with the primer pairs F7-R753, F645-R1150, and F449-RrbcSstart [18,19]. The COI-5P gene was obtained with the primer pair GazF1-GazR1 [20]. The PCR protocol consisted of 5 min at 96 °C for initial denaturation, followed by 35 cycles of 60 s at 94 °C, 60 s at 45 °C, and 90 s at 72 °C, with a final extension for 8 min at 72 °C, and then a soaking cycle at 15 °C. The PCR product was purified with the QIAquick PCR Purification Kit (Qiagen, Hilden, Germany). Commercial sequencing was performed using the Taq Big Dye Terminator Cycle Sequencing Kit (Applied Biosystems, Foster City, CA, USA). The forward and reverse primers used for sequencing were the same as those used for PCR amplification.

Phylogenetic analysis: We obtained the DNA sequences of the *rbc*L and COI-5P genes from the unknown endophytic alga in this study; other sequences were retrieved from GenBank, which included the species from the order

Colaconematales and the order Acrochaetiales (Balliales, Batrachospermales, Entwisleiales, Nema-liales, Palmariales, and Thoreales). All new sequence data generated from this study were deposited in GenBank, with the following accession numbers: MH414967, MW182306-MW182308 (*rbc*L); and MW182302-MW182305 (COI-5P). The sequences were aligned with multiple sequence alignment (MUSCLE) [21] using MEGA v10.1 [22], then checked and trimmed by eye using BioEdit 7.1.3.0 [23]. IQ-TREE software [24] was used to select the best model of evolution, as inferred from the implementation of the Bayesian Information Criterion (BIC) and Akaike Information Criterion (AIC). The best-selected models for the *rbc*L and COI-5P sequences were GTR+F+I+G4 and TIM2+F+I+G4, respectively. The IQ-TREE program was used to perform maximum-likelihood (ML) phylogenetic inference and the results were displayed using FigTree v. 1.4.0 software [25]. Statistical support was assessed with 1000 bootstrap replications [26]. The Bayesian inference (BI) was conducted with Mr Bayes 3.2.7 [27] and followed the best models identified above. Two independent runs of four chains of Markov Chain Monte Carlo (MCMC) were conducted, with one tree sampled every 100 generations for one million generations. The results of the generations were used to build the consensus tree.

3. RESULTS

3.1 Phylogenetic Results

The final *rbc*L alignment included 1202 nucleotides and 68 sequences, including four new *Colaconema* sp. specimens (MH414967, MW182306–MW182308). The final COI- 5P dataset included 582 nucleotides and 57 sequences, including sequences from four new *Colaconema* sp. specimens (MW182302-MW182305). Both ML and BI had virtually congruent topologies, thus, the ML consensus tree was used and the BI values were plotted on the nodes. The unknown species belonged to the order Colaconematales, forming a monophyletic clade with high support (*rbc*L: ML = 80%, BI = 1.00; COI-5P: ML = 92%, BI = 0.99).

In the *rbc*L phylogeny (Figure 1), *C. formosanum* was sister to *C. savianum* (DQ787561) from Korea in a fully supported clade; the genetic distance between *C. formosanum* and *C. savianum* was 7.19–7.83%. For the COI-5P gene (Figure 2), *C. formosanum* was sister to two specimens of *Colaconema* sp. (HQ422955 and HQ422983) from USA with strong support (ML = 91%, BI = 1.00). The genetic distance between *C. formosanum* and *Colaconema* sp. was 14.02–14.19%. In this study, the morphological and molecular analyses supported the taxonomic position of this new filamentous red algal species from Taiwan. We therefore propose the new species, *Colaconema formosanum*.

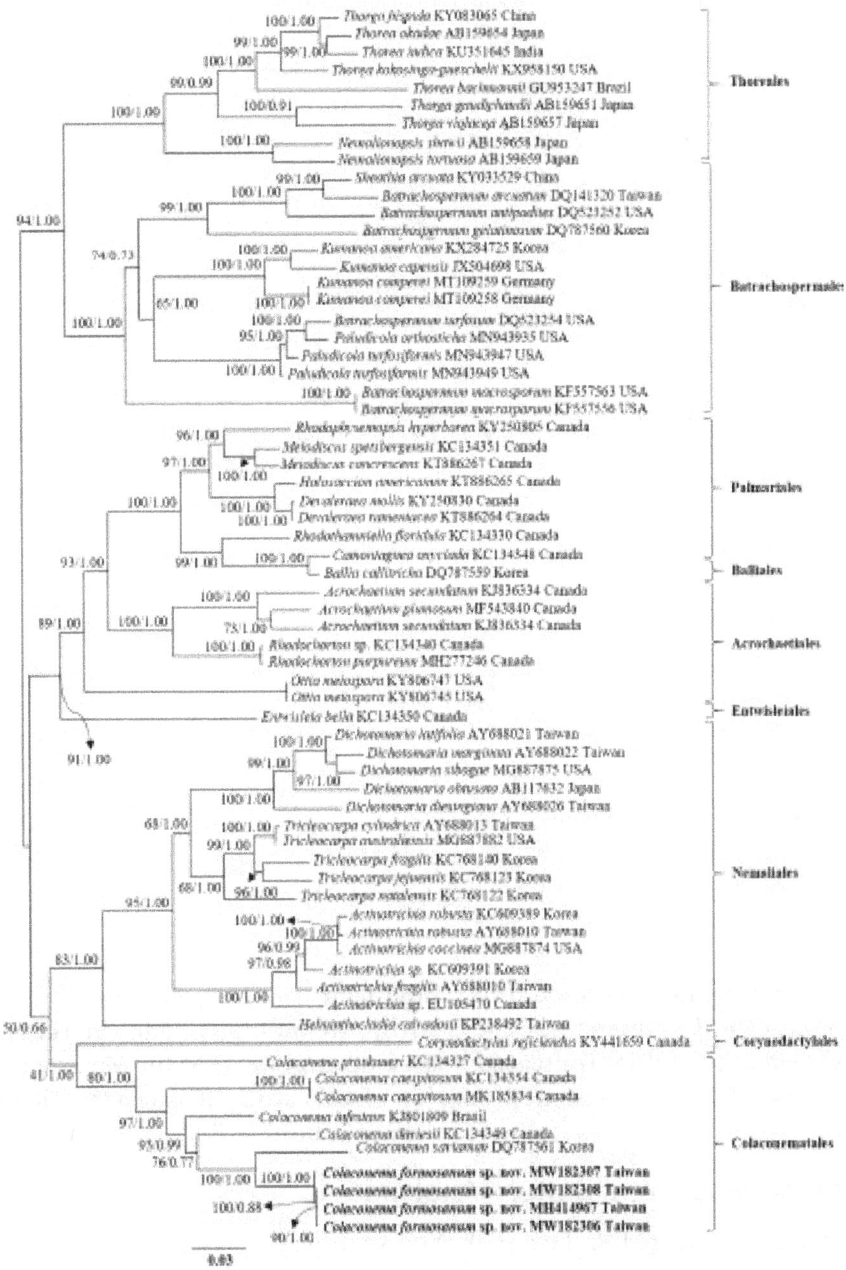

Figure 1. Maximum-likelihood (ML) phylogeny of Nemaliophycidae inferred from partial *rbc*L sequences. Orders within Nemaliophycidae are represented. Values shown at the nodes indicated bootstrap support (BS) for ML and posterior probabilities (PP) for BI. The new species of this study (*Colaconema formosanum* sp. nov.) is shown in bold.

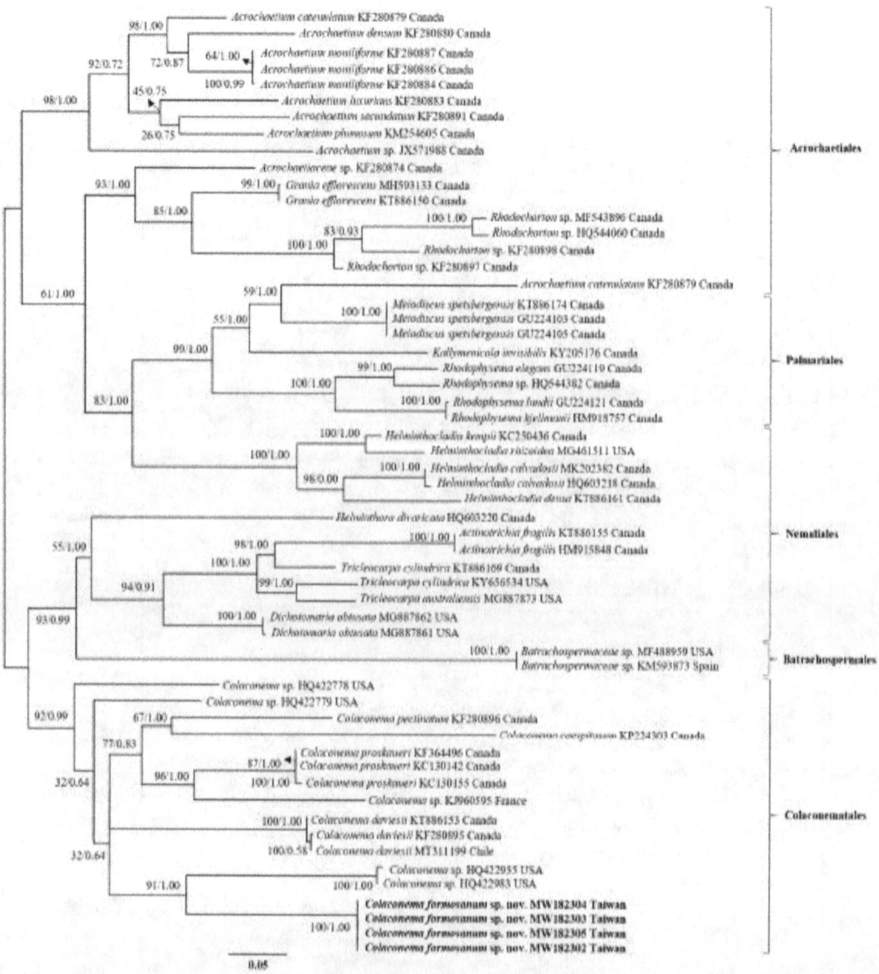

Figure 2. Maximum-likelihood (ML) phylogeny of Nemaliophycidae inferred from partial COI-5P sequences. Orders within Nemaliophycidae are represented. Values shown at the nodes indicated bootstrap support (BS) for ML and posterior probabilities (PP) for BI. The new species of this study (*Colaconema formosanum* sp. nov.) is shown in bold.

3. 2 Morphological and Culture Analyses

Colaconema formosanum M.-C. Lee, H.-Y. Yeh sp. nov.

Diagnosis: Filamentous endophytic thalli infected *Sarcodia suae* between its cortical and subcortical cells. Erect axes forked several times, producing irregular filaments. Oval monospores borned on the terminal cells.

Holotype: Voucher number—NTOU LMC CF2101 (deposited in Algal Cultivation and Biotech Laboratory, Department of Aquaculture, National Taiwan Ocean University, Taiwan), collected by M.-C. Lee & H.-Y. Yeh on 10 July 2014 at Wanli-tong,

Pingtung, endophytic on the red alga *Sarcodia suae* (Figure 3A–D). *rbc*L sequence = GenBank accession MH414967; COI-5P sequence = GenBank accession MW182302.

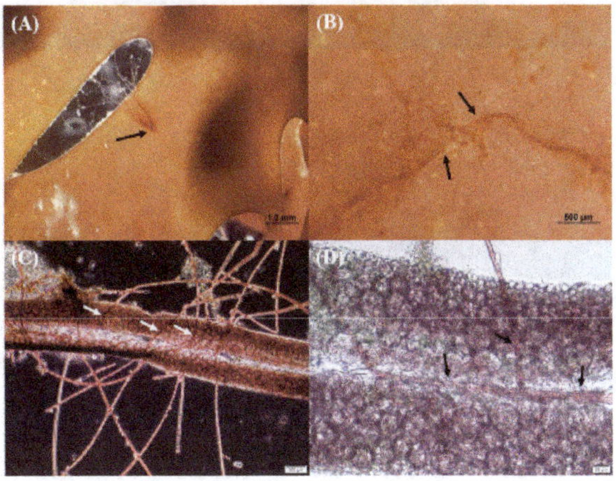

Figure 3. *Colaconema formosanum* sp. nov. (A,B) Thalli endophytic on *Sarcodia suae* (arrows). (C,D) Transversal sections of *S. suae* showing partially endophytic thalli of *C. formosanum* growing between cortical and subcortical cells (arrows).

Isotype: NTOU LMC CF0002- CF0004, deposited in NTOU, Taiwan. *rbc*L sequences = GenBank accession MW182306, MW182307, and MW182308, respectively; COI-5P se- quences = GenBank accession MW182303, MW182304, and MW182305, respectively.

Type locality: Wan-li-tong, Ping-tung, Taiwan (21°59′43″ N, 120°42′16″ E).

Etymology. The specific epithet, *formosanum*, is derived from a Latin and Portuguese noun meaning beautiful, which was a former name of Taiwan. Herein, the meaning is that the species is from Taiwan.

Distribution: The type locality is in the intertidal zone of Ping-Tung City, Taiwan, the species is known only from the type locality.

Habitat: Filamentous thalli endophytic on the marine red alga ***Sarcodia suae.***

3.3 Vegetative Morphology

The thalli formed clusters of red to deep-red filaments, 2.0–2.8 cm long, arising from a basal disc (Figure 4A,B). Erect axes were pseudodichotomous, branching at irregular segments (Figure 4A). Each cylindrical cell was 6.0–7.5 µm wide and 12.5–15 µm long (Figure 4D). Segment length to diameter (L:D) ratios at lower and mid-branch positions and towards apices were 1.6:1.8, 4.0:4.2, and 3.7:3.9, respectively (Figure 4D). Cells with a single pyrenoid, and single parietal or lobed plastids (Figure 4D). The endophytic cells showed the ability to completely penetrate into the cortical and subcortical cells of *S. suae* (Figure 3C,D).

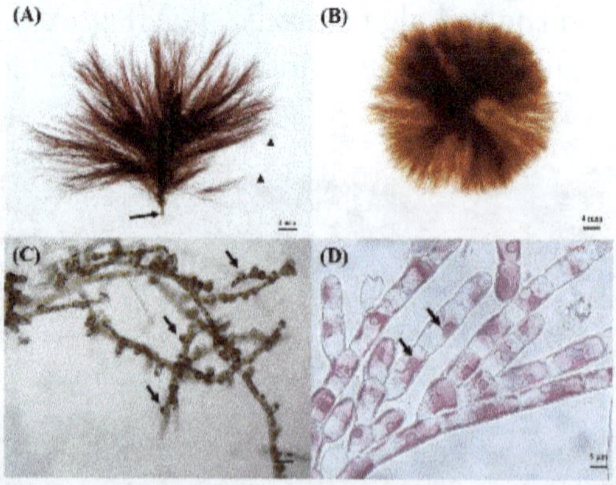

Figure 4. *Colaconema formosanum* sp. nov. (A) General aspect of a tuft showing erect filaments (arrowheads) formed from a basal disc (arrow). (B) Concentric tufts formed under culture conditions. (C) Monosporangia (arrows) formed on prostrate cultured thallus. (D) Detail of erect filaments showing cells with one pyrenoid (arrows).

3.4 Reproductive Morphology

Monosporangia were observed only on the apical cells of the lateral branch, solitary or in clusters. The monosporangia were subspherical, 12–13 μm long and 9.5–10 μm wide (Figure 5A). Upon maturity, the monospores were released from the monosporangia (Figure 5B,C), and some paraphyses could be seen on the apical portion of the monosporan- gia (Figure 5D). The released monospores exhibited a circular shape, 9–10 μm in diameter at the initial stage (Figure 6A). After 24 h, a tube-like cell emerged and elongated gradually (Figure 6B,C). After 2 to 3 days under culture conditions, a second set of segments was observed (Figure 6D). An apical part of the cell released a mucus-like matrix at this stage that engulfed all surrounding filament, and a filamentous growth form was observed (Figure 6E). Male and female structures could not be distinguished.

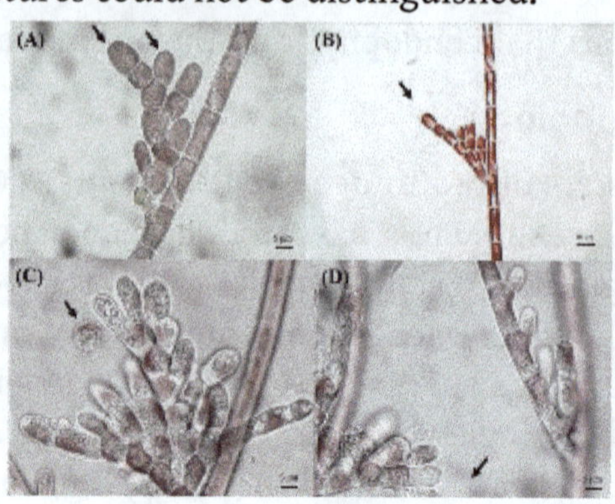

Figure 5. The process of monospore release of *Colaconema formosanum* sp. nov. (A) Mature monosporangia (arrow). (B) After monospore release, only the cell wall

persists (arrow). (C) Monospore released from monosporangia (arrow). (D) When the monospore is released, paraphyses are formed from the monosporangia (arrow).

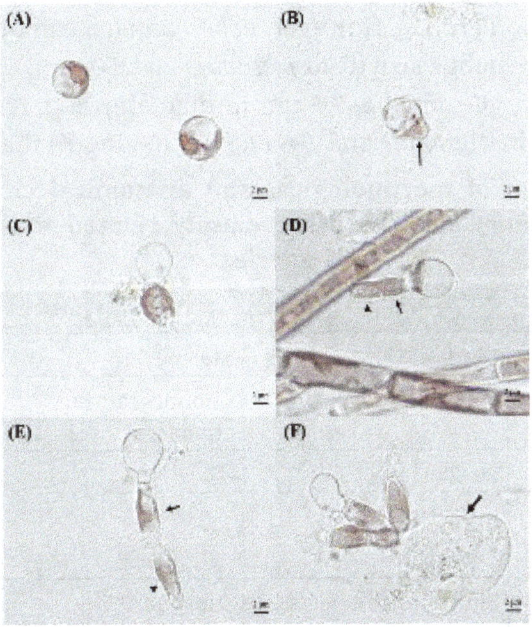

Figure 6. Sporeling development in *Colaconema formosanum* sp. nov. (A) Newly released spores had a concentric circular shape. (B,C) After 24 h, a tube-like cell emerged from the spore and gradually expanded. (D,E) After 2 to 3 days under culture conditions, a second cell could be clearly observed (arrow = first cell, arrow head = second cell). (F) The apical part of the cell in this stage released a mucus-like matrix (arrow).

4. DISCUSSION

Due to the close relationship among species within Nemaliophycidae lines, it is a taxonomic conundrum if only relying on morphology. Even though this has led to a long standing debate on over-relying on morphological attributes within Nemaliophycidae, some features of algae are still worth evaluating [8]. In this study, C. *formosanum* showed some typical features of the *Colaconema* genus, e.g., monosporangia, cylindrical cells, irregularly branched filaments, a single pyrenoid, and single parietal or lobed plastids. In addition, we found that the phycoerythrin of C. *formosanum* was of R-type (un-published data), meanwhile, the germination form was type II, meaning that when the germ tube developed, the protoplast was retained [7], which was similar to C. *daviesii* but is mentioned for the first time in *Colaconema* genus. We recommend tracing this feature in further studies. The morphological evidence supports C. *formosanum* as a new species.

Since few reports investigate the detailed morphological features of *Colaconema* species, only *C. daviesii* and *C. infestans* would be explored in this study, and we mainly focus on cell type, monosporangia, number of pyrenoids, number of plastids and spore germination [12,13]. However, *C. formosanum* can be differentiated from *C. infestans* by larger thallus size (*C. formosanum* 20–250 mm; *C. infestans* 5 mm) and cell dimensions (*C. formosanum* 6–7.5 μm in diameter and 12–15 μm in length; *C. infestans* 2.4–6.4 μm in diameter and 10–17.6 μm in length) (Table 1).

Table 1. Comparison of morphological and anatomical characteristics among *Colaconema formosanum* and the other closely related filamentous *Colaconema* species.

Characters	*C. formosanum*	*C. daviesii*	*C. infestans*
Habit	Marine	Marine	Marine
Host	Sarcodia suae	Chondracanthus chamissoi	Kappaphycus alvarezii
Thallus long (mm)	20–250	-	5
Cell dimensios (μm)	6–7.5 × 12–15	-	2.4–6.4 × 10–17.6
Cell type	Cylindrical	Cylindrical	Cylindrical
Monosporangia	Subspherical, solitary or in clusters	Subspherical	Subspherical
Pyrenoid	Single	-	Single
Plastid	Single parietal	-	Single parietal
Spore (μm)	9–10	-	-
Spore germination	Type II	Type II	-
Phycoerythrin type	R	-	-
Type locality	Ping-Tung, Taiwan	Bio-Bio, Chile	Pitimbu, Brazil
References	This study	[13]	[12]

Although the features above are sufficient to satisfy the taxonomic requirements of the monophyly of *Colaconema* genus and meanwhile show the specific position of the new species [12–14], they still do not accurately confirm the taxonomy. In contrast, it has been demonstrated that the cell dimensions are affected by multiple environmental factors, making it hard to evaluate taxonomy using these features [3,28]. Consequently, molecular phylogenetics is a more reliable method to apply to taxonomy within Nemaliophycidae.

Molecular assisted identification with plastid-encoded *rbc*L and COI-5P have strongly helped in discriminating species of *C. formosanum*, although the information on divergences in *Colaconema* is poorly reported, it is useful to follow other cases, for instance, the *rbc*L divergences between *Halymenia* species are reported to be 0.7–10.6% [29]; the cox1 divergences between *Gelidium* species are reported to be 2.7–5.4% [30]. *C. formosanum* is well segregated from the *C. savianum* (DQ787561) and *Colaconema* sp. (HQ422955 and HQ422983) by pairwise divergences of *rbc*L (7.19–7.83%) and cox1 (14.02–14.19%), respectively. The results illustrate that the interspecific divergences of *C. formosanum* are sufficiently high to warrant *C. formosanum* as a new species.

It is not surprising that the *Colaconema* genus exhibits endophytic behavior, as other examples have been reported worldwide [12–14] (Table 1). Although those cases (for the *Colaconema* genus) only present on macroalgae, evaluating the endophytic phenomenon within other genera at the Nemaliophycidae level suggests that their host specificity may be ecologically determined, and there are reports of attachment to rocks, macroalgae, and some marine animals [31]. This is the first occurrence of an endophytic *Colaconema* species in the red alga *S. suae* in the intertidal zone of Taiwan. In order to reproduce the endophytic phenomenon to exclude the probability of coincidence, in our study we recreated the infection process under artificial culture conditions. When we co-cultured the endophytic *C. formosanum* and the host plant *S. suae* (without pollution), the endophytic phenomenon could be clearly observed under a dissecting microscope after five days of cultivation, indicating that the entire endophytic process occurs rapidly and easily (unpublished data). The host may affect the life phases of an endophytic alga [32]. Interestingly, we found there was no temporal correlation between life phases of *C. formosanum* (endophytic alga) and *S. suae* (host alga), which was confirmed using purified cultures under laboratory conditions and observing the life phases of *C. formosanum*. This result is also supported by previous studies on other endophytic algae, for instance, Correa and McLachlan (1991) isolate *Ulvella operculata* and *U. heteroclada* from *Chondrus crispus* [33]; Gonzalez and Goff (1989) isolate Campylaephora californica and Microcladia coulteri from Egregia menziesii [34].

The probable mechanism, based on a previous study conducted on *C. infestans*, is that the infection changes the biosynthesis pathways of starch enzymes in the Calvin cycle of the host, activating the degradation pathway to produce certain defense compounds. One type of defense compound, uridine diphosphate glucose (UDP-glucose), induces the synthesis of cell wall components instead of floridean starch grains [35]. This phenomenon was also observed in our study; after the host was infected by *C. formosanum*, the growth performance decreased and the algal texture thickened compared with uninfected host samples. However, this observation should be confirmed by further studies.

Some records show that endophytic *Colaconema* species can attach to economically important macroalgae; the initial symptoms are tiny black spots on the surface of the cortical layer of the host, which subsequently become rough tufts. After an outbreak, host cells degrade, forming tiny pores, which allows secondary infection by opportunistic bacteria, with further negative effects on the host and impacts on the aquaculture industry [12,36]. We observed a similar phenomenon in the infection process in this study, but after infection the growth development of the host is unknown, and it would be worthwhile to investigate this further.

In summary, our molecular analysis and morphological observations support the identification of the new species *C. formosanum* This study is the first to describe this alga in Taiwan and report its endophytic habit, attaching to the edible alga *S. suae*. Its endophytic attachment phenomenon and life phases can be reproduced under laboratory conditions. A detailed investigation of the endophytic mechanisms that are generated between *C. formosanum* and *S. suae* and potential applications for this endophytic alga will be necessary topics to explore in future research.

REFERENCES

1. Guiry, G.M. AlgaeBase. World-Wide Electronic Publication, National University of Ireland, Galway. 2021. Available online: https://www.algaebase.org (accessed on 2 March 2021).
2. Lam, D.W.; Verbruggen, H.; Saunders, G.W.; Vis, M.L. Multigene phylogeny of the red algal subclass Nemaliophycidae. *Mol. Phylogenet. Evol.* **2016**, *94*, 730–736. [CrossRef] [PubMed]
3. Soares, L.P.; de Beauclair Guimarães, S.M.P.; Fujii, M.T.; Yoneshigue-Valentin, Y.; Batista, M.G.S.; Yokoya, N.S. *Rhodachlya westii* sp. nov. (Rhodachlyales, Rhodophyta), a new species from Brazil, revealed by an integrative taxonomic approach. *Phycologia* **2020**, *59*, 346–354. [CrossRef]
4. Saunders, G.W.; Hommersand, M.H. Assessing red algal supraordinal diversity and taxonomy in the context of contemporary systematic data. *Am. J. Bot.* **2004**, *91*, 1494–1507. [CrossRef] [PubMed]
5. Harper, J.; Saunders, G. A molecular systematic investigation of the Acrochaetiales (Florideophycidae, Rhodophyta) and related taxa based on nuclear small-subunit ribosomal DNA sequence data. *Eur. J. Phycol.* **1998**, *33*, 221–229. [CrossRef]
6. Woelkerling, W.J. The *Audouinella* (Acrochaetium-Rhodochorton) complex (Rhodophyta): Present perspectives. *Phycologia* **1983**, *22*, 59–92. [CrossRef]
7. Harper, J.T.; Saunders, G.W. A re-classification of the Acrochaetiales based on molecular and morphological data, and establishment of the Colaconematales ord. nov. (Florideophyceae, Rhodophyta). *Eur. J. Phycol.* **2002**, *37*, 463–476. [CrossRef]

8. Garbary, D.; Gabrielson, P. Acrochaetiales (Rhodophyta): Taxonomy and evolution. *Cryptogam. Algol.* **1987**, *8*, 241–252.

9. Batters, E.A.L. On some New British Marine Algae. *Ann. Bot.* **1895**, *9*, 307–321. [CrossRef]

10. Saunders, G.W.; Jackson, C.; Salomaki, E. Phylogenetic analyses of transcriptome data resolve familial assignments for genera of the red-algal Acrochaetiales-*Palmariales* Complex (Nemaliophycidae). *Mol. Phylogenet. Evol.* **2018**, *119*, 151–159. [CrossRef]

11. Skriptsova, A.; Kalita, T. A re-evaluation of *Palmaria* (Palmariaceae, Rhodophyta) in the North-West Pacific. *Eur. J. Phycol.* **2020**, *55*, 266–274. [CrossRef]

12. Araújo, P.G.; Araújo, P.G.; Schmidt Éder, C.; Kreusch, M.G.; Kano, C.H.; Guimarães, S.M.; Bouzon, Z.L.; Fujii, M.; Yokoya, N.S. Ultrastructural, morphological, and molecular characterization of *Colaconema infestans* (Colaconematales, Rhodophyta) and its host *Kappaphycus alvarezii* (Gigartinales, Rhodophyta) cultivated in the Brazilian tropical region. *Environ. Boil. Fishes* **2014**, *26*, 1953–1961. [CrossRef]

13. Montoya, V.; Meynard, A.; Contreras-Porcia, L.; Contador, C.B. Molecular identification, growth, and reproduction of *Colaconema daviesii* (Rhodophyta; Colaconematales) endophyte of the edible red seaweed Chondracanthus chamissoi. *Environ. Boil. Fishes* **2020**, *32*, 3533–3542. [CrossRef]

14. Wynne, M.; Schneider, C. *Colaconema basiramosum* sp. nov (Colaconemataceae, Rhodophyta) from the Sultanate of Oman, Northern Arabian Sea. *Cryptogam. Algol.* **2008**, *29*, 69–80.

15. Shih, C.-C.; Hwang, H.-R.; Chang, C.-I.; Su, H.-M.; Chen, P.-C.; Kuo, H.-M.; Li, P.-J.; Tsui, K.-H.; Lin, Y.-C.; Huang, S.-Y.; et al. Anti-Inflammatory and Antinociceptive Effects of Ethyl Acetate Fraction of an Edible Red Macroalgae *Sarcodia ceylanica*. *Int. J. Mol. Sci.* **2017**, *18*, 2437. [CrossRef] [PubMed]

16. Rodríguez-Prieto, C.; De Clerck, O.; Kitayama, T.; Lin, S.-M. Systematic revision of the widespread species *Sarcodia ceylanica* (Sarcodiaceae, Rhodophyta) in the Indo-Pacific Oceans, including *S. suiae* sp. nov. *Phycologia* **2017**, *56*, 63–76. [CrossRef]

17. Provasoli, L. Media and Prospects for the Cultivation of Marine Algae. In Proceedings of the US-Japan Conference, Hakone, Japan, 12–15 September 1968; pp. 63–75.

18. Freshwater, D.W.; Rueness, J. Phylogenetic relationships of some European *Gelidium* (Gelidiales, Rhodophyta) species, based on rbcL nucleotide sequence analysis. *Phycologia* **1994**, *33*, 187–194. [CrossRef]

19. Lin, S.-M.; Fredericq, S.; Hommersand, M.H. Systematics of the Delesseriaceae (Ceramiales, Rhodophyta) based on large subunit rDNA and rbcl sequences, including the phycodryoideae, subfam. Nov. *J. Phycol.* **2001**, *37*, 881–899. [CrossRef]
20. Saunders, G.W. Applying DNA barcoding to red macroalgae: A preliminary appraisal holds promise for future applications. Philos. Trans. R. Soc. B Biol. Sci. **2005**, 360, 1879–1888. [CrossRef]
21. Edgar, R.C. MUSCLE: Multiple sequence alignment with high accuracy and high throughput. *Nucleic Acids Res.* **2004**, *32*, 1792–1797. [CrossRef]
22. Kumar, S.; Stecher, G.; Li, M.; Knyaz, C.; Tamura, K. MEGA X: Molecular Evolutionary Genetics Analysis across Computing Platforms. *Mol. Biol. Evol.* **2018**, *35*, 1547–1549. [CrossRef]
23. Hall, T.A. BioEdit: A user-friendly biological sequence alignment editor and analysis program for Windows 95/98/NT. *Nucleic Acids Symp. Ser.* **1999**, *41*, 95–98.
24. Trifinopoulos, J.; Nguyen, L.-T.; Von Haeseler, A.; Minh, B.Q. W-IQ-TREE: A fast online phylogenetic tool for maximum likelihood analysis. *Nucleic Acids Res.* **2016**, *44*, W232–W235. [CrossRef]
25. Rambaut, A. FigTree. Version 1.4.2 (Inst. Evol. Biol., Univ. Edinburgh, 2014). Available online: http://tree.bio.ed.ac.uk/software/ figtree/(accessed on 1 February 2021).
26. Nguyen, L.-T.; Schmidt, H.A.; Von Haeseler, A.; Minh, B.Q. IQ-TREE: A Fast and Effective Stochastic Algorithm for Estimating Maximum-Likelihood Phylogenies. *Mol. Biol. Evol.* **2015**, *32*, 268–274. [CrossRef] [PubMed]
27. Ronquist, F.; Teslenko, M.; Van Der Mark, P.; Ayres, D.L.; Darling, A.; Hoehna, S.; Larget, B.; Liu, L.; Suchard, M.A.; Huelsenbeck, J.P. MrBayes 3.2: Efficient Bayesian Phylogenetic Inference and Model Choice Across a Large Model Space. *Syst. Biol.* **2012**, *61*, 539–542. [CrossRef] [PubMed]
28. Garbary, D. The Effects of Temperature on the Growth and Morphology of Some *Audouinella* Spp. (Acrochaetiaceae, Rhodophyta). *Bot. Mar.* **1979**, *22*, 493–498. [CrossRef]
29. Tan, P.-L.; Lim, P.-E.; Lin, S.-M.; Phang, S.-M. *Halymenia johorensis* sp. nov. (Halymeniaceae, Rhodophyta), a new foliose red algal species from Malaysia. *Environ. Boil. Fishes* **2018**, *30*, 187–195. [CrossRef]
30. Boo, G.H.; Kim, K.M. A new species of marine algae from Korea based on morphology and molecular data: *Gelidium palmatum* sp. nov. (Gelidiales, Rhodophyta). *ALGAE* **2020**, *35*, 33–43. [CrossRef]

31. Bentall, G.B.; Rosen, B.H.; Kunz, J.M.; Miller, M.A.; Saunders, G.W.; Laroche, N.L. Characterization of the putatively introduced red alga *Acrochaetium secundatum* (Acrochaetiales, Rhodophyta) growing epizoically on the pelage of *southern sea otters* (Enhydra lutris nereis). *Mar. Mammal Sci.* **2015**, *32*, 753–764. [CrossRef]

32. Andrews, J.H. Pathology of seaweeds: Current status and future prospects. *Cell. Mol. Life Sci.* **1979**, *35*, 429. [CrossRef]

33. Correa, J.; McLachlan, J. Endophytic algae of *Chondrus crispus* (Rhodophyta). IV Effects on the host following infections by Acrochaete operculata and A. heteroclada (Chlorophyta). *Mar. Ecol. Prog. Ser.* **1992**, *81*, 73–87. [CrossRef]

34. González, M.A.; Goff, L.J. The red algal ekpiphytes microcladia coulteri and m. californica (rhodophyceae, ceramiaceae). *J. Phycol.* **1989**, *25*, 558–567. [CrossRef]

35. Bouzon, Z.L.; Ferreira, E.C.; dos Santos, R.W.; Scherner, F.; Horta, P.A.; Maraschin, M.; Schmidt, E.C. Influences of cadmium on fine structure and metabolism of *Hypnea musciformis* (Rhodophyta, Gigartinales) cultivated in vitro. *Protoplasma* **2011**, *249*, 637–650. [CrossRef] [PubMed]

36. Vairappan, C.S.; Chung, C.S.; Hurtado, A.Q.; Soya, F.E.; Lhonneur, G.B.; Critchley, A. Distribution and symptoms of epiphyte infection in major carrageenophyte-producing farms. In Proceedings of the 19th International Seaweed Symposium, Kobe, Japan, 26–31 March 2007;

37. Borowitzka, M.A., Critchley, A.T., Kraan, S., Peters, A., Sjøtun, K., Notoya, M., Eds.; Springer: Dordrecht, The Netherlands, 2009; pp. 27–33.

CHAPTER-14

DETECTION OF THE BENTHIC DINOFLAGELLATES, *OSTREOPSIS* CF. *OVATE* AND *AMPHIDINIUM MASSARTII* (DINOPHYCEAE), USING LOOP-MEDIATED ISOTHERMAL AMPLIFICATION

1. INTRODUCTION

The genus *Ostreopsis* Johannes Schmidt and most of the species of *Amphidinium* Claparède & J. Lachmann are benthic dinoflagellates that grow on macrophytes or are attached to sand or coral rubble [1–3]. Their occurrence has been generally reported in tropical and subtropical seas [4,5]. The global occurrence of some species in these genera has significantly increased over the last decade and is expected to expand to temperate regions. The main harmful effects of some species of benthic dinoflagellates are related to the fact that they not only affect marine life and the aquaculture industry, but also pose a threat to human health [6–10]. *Ostreopsis* species have a particularly rich mucilaginous matrix, and some of them have thus far been reported to produce several toxins [1,11–15]. Moreover, *Ostreopsis* species are potentially toxic and can affect marine organisms and humans through the food web [16]. Their toxins can cause severe irritation to human skin and respiratory problems through aerosolization. They can also cause vomiting, kidney problems, and even death in severe cases [17,18].

Over the past several decades, blooms of benthic dinoflagellates have been observed in temperate to tropical coastal waters, in both the southern and northern hemispheres, whereas the proliferation of *Ostreopsis* cf. *ovata* was found in temperate regions during the summer [5]. The expansion of toxin-producing

Ostreopsis spp. to temperate regions can potentially occur due to ballast water discharge by cargo ships, and, mainly due to the marginal dispersal associated with global warming, can induce bloom formations [19].

The cosmopolitan dinoflagellate genus *Amphidinium* has been found in pelagic and mainly in benthic environments with frequent occurrence [20–23]. Some *Amphidinium* species, particularly *A. caterae*, have been known to produce a number of bioactive com- pounds with cytotoxic or hemolytic activity to marine organisms [24–28]. Moreover, the cytotoxicity of *A.* cf. *massartii* affecting *Artemia salina* mortality has been revealed [29].

Several benthic dinoflagellate species, including *Ostreopsis*, have been reported as potential causative agents of toxic poisoning in Korean coastal waters [30–35]. Recently, a rapid increase in the *O.* cf. *ovata* biomass in Jeju coastal waters has been reported [36]. Most of these reports on the occurrence of benthic dinoflagellates in Korean coastal waters were based on conventional microscopic analysis methods [37–40]. Because blooms in coastal and oceanic waters with negative impacts on environmental health have the possibility of occurrence, several toxigenic dinoflagellates belonging to the genera *Amphidinium*, *Coolia*, *Gambierdiscus*, and *Ostreopsis* (including *O.* cf. *Ovate* and *A.* cf. *massartii*) have been seriously considered as harmful organism candidates in Korea [41].

Following the development of molecular techniques, a number of modifications of polymerase chain reaction (PCR) methods have been established [42,43]. However, whereas these PCR-based assays have provided a reliable, sensitive, and specific tool to detect potentially harmful dinoflagellates, their economic limitations such as dependence on an expensive apparatus, practical limitations such as low amplification efficiencies, and long reaction times ultimately restrict their widespread application. Therefore, the development of rapid, simple, and cost-effective detection methods is still necessary to effectively detect harmful dinoflagellates.

Since Loop-mediated Isothermal Amplification (LAMP) was first introduced by Notomi et al. in 2000 [44], more than 100 LAMP detection methods have been developed for animal pathogens, including humans, and are ideally used due to the advantages of low cost and high sensitivity [45–47]. Because LAMP is performed under isothermal conditions, this method facilitates the amplification of only a few copies of initial DNA to obtain approximately 10^9 copies in less than 1 h, which can be visualized after the reaction using SYBR Green dye. The development of a LAMP assay, therefore, enhances the detection of various dinoflagellate species [48,49].

Owing to the recent increase in the abundance of benthic dinoflagellates in Korean coastal waters, their rapid detection and extensive monitoring are required. In this study, we developed a highly specific LAMP assay for the sensitive detection of two species of benthic dinoflagellates, *O.* cf. *ovata* and *A. massartii*. Moreover, we established the proper LAMP conditions for each species and applied them to the field samples to verify the sensitivity.

2. MATERIALS AND METHODS

2.1 Sampling Site and Establishment of Dinoflagellate Strains

Macrophytes samples were collected by scuba divers in May 2018 within a water depth of 10 m at Seongsan, Jeju Island (33°27.35′ N, 126°56.01′ E). The seawater temperature and salinity were recorded as 17.4°C and 33.2, respectively, at that time. The collected macrophytes were transferred into a 1 L bottle, which was filled with filtered seawater, and then shaken vigorously to detach the attached dinoflagellates. The samples were filtered by a 100 μm mesh to separate the macrophyte from seawater, and then transferred to the laboratory.

To establish a single cell strain, 5 mL of the sample was placed in a six-well plate, and single-cell isolation was performed under a dissecting microscope (SZX10, Olympus, Tokyo, Japan). After the clonal cultures of *O.* cf. *ovata* and *A. massartii* were established, the strains were transferred into 30 mL flasks containing fresh f/2-Si medium. The cultures were placed under white fluorescent lights at 22°C with a continuous illumination of 20 $\mu E \cdot m^{-2} \cdot s^{-1}$.

2.2 DNA Extraction and Species Identification

The DNA sequences of these cells were analysed when the concentration of each strain was more than 10^3 cells mL^{-1}. The dense culture (10 mL) was centrifuged at 10,000× *g* for 3 min at room temperature and the pellet was used for DNA extraction. Genomic DNA was extracted using an AccuPrep Genomic DNA Extraction Kit (BIONEER, Daejeon, Korea). The quality and purity of the DNA were assessed using agarose gel electrophoresis and spectro-photometry. For species identification, the SSU, ITS1, 5.8S, ITS2, and LSU rDNA sequences were amplified using universal eukaryotic primers [50,51]; the obtained DNA sequences were confirmed using BLAST (https://blast.ncbi.nlm.nih.gov/Blast.cgi, NCBI (accessed on 3 September 2018). The rDNA sequences of the two species were identical to those of the Korean strain (*Ostreopsis* cf. *ovata* [52]; *Amphidinium massartii* [53]).

2.3 Construction of LAMP Primers

Primers for LAMP assay were designed based on the internal transcribed spacer region (ITS gene) of *O.* cf. *ovata* (HE793379.2) and the D1–D6 region of large subunit rRNA (LSU gene) of *A. massartii* (AY455670.1) using Primer Explorer V5 (http://primerexplorer.jp/lampv5e/index.html (accessed on 17 September 2018) software. A total of six distinct sequences (B1, B2, B3, F1, F2, and F3) in the target DNA were designed for LAMP assay (B3, F3, backward inner primer (BIP), and forward inner primer (FIP)). Primer details are listed in Table 1. The primer sequences and their respective binding sites are shown in Figure 1. Although LAMP reaction could be accelated using additional primers (termed loop primers), no suitable loop primer was found within the target gene. The LAMP reactions of the designed primer sets were confirmed as typical ladder-like patterns by gel electrophoresis as well as direct colour change by SYBR Green dye [44].

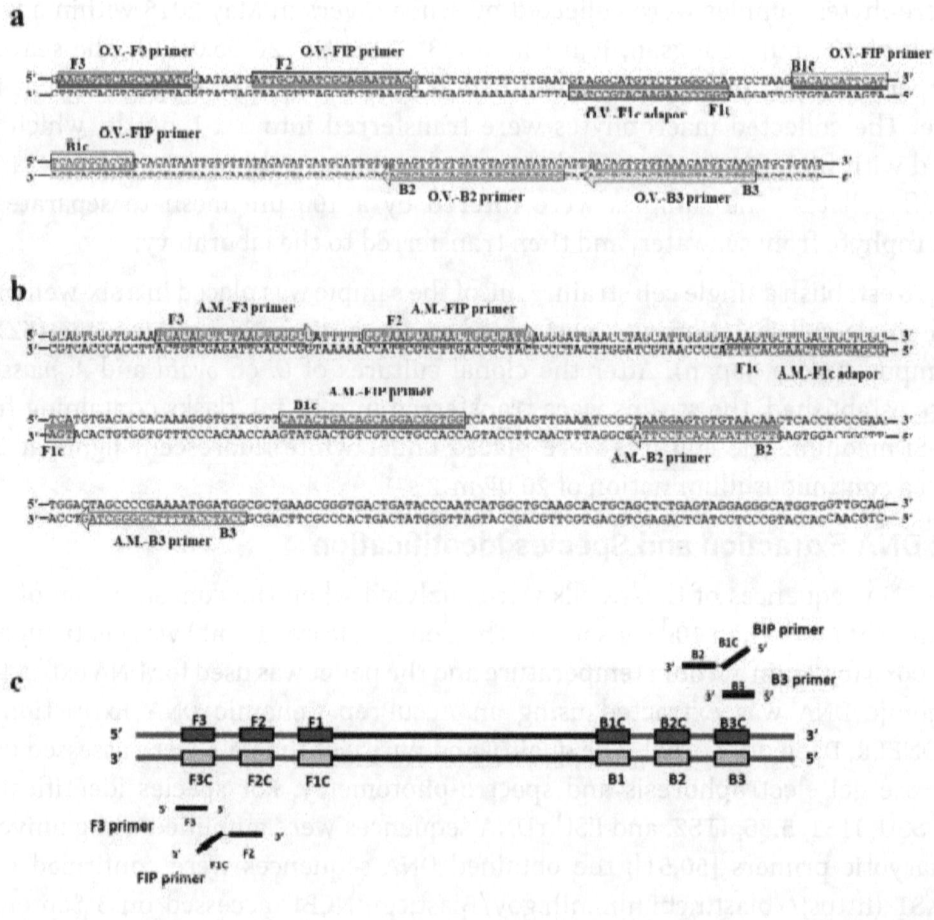

Figure 1. Nucleotide sequence used to design the primers for loop-mediated isothermal amplification (LAMP). The square boxes indicate the recognition sequences of the primers. The right arrow indicates that a sense sequence was used for the primers. The left arrow indicates that a complementary sequence was used for the primer. (a) Primer sequences for *Ostreopsis* cf. *ovata.* (b) Primer sequences for *Amphidinium massartii.* (c) Schematic representation of the primers used in this study. Construction of two outer (F3 and B3) and two inner (forward inner primer (FIP) and backward inner primer (BIP)) primers for loop-mediated isothermal amplification (LAMP).

Table 1. Oligonucleotide primers developed for detecting *Ostreopsis* cf. *ovata* and *Amphidinium* massartii using loop-mediated isothermal amplification (LAMP), PCR, and qPCR.

Species Name	LAMP	Primers	Sequence (5_-3_)
Ostreopsis cf. ovata	PCR	F3	AAGAGTGCAGCCAAATGC
		B3	GTTGACATGTTTACACACTGTA
		FIP (F1c-F2)	TGGCCCAAGAACATGCCTACATTGCAAATCG-CAGAATTACG
		BIP (B1c-B2)	GACATCATTCATTCAGTGCACCAGTATATACTA-CATCACACACTCAC
		Forward	TGATGTGTACAACTCCCTT
		Reverse	GAATGATGTCCTTAGGAATGG
	qPCR	Forward	GGCCATTCCTAAGGACATCA
		Reverse	TGGCCATATACAGCATGTTGAC
		Probe	6-FAM-ATCATGCATTGTGTGAGTGTGTGATGT-BHQ-1
Amphidinium massartii	LAMP	F3	TGACAGCTCTAAGTGGGCC
		B3	CCATCCATTTTCGGGGCTAG
		FIP (F1c-F2)	TGAGCGAGCAGTCAAGCACTTTGGTAAGCAGAACT-GGCGATG
		BIP (B1c-B2)	CATACTGACAGCAGGACGGTGGATTCGGCAGGT-GAGTTGTT
	PCR	Forward	AGTGGGTGGAATGACAGCT
		Reverse	CATCCATTTTCGGGGCTAGT

2.4 Optimization of LAMP Reaction Conditions

To determine the optimal conditions of LAMP assay for detecting *O.* cf. *ovata* and *A. massartii*, experiments were performed in which the different variables known to affect this assay, such as reaction times, temperatures, and buffer composition concentrations (dNTP, *Bst* polymerase, and $MgSO_4$), were altered (Table 2). The reaction conditions and buffer compositions were varied to include ranges of 0–16 units of *Bst* DNA polymerase, 0–18 mM of $MgSO_4$, 0–4.5 mM of dNTP each, 54–64 C of temperatures, and 15–90 min of reaction time. After that, the LAMP reaction mixture contained 20 mM tris-HCl, 50 mM KCl, 10 mM $(NH_4)_2SO_4$, 1.0 μM of each FIP and BIP, 0.4 μM of each of the outer F3 and B3 primers, and 4 μL of genomic DNA extracted from 100 cells using syringe filter set, which resulted in a final reaction volume of 20 μL.

Table 2. Conditions and tested ranges of each factor for optimizing LAMP.

Conditions	Tested Ranges					
Temperature (°C)	54	56	58	60	62	62
Reaction Time (min.)	15	30	45	60	75	90
dNTP (mM)	0	0.5	1.0	1.8	2.5	45
Bst DNA polymerase (units)	0	2	4	8		
MgSO4 (mM)	0 2	4 6	8 10	12	14	16

The detectable ranges for the species (*O.* cf. *ovata* or *A. massartii*) included the con- centraions of compositon in the commercial LAMP mix (2× LAMP Maeter mix, NEB E1700L): 8 U *Bst* polymerase, 1.4 mM dNTPs, and 8 mM MgSO4. Therefore, the following tests (sensitivity, specific-specificity, and field samples application) were performed with the commercial LAMP mix, at 62°C in *O.* cf. *ovata* and 60°C *A. massartii* for 60 min. The reaction temperature was determined to the median between the minimum reaction temperature and the manufacturer's recommended temperature.

By adding 1 μL of fluorescent dye SYBR Green I (×1000, Thermo Fisher Scientific, Waltham, MA, USA) after completing the LAMP reaction, the positive reactions could be confirmed with the naked eye. The colour of positive reactions turned orange to yellow-green, while negative reactions remained orange. Moreover, by performing the gel electrophoresis, the LAMP products could be double-checked. The positive sample of LAMP showed a band in the ladder pattern, while the negative sample had no band in the gel. The photographs of the agarose gel were taken using an Azure c200 gel imaging workstation (Azure Biosystems, Dublin, CA, USA).

2.5 *Sensitivity of LAMP*

2.5.1 Sensitivity Test from Extracted DNA

The sensitivity tests were evaluated using DNA templates extracted from the clonal cultures of *O.* cf. *ovata* or *A. massartii* each (3×10^3 cells of *O.* cf. *ovata* and 10^4 cells of *A. massartii*). The cells were concentrated with GF/C filters (Whatman® glass microfiber filters, Little Chalfont, UK) using a syringe filter, and then the genomic DNAs were extracted using an AccuPrep Genomic DNA Extraction Kit (BIONEER, Daejeon, Korea). The extracted DNA was diluted serially.

2.5.2 Sensitivity Test from 10 Cells Directly

To evaluate the detection limit more precisely, we used the cells isolated from the clonal cultures as the DNA templates directly without the column-based DNA extraction processing. Ten cells were isolated each from the culture of *O.* cf. *ovata* or *A. massartii* using a sterile micropipette under a dissecting microscope (SZX10,

Olympus, Tokyo, Japan). Individual cells were transferred to sterile seawater to remove any contaminants and then suspended in approximately 1 µL of TE buffer. The ten cells were lysed by freeze-thawing process and diluted ten-fold serially, by adding sterile distilled water (Invitrogen, Carlsbad, CA, USA), from 1 to 10^{-4}. All samples were frozen at -20 °C in preparation for the test.

2.5.3 Comparison of the Detection Sensitivity with Other Molecular Assays

We carried out the conventional PCR and qPCR (quantitative PCR) for comparison with LAMP. The PCR was performed on a Mastercycler Nexus (Eppendorf, Hamburg, Germany) in a 20 µL PCR mixture containing 4 µL of HiPi PCR premix, 1 µL of 10 µM each forward/reverse primers, 10 µL of sterile distilled water, and 4 µL of DNA template. PCR conditions of *O.* cf. *ovata* and *A. massartii* were as follows: an initial denaturation step of 95°C for 15 min, followed by 35 cycles of denaturation at 94°C for 30 s, and then an annealing step at 55°C for 30 s, and elongation at 72°C for 50 s, followed by a final extension step 72°C for 7 min. The PCR products were analysed on a 1.5% agarose gel using gel electrophoresis.

Table 3. List of dinoflagellate species used to confirmed the species specificity of the two sets of LAMP primers (*O.* cf. *ovata* and *A. massartii*).

Species Name	Strain	Remarks
Alexandrium tamarense	S041118-JSP	Busan, Korea
Amphidinium carterae	CCMP1314	USA
Coolia malayensis	JSGP-CM	Seoguipo, Korea
Gambierdiscus jejuensis	JSGP201117-GJ	Seoguipo, Korea
Gymnodinium aureolum	GASH1103	Kunsan, Korea
Heterocapsasteinii	SMS-MSJ	Masan bay, Korea
Ostreopsis lenticularis	JSGP201117-OL	Seoguipo, Korea
Prorocentrum minimum	WKS-BJH	Kunsan, Korea
Prorocentrum koreanum	JSS-KEJ	Jeju, Korea
Scrippsiellaacuminata	US1-G6	Ulsan, Korea
Symbiodinium voratum	JSGP201117-SV	Seoguipo, Korea
Ostreopsis cf. ovata	JSS200917-OO	Seongsan, Korea
Amphidinium massartii	J-MSJ-AM	Jeju, Korea

Quantitative PCR (qPCR) was performed in duplicates with a PCR max Eco 48 real-time PCR system (PCR max, Stone, UK) using qPCRBIO probe Mix No-ROX (PCR biosystms, London, England) following the manufacturer's guidelines: 10 µL of 2× qPCRBIO Probe Mix, 1 µL of 10 µM each of forward/reverse primers, 0.5 µL of 10 μM probe labelled at the 5′ and 3′ ends with the fluorescent dye 6-FAM and BHQ-1, 4.5 μL of distilled water, and 3 μL of DNA template. The primers and TaqMan probes (coupled to for qPCR were also designed to amplify the ITS gene (Table 3). Thermal cycling was performed under the following conditions: 95°C for 3 min of initial denaturation, then 40 cycles of amplification at 95°C for 10 s and 60°C for 30 s. The standard templates (3×10^3 cells of *O.* cf. *ovata*) were concentrated with GF/C filters using a syringe filter. Following DNA extraction and qPCR methods were as described above. The eluted genomic DNA was diluted to the equivalent of 1000, 300, 100, 30, and 10 cells. In the application of the field samples, the relatively quantified cell concentrations were calculated according to Park et al. [36].

2.6 Confirmation of Species-Specific LAMP Primers

To confirm species specificity and exclude the possibility of false positives about the two sets of LAMP primers (*O.* cf. *ovata* and *A. massartii*), DNAs of eleven dinoflagellate species (*Alexandrium tamarense* (Lebor) Balech; *Amphidinium carterae* Hulbert; *Coolia malayen- sis* Leaw, P.-T. Lim & Usup; *Gambierdiscus jejuensis* S. H. Jang & H. J. Jeong; *Gymnodinium aureolum* (Hulburt) G. Hansen; *Heterocapsa steinii* Tillmann, Gottschling, Hoppenrath, Kus- ber & Elbrächter; *Ostreopsis lenticularis* Y. Fukuyo; *Prorocentrum minimum* (Pavillard) Schiller; *Prorocentrum koreanum* M.-S. Han, S. Y. Cho & P. Wang; *Scrippsiella acuminata* (Ehrenberg) Kretschmann, Elbrächter, Zinssmeister, S. Soehner, Kirsch, Kusber & Gottschling); and *Symbiodinium voratum* Jeong, Lee, Kang, LaJeunesse) were extracted and used as a template for the LAMP reaction (Table 3). All strains, mentioned above, have been established and maintained in our cell culture laboratory (at 22°C under continuos illumination of 20 $\mu E \cdot m^{-2} \cdot s^{-1}$) of the Advanced Institute of Convergence Technology (AICT, Suwon, Korea), excluding *Amphidinium carterae* strain (CCMP1314). Two thousand cells from each species were isolated from the dense culture media of the dinoflagellate strains and mixed in a tube. The mixture was concentrated with GF/C filters using a syringe filter. Then, the genomic DNA was extracted using an AccuPrep Genomic DNA extraction kit (Bioneer, Daejeon, Korea). The positive mixture including either *O.* cf. *ovata* or *A. massartii* and the negative mixture without *O.* cf. *ovata* and *A. massartii* were used as the controls. The LAMP reaction was conducted at 62°C in *O.* cf. *ovata* and 60°C in *A. massartii* for 60 min.

All DNA templates tested in this work were from fresh samples only that had not been stored for more than two days. The amplification results were obtained following electrophoresis, and gels were stained to verify specificity.

2.7 Testing of Field Samples

To test the *O.* cf. *ovata* LAMP primers using field samples, four different macroalgal species were randomly collected from two sampling sites on Jeju Island (Sasu and

Seongsan) in 2019 (January, March, and June) (Figure 2). Macroalgae living within 10 m of water depth were collected by scuba divers and pooled in a 1 L bottle with ambient seawater. The bottle was shaken vigorously to detach the benthic dinoflagellates. After that, macroalgae were frozen on dry ice immediately and the water sample was filtered using a 100 μm mesh to remove macroalgal particles and zooplankton. Each water sample (50 mL) was concentrated with a GF/C filter using a syringe filter. The GF/C filter was placed in a 1.5 mL tube and kept frozen immediately until DNA extraction. Genomic DNAs were extracted with a Bioneer AccuPrep® Genomic DNA Extraction kit, and a LAMP assay was performed. The LAMP reaction was conducted at 62°C for 60 min.

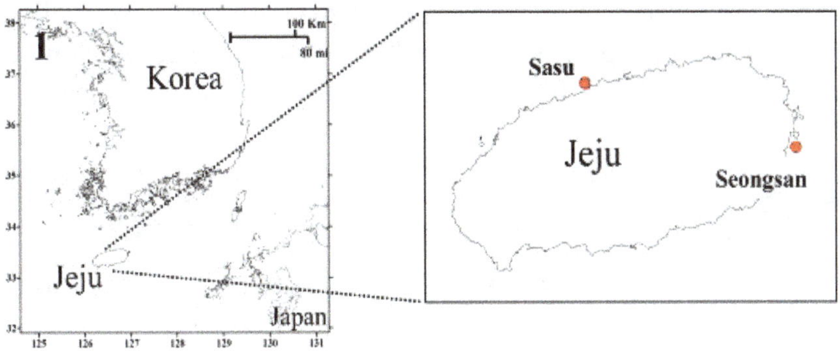

Figure 2. Map of Korea and Jeju Island indicating the sampling sites.

To compare microscopic detection, 300 mL of water sample was fixed with formalin (final concentration, 1%). The fixed samples were concentrated for microscopic observation. Fixed samples were left overnight to allow the cells to sink, and the supernatant was removed to adjust the final volume to 50 mL. A solution of Calcofluor (Sigma-Aldrich, St. Louis, MO, USA) was added at a final concentration of 10 μg mL⁻1 for 1–2 min in the dark before observation. Then, 1 mL samples were observed in a Sedgewick-Rafter chamber (SPI Supplies, West Chester, PA, USA) at 100× magnification using an epifluorescence microscope (BX 53, Olympus, Tokyo, Japan) to confirm the presence of cells. For the same reason, dense plankton samples were collected at each sampling sites by towing a 20 μm mesh sized plankton net along the water column for 1 min. Collected samples were fixed with formalin and observed with light and epifluorescence microscopes (BX 53, Olympus, Tokyo, Japan).

3. RESULTS

3.1 Optimization of LAMP Conditions

LAMP reaction in *O.* cf. *ovata* required at least 45 min and occurred at a temperature of 56°C or higher. We confirmed that LAMP reactions occurred even at two units of *Bst* polymerase per reaction tube. To find an optimal dNTP concentration, each 0–4.5 mM of dNTPs were tested in 20 μL of the reaction mixture. LAMP reaction occurred in 1.0–2.5 mM dNTPs, but it did not occur in 0.5 and 4.5 mM dNTPs. In the concentration test of MgSO4 per reaction tube, LAMP reaction occurred from 6 to 16 mM of MgSO4 (Figure 3a). LAMP reaction in *A. massartii* showed that the reaction was initiated at least after 30 min and at all temperatures from 54 to 64 °C. The required concentration of dNTP per reaction tube was from 1.0 (very weak) to 2.5 mM. Compared to what was confirmed in *O.* cf. *ovata* LAMP reactions, LAMP reactions in *A. massartii* were in a narrow range of MgSO4 concentration, from 8 to 14 uM per reaction tube (Figure 3b). Too low or too high dNTP and MgSO4 concentrations resulted in false-negative results in both *O.* cf. *ovata* and *A. massartii.* These composition (*Bst* polymerase, dNTPs, MgSO4) concentration tests were performed at 62°C in *O.* cf. *ovata* and 60°C *A. massartii* for 60 min.

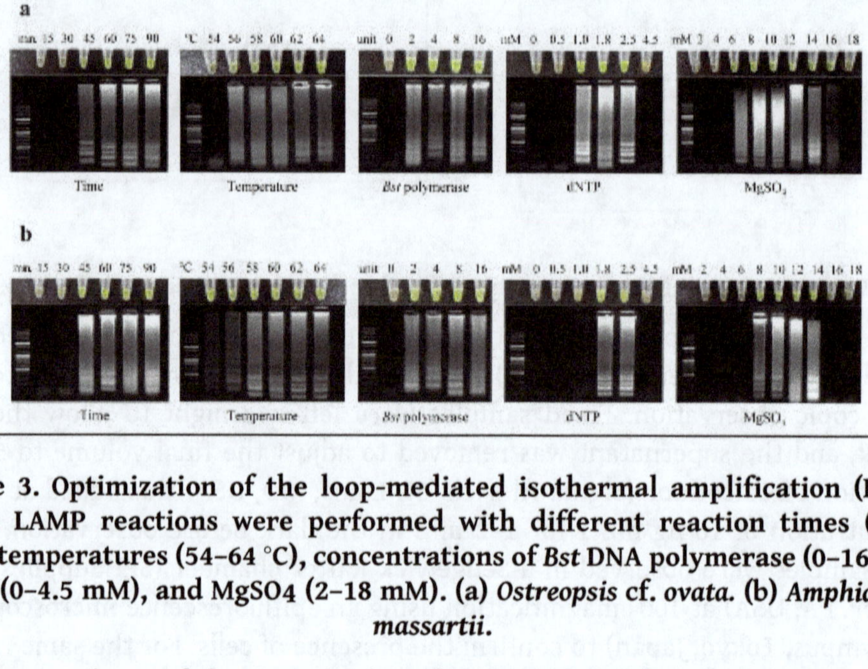

Figure 3. Optimization of the loop-mediated isothermal amplification (LAMP) assay. LAMP reactions were performed with different reaction times (15-90 min), temperatures (54-64 °C), concentrations of *Bst* DNA polymerase (0–16 unit), dNTP (0–4.5 mM), and MgSO4 (2–18 mM). (a) *Ostreopsis* cf. *ovata.* (b) *Amphidinium massartii.*

3.2 Comparison of the Sensitivity of LAMP and Other Molecular Assays

3.2.1 *Ostreopsis.* cf. *ovata*

When tested with extracted DNA from 3 × 10^3 cells concentrated with GF/C filter, the detection limit of LAMP was 10 cells (Figure 4a). Meanwhile, when we tested using lysed DNA from 10 cells isolated directly, sensitivity was increased as 0.1 cell of *O.* cf. *ovata* (Figure 4b). The sensitivity was higher when performing LAMP with isolated cells than with extracted DNA. The sensitivity of qPCR was similar to the LAMP result while the PCR was ten-fold more sensitive than the LAMP assay

(Figure 4a). The qPCR was only performed for O. cf. *ovata* as suitable primers and probes, but could not be designed for *A. massarti.*

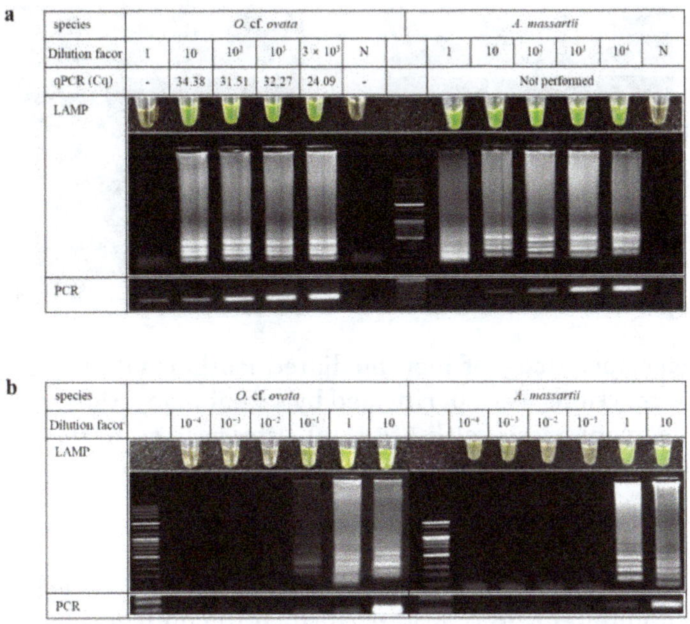

Figure 4. Comparison of the detection sensitivity with loop-mediated isothermal amplification (LAMP) and conventional PCR. (a) Samples in each line were serially diluted (from 3 × 10^3 cells in O. cf. *ovata* and 10^4 cells in *A. massartii*) from genomic DNA which was extracted from the sample concentrated with GF/C filter. (b) Samples in each line were serially diluted from directly isolated ten cells without the column-based DNA extraction processing. cq = quantification cycle.

3.2.2 Amphidinium Massartii

The detecton limit of *A. massartii* was 1 cell in both DNA extracted from 10^4 cells concentrated with GF/C filter (Figure 4a) and DNA lysed from 10 cells isolated directly (Figure 4b). The PCR detection limit of *A. massartii* was similar to that of the LAMP results.

3.3 Confirmation of Species-Specific LAMP Primers

By performing LAMP reactions on the DNA mixture containing eleven strains of dinoflagellates (*A. tamarense, A. carterae, C. malayensis, G. jejuensis, G. aureolum, H. steinii, O. lenticularis, P. minimum, P. koreanum, S. acuminata,* and *S. voratum*), we confirmed that the positive LAMP reactions were only observed for target species and positive control mixtures containing either O. cf. *ovata* or *A. massartii,* while no products were amplified in both negative control and negative control mixture (Figure 5). Therefore, the results showed that the designed LAMP primers for the two species react specifically.

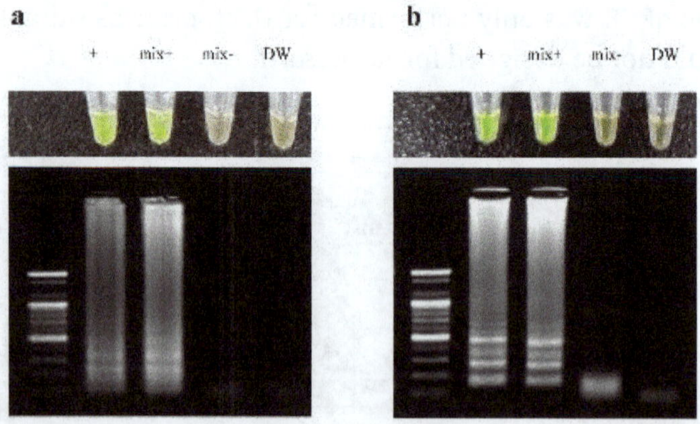

Figure 5. Species-specificity of loop-mediated isothermal amplification (LAMP) primers. LAMP reactions were performed by visual inspection with diluted SYBR Green I and electrophoresis with EtBr. (a) *Ostreopsis* cf. *ovata.* (b) *Amphidinium massartii.*

3.4. Application to Field Samples

The LAMP assay for *O.* cf. *ovata* was applied using the field samples collected from Jeju, Korea. PCR and qPCR were also performed to confirm the accuracy and the reliability of LAMP results. The result of LAMP showed that *O.* cf. *ovata* was detected in both Sasu and Seongsan in January and April, but detected only in Seongsan in June. Specifically, the positive LAMP results of *O.* cf. *ovata* were obtained only in samples estimated to be more than 10 cells mL^{-}1 in qPCR assay, while PCR results showed that they were amplified even at less than 10 cells mL^{-}1. In the microscopy examination results, *O.* cf. *ovata* was not observed in some of the samples (Seongsan no.4 in Jan. and no.1, 2 in Jun.) in which it was confirmed by all molecular assays, including LAMP (Figure 6).

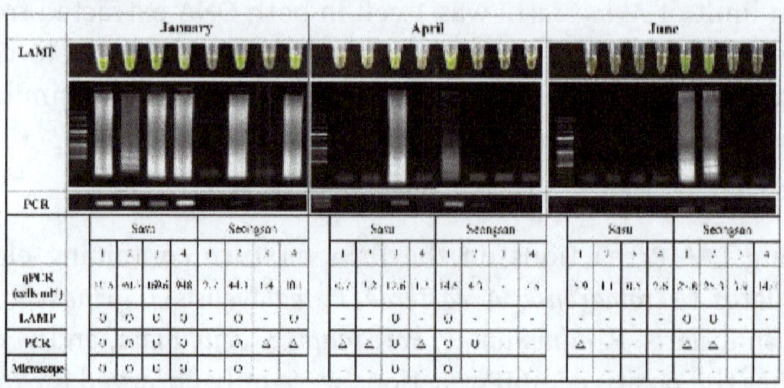

Figure 6. Field sample detection using loop-mediated isothermal amplification (LAMP) primers. Detection of each species was confirmed using a microscope, LAMP, PCR, and qPCR in seawater samples from Sasu and Seongsan of Jeju Island, 2019. The numbers (1 to 4) in the table indicate the randomly collected macrophytes (unidentified).

4. DISCUSSION

Traditional PCR and qPCR assays using *Taq* polymerase have already been used to detect or quantitatively evaluate many dinoflagellates with high sensitivity and specificity. However, these assays require special and expensive devices, such as a thermal cycler. LAMP assay using *Bst* polymerase is a simple and fast molecular diagnostic tool that does not require an expensive thermocycler because *Bst* polymerase amplifies nucleic acids under isothermal conditions [54]. Until now, attempts have been made to apply LAMP to some dinoflagellates related to harmful algal blooms, but no studies have been conducted on the early detection of benthic dinoflagellates. Novel and highly specific LAMP assays for the sensitive detection of benthic dinoflagellates *O.* cf. *ovata* and *A. massartii* were established in this study. Moreover, another purpose of this work was to evaluate the detection capabilities of LAMP compared to other molecular assays. The LAMP assay was sensitive enough to detect for *O.* cf. *ovata* and *A. massartii*, similar to both PCR and qPCR results.

The LAMP assay provides several advantages over a conventional PCR assay in that it can amplify target DNA sequence under isothermal conditions faster (≤1 h) and does not require the use of sophisticated or expensive equipment. Thus, in laboratories or isolated areas where equipment is minimal, the entire reaction process can still be performed using only a heat block or temperature-controlled water tank. However, the most substantial feature is the ability to visually detect amplification through the addition of fluorescent dyes such as SYBR Green I [55], making it suitable for implementation in rapid field trials.

Recently, LAMP has been applied in the detection of marine dinoflagellates such as *Alexandrium* sp., *Karenia mikimotoi*, and *Prorocentrum* sp., and many studies have been conducted for the rapid detection of harmful dinoflagellates [56–58]. To date, most LAMP methods for detecting dinoflagellates have been limited to planktonic dinoflagellates. In the present study, we successfully detected benthic dinoflagellates and confirmed that the detection threshold of LAMP was similar to that of PCR and qPCR assays. When the LAMP method was used, we completed the analysis and obtained the results in situ with high accuracy within 60 min to detect the occurrence of *O.* cf. *ovata* and *A. massartii*. Generally, LAMP positive products could be confirmed in the ladder pattern band through the gel electrophoresis. However, we found that the samples kept frozen for a long time (more than a month) or repeated freezing-thawing several times caused false-positive results with a smear band rather than a ladder pattern. This is why we only used the fresh samples to avoid false-positive results.

In the field samples detached from macrophytes, we observed not only *O.* cf. *ovata* and *A. massartii* but also diverse benthic dinoflagellates (such as *Ostreopsis lenticularis*, *Coolia* spp., *Gambierdiscus* spp., *Prorocentrum* spp.) with high abundances, but there was no LAMP reaction from the samples without *O.* cf. *ovata* and *A. massartii*. The LAMP sets in this study have been proven to react only with *O.* cf. *ovata* or *A. massartii*.

5. CONCLUSIONS

In this study, we developed a LAMP method for in situ and precise detection of benthic dinoflagellate species, *O.* cf. *ovata* and *A. massartii*. This provides a useful detection technique for field research, owing to the use of simple reaction conditions and inexpensive equipment and the lack of a need for an expert phycologist for algal identification by microscopy. Thus, the LAMP method can be applied to monitor the occurrence and distribution of benthic dinoflagellates in large-scale environments such as the coastal area of Korea.

REFERENCES

1. Mangialajo, L.; Chiantore, M.; Cattaneo-Vietti, R. Loss of fucoid algae along a gradient of urbanisation, and structure of benthic assemblages. *Mar. Ecol. Prog. Ser.* **2008**, *358*, 63–74. [CrossRef]
2. Shah, M.M.R.; An, S.-J.; Lee, J.-B. Seasonal abundance of epiphytic dinoflagellates around coastal waters of Jeju Island, Korea. *J. Mar. Sci. Technol.* **2013**, *21*, 156–165. [CrossRef]
3. Accoroni, S.; Romagnoli, T.; Penna, A.; Capellacci, S.; Ciminiello, P.; Dell'Aversano, C.; Tartaglione, L.; Abboud-Abi Saab, M.; Giussani, V.; Asnaghi, V.; et al. *Ostreopsis fattorussoi* sp. nov. (Dinophyceae), a new benthic toxic *Ostreopsis* species from the eastern Mediterranean Sea. *J. Phycol.* **2016**, *52*, 1064–1084. [CrossRef] [PubMed]
4. Murray, S.; Patterson, D. The benthic dinoflagellate genus *Amphidinium* in south-eastern Australian waters, including three new species. *Eur. J. Phycol.* **2002**, *37*, 279–298. [CrossRef]
5. Rhodes, L. World-wide occurrence of the toxic dinoflagellate genus *Ostreopsis* Schmidt. *Toxicon* **2011**, *57*, 400–407. [CrossRef] [PubMed]
6. Delgado, G.; Lechuga-Devéze, C.H.; Popowski, G.; Troccoli, L.; Salinas, C.A. Epiphytic dinoflagellates associated with ciguatera in the northwestern coast of Cuba. *Rev. Biol. Trop.* **2006**, *54*, 299–310. [CrossRef]
7. Friedman, M.A.; Fleming, L.E.; Fernandez, M.; Bienfang, P.; Schrank, K.; Dickey, R.; Bottein, M.-Y.; Backer, L.; Ayyar, R.; Weisman, R.; et al. Ciguatera fish poisoning: Treatment, prevention and management. *Mar. Drugs* **2008**, *6*, 456–479. [CrossRef]

8. Pagliara, P.; Caroppo, C. Toxicity assessment of *Amphidinium carterae*, *Coolia* cfr. *monotis* and *Ostreopsis* cfr. *ovata* (Dinophyta) isolated from the northern Ionian Sea (Mediterranean Sea). *Toxicon* **2012**, *60*, 1203–1214. [CrossRef]

9. Berdalet, E.; Fleming, L.E.; Gowen, R.; Davidson, K.; Hess, P.; Backer, L.C.; Moore, S.K.; Hoagland, P.; Enevoldsen, H. Marine harmful algal blooms, human health and wellbeing: Challenges and opportunities in the 21st century. *J. Mar. Biol. Assoc. UK* **2016**, *96*, 61–91. [CrossRef]

10. Berdalet, E.; Tester, P.A.; Chinain, M.; Fraga, S.; Lemée, R.; Litaker, W.; Penna, A.; Usup, G.; Vila, M.; Zingone, A. Harmful algal blooms in benthic systems: Recent progress and future research. *Oceanography* **2017**, *30*, 36–45. [CrossRef]

11. Lenoir, S.; Ten-Hage, L.; Turquet, J.; Quod, J.-P.; Bernard, C.; Hennion, M.-C. First evidence of palytoxin analogues from an *Ostreopsis mascarenensis* (Dinophyceae) benthic bloom in southwestern Indian Ocean. *J. Phycol.* **2004**, *40*, 1042–1051. [CrossRef]

12. Mohammad-Noor, N.; Daugbjerg, N.; Moestrup, Ø.; Anton, A. Marine epibenthic dinoflagellates from Malaysia—A study of live cultures and preserved samples based on light and scanning electron microscopy. *Nord. J. Bot.* **2007**, *24*, 629–690. [CrossRef]

13. Ciminiello, P.; Dell!Aversano, C.; Iacovo, E.; Fattorusso, E.; Forino, M.; Tartaglione, L.; Yasumoto, T.; Battocchi, C.; Giacobbe, M.; Amorim, A.; et al. Investigation of toxin profile of Mediterranean and Atlantic strains of *Ostreopsis* cf. *siamensis* Dinophyceae) by liquid chromatography–high resolution mass spectrometry. *Harmful Algae* **2013**, *23*, 19–27. [CrossRef]

14. Ben-Gharbia, H.; Yahia, O.K.; Amzil, Z.; Chomérat, N.; Abadie, E.; Masseret, E.; Sibat, M.; Triki, Z.H.; Nouri, H.; Laabir, M. Toxicity and growth assessments of three thermophilic benthic Dinoflagellates (*Ostreopsis* cf. *ovata*, *Prorocentrum lima* and *Coolia monotis*) developing in the Southern Mediterranean Basin. *Toxins* **2016**, *8*, 297. [CrossRef] [PubMed]

15. Lassus, P.; Chomérat, N.; Hess, P.; Nézan, E. *Toxic and Harmful Microalgae of the World Ocean/Micro-Algues Toxiques et Nuisibles de l'Océan Mondial*; International Society for the Study of Harmful Algae, United Nations Educational, Scientific and Cultural Organisation: Copenhagen, Denmark, 2016; pp. 111–170, ISBN 978-87-990827-6-6.

16. Aligizaki, K.; Katikou, P.; Nikolaidis, G.; Panou, A. First episode of shellfish contamination by palytoxin-like compounds from *Ostreopsis* species (Aegean Sea, Greece). *Toxicon* **2008**, *51*, 418–427. [CrossRef] [PubMed]

17. Deeds, J.R.; Schwartz, M.D. Human risk associated with palytoxin exposure. *Toxicon* **2010**, *56*, 150–162. [CrossRef]

18. Tester, P.A.; Litaker, R.W.; Berdalet, E. Climate change and harmful benthic microalgae. *Harmful Algae* **2020**, *91*, 101655. [CrossRef]
19. Shears, N.T.; Ross, P.M. Blooms of benthic dinoflagellates of the genus *Ostreopsis*; an increasing and ecologically important phenomenon on temperate reefs in New Zealand and worldwide. *Harmful Algae* **2009**, *8*, 916–925. [CrossRef]
20. López-Flores, R.; Boix, D.; Badosa, A.; Brucet, S.; Quintana, X.D. Is Mirtox toxicity related to potentially harmful algae proliferation in Mediterranean salt marshes? *Limnetica* **2010**, *29*, 257–268. [CrossRef]
21. Murray, S.A.; Kohli, G.S.; Farrell, H.; Spiers, Z.B.; Place, A.R.; Dorantes-Aranda, J.J.; Ruszczyk, J. A fish kii associated with a bloom of *Amphidinium carterae* in a coastal lagoon in Sydney, Australia. *Harmful Algae* **2015**, *49*, 19–28. [CrossRef]
22. Yong, H.L.; Mustapa, N.I.; Lee, L.K.; Lim, Z.F.; Tan, T.H.; Usup, G.; Gu, H.; Litaker, R.W.; Tester, P.; Lim, P.T.; et al. Habitat complexity affects benthic harmful dinoflagellate assemblages in the fringing reef of Rawa Island, Malaysia. *Harmful Algae* **2018**, *78*, 56–68. [CrossRef]
23. Lee, L.K.; Lim, Z.F.; Gu, H.; Chan, L.L.; Litaker, R.W.; Tester, P.A.; Leaw, C.P.; Lim, P.T. Effects of substratum and depth on benthic harmful dinoflagellate assemblages. *Sci. Rep.* **2020**, *10*, 11251. [CrossRef]
24. Kobayashi, J.; Ishibashi, M.; Nakamura, H.; Ohizumi, Y. Amphidinolide-A, a novel antineoplastic macrolide from the marine dinoflagellate *Amphidinium* sp. *Tetrahedron Lett.* **1986**, *27*, 5755–5758. [CrossRef]
25. Yasumoto, T.; Seino, N.; Murakami, Y.; Murata, M. Toxins produced by benthic dinoflagellates. *Biol. Bull.* **1987**, *172*, 129–131. [CrossRef]
26. Kobayashi, J.; Ishibashi, M.; Wälchli, M.R.; Nakamura, H.; Hirata, Y.; Sasaki, T.; Ohizumi, Y. Amphidinolide C: The first 25- membered macrocyclic lactone with potent antineoplastic activity from the cultured dinoflagellte *Amphidinium* sp. *J. Am. Chem. Soc.* **1988**, *110*, 490–494. [CrossRef]
27. Murray, S.A.; Garby, T.; Hoppenrath, M.; Neilan, B.A. Genetic diversity, morphological uniformity and polyketide production in dinoflagellates (*Amphidinium*, Dinoflagellata). *PLoS ONE* **2012**, *7*, e38253. [CrossRef]
28. Moreira-González, A.R.; Fernandes, L.F.; Uchida, H.; Uesugi, A.; Suzuki, T.; Chomérat, N.; Bilien, G.; Pereira, T.A.; Mafra, L.L., Jr. Morphology, growth, toxin production, and toxicity of cultured marine benthic dinoflagellates from Brazil and Cuba. *J. Appl. Phycol.* **2019**, *31*, 3699–3719. [CrossRef]
29. Karafas, S.; Teng, S.T.; Leaw, C.P.; Alves-de-Souza, C.A. An evaluation of the genus *Amphidinium* (Dinophyceae) combining evidence from morphology, phylogenetics, and toxin production, with the introduction of six novel species. *Harmful Algae* **2017**, *68*, 128–151. [CrossRef] [PubMed]

30. Kim, H.S.; Yih, W.; Kim, J.H.; Myung, G.; Jeong, H.J. Abundance of epiphytic dinoflagellates from coastal waters off Jeju Island, Korea during autumn 2009. *Ocean Sci. J.* **2011**, *46*, 205–209. [CrossRef]

31. Hwang, B.S.; Yoon, E.Y.; Kim, H.S.; Yih, W.; Park, J.Y.; Jeong, H.J.; Rho, J.-R. Ostreol A: A new cytotoxic compound isolated from the epiphytic dinoflagellate *Ostreopsis* cf. *ovata* from the coastal waters of Jeju Island, Korea. *Bioorg. Med. Chem. Lett.* **2013**, *23*, 3023–3027. [CrossRef]

32. Lee, K.-W.; Kang, J.-H.; Baek, S.-H.; Choi, Y.-U.; Lee, D.-W.; Park, H.-S. Toxicity of the dinoflagellate *Gambierdiscus* sp. toward the marine copepod *Tigriopus japonicus*. *Harmful Algae* **2014**, *37*, 62–67. [CrossRef]

33. Yang, A.R.; Lee, S.; Yoo, Y.D.; Kim, H.S.; Jeong, E.J.; Rho, J.-R. Limaol: A polyketide from the benthic marine dinoflagellate Prorocentrum lima. J. Nat. Prod. **2017**, 80, 1688–1692. [CrossRef]

34. Lee, S.; Yang, A.R.; Yoo, Y.D.; Jeong, E.J.; Rho, J.-R. Relative configurational assignment of 4-hydroxyprorocentrolide and prorocentrolide C isolated from a benthic Dinoflagellate (*Prorocentrum lima*). *J. Nat. Prod.* **2019**, *82*, 1034–1039. [CrossRef]

35. Lim, A.S.; Jeong, H.J. Benthic dinoflagellates in Korean waters. *Algae* **2021**, *36*, 91–109. [CrossRef]

36. Park, J.; Hwang, J.; Hyung, J.; Yoon, E.Y. Temporal and spatial distribution of the toxic epiphytic dinoflagellate *Ostreopsis* cf. *ovate* in the coastal waters off Jeju Island, Korea. *Sustainability* **2020**, *12*, 5864. [CrossRef]

37. Baek, S.H. First report for appearance and distribution patterns of the epiphytic dinoflagellates in the Korean Peninsula. *Korean J. Environ. Biol.* **2012**, *30*, 355–361. [CrossRef]

38. Jeong, H.J.; Lim, A.S.; Jang, S.H.; Yih, W.H.; Kang, N.S.; Lee, S.Y.; Yoo, Y.D.; Kim, H.S. First report of the epiphytic dinoflagellate *Gambierdiscus caribaeus* in the temperate waters off Jeju Island, Korea: Morphology and molecular characterization. *J. Eukaryot. Microbiol.* **2012**, *59*, 637–650. [CrossRef]

39. Shah, M.M.R.; An, S.-J.; Lee, J.B. Occurrence of sand-dwelling and epiphytic dinoflagellates including potentially toxic species along the coast of Jeju Island, Korea. *J. Fish. Aquat. Sci.* **2014**, *9*, 141–156. [CrossRef]

40. Kang, S.-M.; Lee, J.-B. New records of genus *Dinophysis, Gonyaulax, Amphidinium, Heterocapsa* (Dinophyceae) from Korean waters. *Korean J. Environ. Biol.* **2018**, *36*, 260–270. [CrossRef]

41. Ministry of Oceans and Fisheries. *Improvement of Management Strategies on Marine Ecosystem Disturbing and Harmful Organisms (MoMHO)*; Advanced Institute of Convergence Technology: Suwon, Korea, 2021; p. 751.

42. Bolch, C.J.; de Salas, M.F. A review of the molecular evidence for ballast water introduction of the toxic dinoflagellates *Gymnodinium catenatum* and the *Alexandrium* "*tamarensis* complex" to Australasia. *Harmful Algae* **2007**, *6*, 465–485. [CrossRef]

43. Battocchi, C.; Totti, C.; Vila, M.; Masó, M.; Capellacci, S.; Accoroni, S.; Reñé, A.; Scardi, M.; Penna, A. Monitoring toxic microalgae *Ostreopsis* (dinoflagellate) species in coastal waters of the Mediterranean Sea using molecular PCR-based assay combined with light microscopy. *Mar. Pollut. Bull.* **2010**, *60*, 1074–1084. [CrossRef]

44. Notomi, T.; Okayama, H.; Masubuchi, H.; Yonekawa, T.; Watanabe, K.; Amino, N.; Hase, T. Loop-mediated isothermal amplifica- tion of DNA. *Nucleic Acids Res.* **2000**, *28*, e63. [CrossRef]

45. Karanis, P.; Ongerth, J. LAMP—A powerful and flexible tool for monitoring microbial pathogens. *Trends Parasitol.* **2009**, *25*, 498–499. [CrossRef]

46. Deb, R.; Chakraborty, S. Trends in veterinary diagnostics. *J. Vet. Sci. Technol.* **2012**, *3*, 1. [CrossRef]

47. Li, Y.; Fan, P.; Zhou, S.; Zhang, L. Loop-mediated isothermal amplification (LAMP): A novel rapid detection platform for pathogens. *Microb. Pathog.* **2017**, *107*, 54–61. [CrossRef]

48. Wang, L.; Li, L.; Alam, M.; Geng, Y.; Li, Z.; Yamasaki, S.; Shi, L. Loop-mediated isothermal amplification method for rapid detection of the toxic dinoflagellate *Alexandrium*, which causes algal blooms and poisoning of shellfish. *FEMS Microbiol. Lett.* **2008**, *282*, 15–21. [CrossRef]

49. Picón-Camacho, S.M.; Thompson, W.P.; Blaylock, R.B.; Lotz, J.M. Development of a rapid assay to detect the dinoflagellate *Amyloodinium ocellatum* using loop-mediated isothermal amplification (LAMP). *Vet. Parasitol.* **2013**, *196*, 265–271. [CrossRef]

50. Medlin, L.; Elwood, H.J.; Stickel, S.; Sogin, M.L. The characterization of enzymatically amplified eukaryotic 16S-like rRNA-coding regions. *Gene* **1988**, *71*, 491–499. [CrossRef]

51. Litaker, R.W.; Vandersea, M.W.; Kibler, S.R.; Reece, K.S.; Stokes, N.A.; Steidinger, K.A.; Millie, D.F.; Bendis, B.J.; Pigg, R.J.; Tester, P.A. Identification of *Pfiesteria piscicida* (Dinophyceae) and *Pfiesteria*-like organisms using internal transcribed spacer-specific PCR assays. *J. Phycol.* **2003**, *39*, 754–761. [CrossRef]

52. Kang, N.S.; Jeong, H.J.; Lee, S.Y.; Lim, A.S.; Lee, M.J.; Kim, H.S.; Yih, W. Morphology and molecular characterization of the epiphytic benthic dinoflagellate *Ostreopsis* cf. ovata in the temperate waters off Jeju Island, Korea. *Harmful Algae* **2013**, *27*, 98–112. [CrossRef]

53. Lee, K.H.; Jeong, H.J.; Park, K.; Kang, N.S.; Yoo, Y.D.; Lee, M.J.; Lee, J.; Lee, S.; Kim, T.; Kim, H.S.; et al. Morphology and molecular characterization of the epiphytic dinoflagellate *Amphidinium massartii*, isolated from the temperate waters off Jeju Island, Korea. *Algae* **2013**, *28*, 213–231. [CrossRef]

54. Fernandez-Soto, P.; Mvoulouga, P.O.; Akue, J.P.; Abán, J.L.; Santiago, B.V.; Sánchez, M.C.; Muro, A. Development of a highly sensitive loop-mediated isothermal amplification (LAMP) method for the detection of *Loa loa*. *PLoS ONE* **2014**, *9*, e94664. [CrossRef]

55. Hendel, R.C.; Patel, M.R.; Kramer, C.M.; Poon, M.; Carr, J.C.; Gerstad, N.A.; Gillam, L.D.; Hodgson, J.M.; Kim, R.J.; Lesser, J.R. ACCF/ACR/SCCT/SCMR/ASNC/NASCI/SCAI/SIR 2006 appropriateness criteria for cardiac computed tomography and cardiac magnetic resonance imaging/A report of the American college of cardiology foundation quality strategic directions committee appropriateness criteria working group, American college of radiology, society of cardiovascular computed tomography, Society for cardiovascular magnetic resonance, American society of nuclear cardiology, North American society for cardiac imaging, Society for cardiovascular angiography and interventions, and Society of interventional radiology. *J. Am. Coll. Cardiol.* **2006**, *48*, 1475–1497. [CrossRef]

56. Zhang, F.; Ma, L.; Xu, Z.; Zheng, J.; Shi, Y.; Lu, Y.; Miao, Y. Sensitive and rapid detection of *Karenia mikimotoi* (Dinophyceae) by loop-mediated isothermal amplification. *Harmful Algae* **2009**, *8*, 839–842. [CrossRef]

57. Nagai, S.; Itakura, S. Specific detection of the toxic dinoflagellates *Alexandrium tamarense* and *Alexandrium catenella* from single vegetative cells by a loop-mediated isothermal amplification method. *Mar. Genom.* **2012**, *7*, 43–49. [CrossRef]

58. Zhang, F.; Shi, Y.; Jiang, K.; Song, W.; Ma, C.; Xu, Z.; Ma, L. Rapid detection and quantification of *Prorocentrum minimum* by loop-mediated isothermal amplification and real-time fluorescence quantitative PCR. *J. Appl. Phycol.* **2014**, *26*, 1379–1388. [CrossRef]

CHAPTER-15

NUMERICAL STUDY ON THE MASSIVE OUTBREAK OF THE *ULVA PROLIFERA* GREEN TIDES IN THE SOUTH-WESTERN YELLOW SEA IN 2021

1. INTRODUCTION

The Yellow Sea, surrounded by Mainland China and the Korean Peninsula, is a characteristic continental shelf sea (Figure 1). It plays an important role in the environment of China and the Korean Peninsula. The bathymetry of the Yellow Sea is generally shallow and complex. The south-western Yellow Sea is listed as a marginal sea with various contributions from land, rivers, and tides, which has been the focus of multi-disciplinary research in recent years.

Over the past 15 years (from 2007 to 2021), successively occurrences of *Ulva prolifera* green tides have become a striking recurrent phenomenon in the south-western Yellow Sea. The average distribution area and the cover area were reported to be 37,000 km^2 and 450 km^2, respectively, in recent ten years. The general consensus is that the green tide originates from the coast of Subei in late spring. It migrates northward with the ocean current [1–3] and lands on the Lunan coast over long-distance migration every summer. The *Ulva prolifera* green tides have had detrimental effects on the local marine environment and ecosystem [4]. According to the previous studies, *Ulva prolifera* has wide adaptability to temperature and salinity [5–7]. The freshwater from the Changjiang River and local rivers provide abundant nutrients into the south-western Yellow Sea and is also suitable for the growth of *Ulva prolifera*. Dissolved inorganic nitrogen (DIN) has been suggested to play a crucial role in the bloom of *Ulva prolifera* [8–11]. Chen et al. [12] found that the nutrient distribution and structure in the Jiangsu coast was affected by the land-based load, and the ratio of nitrogen to phosphorus could affect the bloom of *Ulva prolifera* significantly. Wang et al. [13] suggested the phosphate limitation on the initial growth of *Ulva prolifera* seedling can occur in the southern Yellow Sea. Later, Sun et al. [14] confirmed that the DIN was the most critical nutrient controlling the magnitude and time of the green tide rather than phosphorus, based on a dynamic growth model. According to the observations, the DIN concentration in the south-western Yellow Sea is lowest in the winter and reaches its highest value in April [7,11,15], which implies that overwintering banks of nutrients might be an important population trigger of *Ulva prolifera* [16,17].

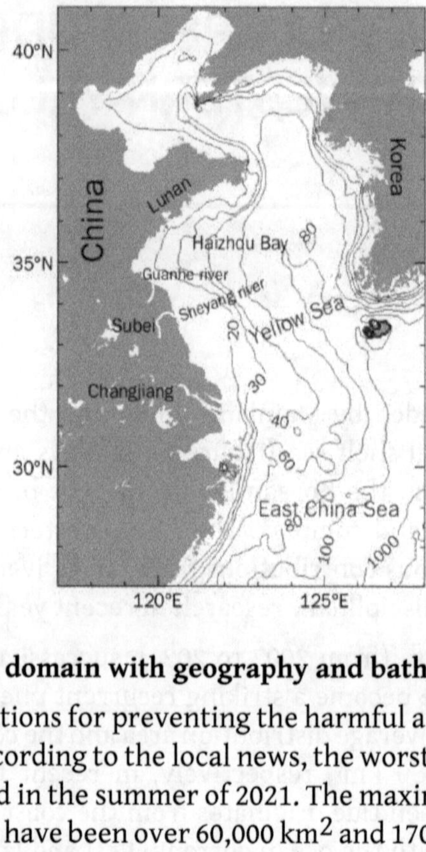

Figure 1. Model domain with geography and bathymetry (unit: m).

However, effective solutions for preventing the harmful algal bloom have not been found so far [17,18]. According to the local news, the worst *Ulva prolifera* green tides in last 15 years occurred in the summer of 2021. The maximal distribution area and the maximal cover area have been over 60,000 km^2 and 1700 km^2, respectively. The amount of the *Ulva prolifera* was about 2.3 times that of 2013, which was reported to be the worst year previously.

Thus, based on observational data and numerical modelling, the hydrographical features of the south-western Yellow Sea from the winter 2020–2021 to summer 2021 were compared with the long-term averaged status. Section 2 comprises a description of the methodology and data used in this study. Section 3 addresses the model results about the inter-annual variation of the green tides. Section 4 is a summary of the study.

2. DATA AND METHODOLOGY

2.1 Observation Data

The water depth in the nearshore region of Subei is quite shallow, and the observation data are relatively scarce. Supported by the Project on Coastal Investigation and Research (i.e., Project 908) of China, four cruises in four seasons were conducted during 2006 and 2007, mostly covering the Subei coast region. The measured salinity data were reported by Zhang et al. [19]. The digitalized isohaline was compared with the simulated sea surface salinity.

Merged satellite and in situ global daily sea surface temperatures (MGDSSTs) of the Japan Meteorological Agency, with a resolution of 0.25° × 0.25°, were used to validate the simulated sea surface temperatures of the present model.

Furthermore, in late April of 2021, four GPS-tracked ARGOS surface floats were deployed in the south-western Yellow Sea to measure the currents off the Subei coast. The floats were released along the edge of the Subei Bank to track the movement of the Subei coastal current. The trajectories of four surface ARGOS drifters were used to check the simulated circulation.

2.2 Numerical Modeling

Based on the Princeton Ocean Model (POM), we established a high-resolution regional circulation model over the Bohai Sea, the Yellow Sea, and a part of the East China Sea. The model, with a horizontal resolution of 1/12° × 1/12° , covered (Figure 1) the domain of 27° N–41° N, 117° E–128° E. The vertical sigma coordinates were layered in the following proportions from top to bottom: 0.000, -0.003, -0.006, -0.013, -0.025,-0.050, -0.100, -0.200, -0.300, -0.400, -0.500, -0.600, -0.700, -0.800, -0.900, and-1.00. The model topography was based on the $1' \times 1'$ Shuttle Radar Topography Mission (SRTM) database. The minimum and maximum depths in this model were set to 10 m and 4000 m, respectively.

The boundary conditions were determined by the simulated results of a 1/12° × 1/15° Eastern Asian Marginal Seas data assimilation model [20]. The monthly mean results from 2001 to 2020 were averaged as the climatologically open boundary of the present study. The eight main tidal components (M2, S2, K1, O1, N2, P1, K2, and Q1) also were considered from the open boundary. The tidal harmonic constants were decided by the results of the NAO.99b model [21].

The surface wind stress was determined by the monthly forcing of the ERA5 dataset with a high horizontal resolution of 0.25° × 0.25°. The long-term averaged values were calculated during the same 20-year period described in the previous paragraph as the climatologically forcings. The surface heat flux was calculated using a bulk formula [22]. The net heat flux was expressed as the sum of shortwave radiation, long wave radiation, sensible heat flux, and latent heat flux, and all of these components followed the empirical formulas of Hirose et al. [20]. The penetration of the shortwave radiation was also considered, and the water quality was type II [23].

The freshwater flux was estimated with the precipitation (P) and evaporation (E) as well as the river runoffs (R), as P + R - E. Precipitation (P) data were also obtained from the long-term averaged monthly ERA5 dataset. The corresponding evaporation (E) was obtained when calculating the latent heat flux as mentioned above. The amounts of the freshwater from Changjiang River were retrieved from the climatologically monthly transports at Datong station, which are reported by the Chinese River Sediment Bulletin. The multi-year mean discharges from Subei local rivers were retrieved from the Jiangsu Province Water Resources Bulletin, which are represented by the Guanhe River and the Sheyang River in this study [24]. The discharges of the Guanhe River and Sheyang River and their seasonal variations were also allotted based on the work of Yang [24]. There were no other relaxations for temperature and salinity in the present model.

The model was first integrated with the climatological fields for 4 years. An additional two-year calculation was conducted due to the overwinter effects, using the above last month's status for the restart conditions. The experiment that used the climatological meteorological forcing and boundary conditions was named Exp.C (referring to climatological meteorological forcing). To investigate the environmental features in the winter 2020–2012, a comparative experiment was designed (Table 1). The realistic monthly meteorological forcing and boundary conditions were employed and named Exp.R (referring to realistic forcings). Exp.R was established under the daily meteorological forcing from January 2020 to June 2021, and the Changjiang River discharges were also changed to the realistic monthly discharges. The values and their comparisons to meteorological values during the interested period are listed in Table 2.

Table 1. List of experiments.

Experiments	Discharges	Forcings and Boundary Conditions
Exp.C	Climatological monthly discharges of Changjiang River and Subei local rivers	Climatological meteorological forcings and boundary conditions
Exp.R	Realistic monthly discharge of Changjiang River and climatological discharge of Subei local rivers	Realistic daily meteorological forcings and boundary conditions

Table 2. Changjiang discharges (108 m3).

Time	December 2020	January 2021	February 2021	March 2021	April 2021	May 2021	June 2021
Monthly Changjiang discharge (related to meteorological values)	436.6 (+10.5%)	407.1 (+31.4%)	326.6 (+9.3%)	530.3 (+16.9%)	686.9 (+10.7%)	1122 (+24.6%)	1216 (+15.8%)

2.3 Passive Tracer

Salinity distribution can present the pathway of the Changjiang River qualitatively. To quantitatively discuss the roles of the Changjiang River discharges, the passive tracer was continuously released in the numerical experiments from December to the following June at the Changjiang River estuary. The initial value of the passive tracer was 0, which means the nutrient content from the Changjiang River was 0 at the beginning. The tracer concentration at the estuary was set to 1, which was dimensionless and represented the nutrient concentrations inputted from the Changjiang River. According to Zhang et al. [11], the area south of 35° N and west of 122° E is one in which the *Ulva prolifera* was rapidly developed. Therefore, the persisted tracer content in the upper layer (within 5 m depth) of the region 32–35° N, 119–122° E was also calculated, following Formula (1):

$$T_c = \int c\, dv \qquad (1)$$

where T_c is the content of tracer in the upper layer, and c is the tracer's concentration.

3. RESULTS

3.1 Model Validations

The simulated sea surface temperatures (SSTs) in Exp.C were compared with MGDSSTs. The basic features of the SST in summer and winter were well-represented (Figure 2). Our model successfully reproduced the surface cold patch off the Subei coast in the summer and the intrusion of the high temperature water in the winter. There were slightly larger differences in the near shore region (shallower than 10 m) between the simulated SST and MGDSSTs. This may have been due to the lower horizontal resolution of MGDSSTs. Furthermore, shown in Figure 3, the root-mean-square errors between simulated SST and MGDSSTs were 0.99 °C in the summer and 1.15 °C in the winter after interpolating the MGDSSTS to the model grids.

The digitalized isohalines in the south-western Yellow Sea in each seasons were over- laid on the simulated results of Exp.C (Figure 4). It should be noted that the digitalized isohalines were based on observations during a particular period. The simulated salinities represent the climatological distributions. Therefore, more attention should be paid to the pattern of simulated isohalines instead of the reproduction of isolated low-salinity water patches. The low-salinity water (<30 PSU) of Changjiang diluted water occupies most of the southwest Yellow and East China Seas in the summer (Figure 4c). Though with a significantly reduced range, it retreats to the coast with large horizontal gradients in other seasons. The 30 and 31 PSU isohalines in individual seasons were also found to be in good agreement with observations. All the characteristics, including the seasonal evolutions of the isohalines (Figure 4a–d), were well-simulated by Exp.C, implying that the present model is able to capture the essential mass transportations in this area.

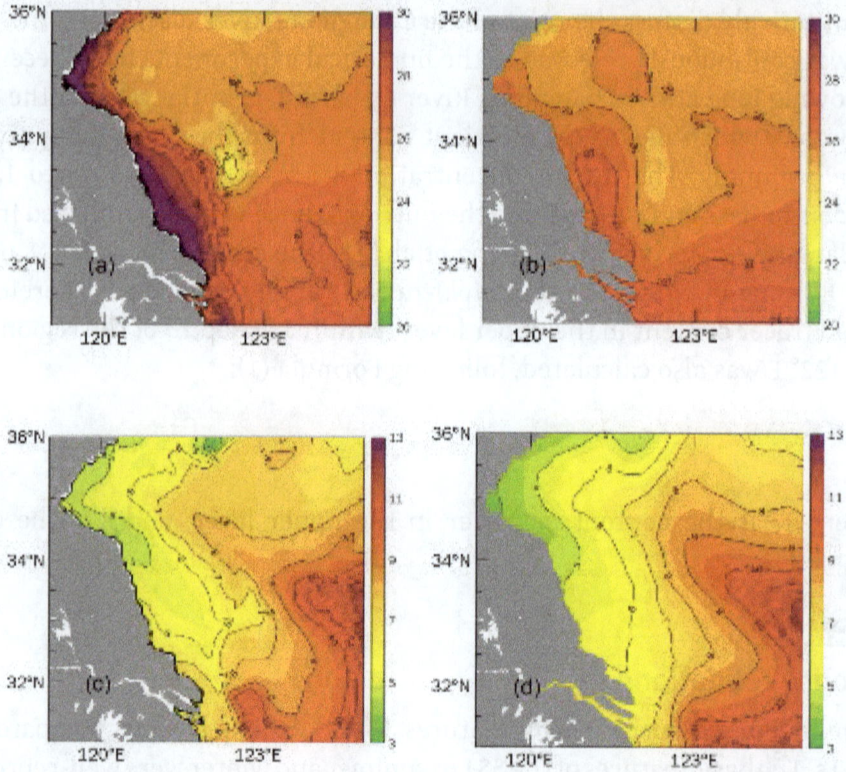

Figure 2. Simulated summer sea surface temperature of (a) Exp. C and (b) MGDSSTs; simulated winter sea surface temperature of (c) Exp.C and (d) MGDSSTs (unit: °C).

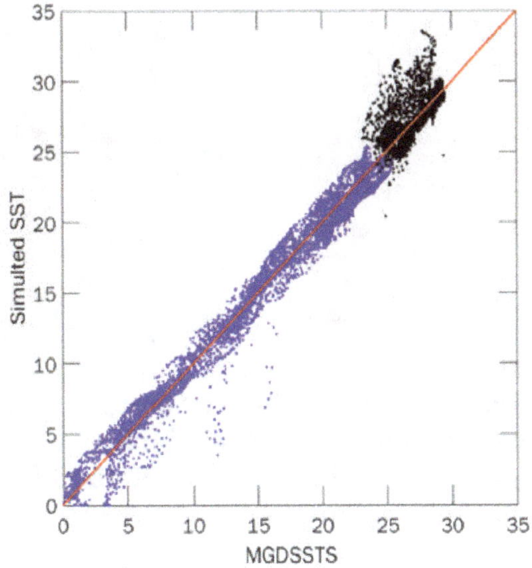

Figure 3. Validations of simulated surface temperature with MGDSSTs. Black and blue dots represent summer and winter values, respectively (unit: °C).

Four ARGOS surface drifts (Figure 5a) were deployed off the Subei coast since late April of 2021. For comparison, in the realistic case of Exp.R, eleven modeled drifters (Figure 5b) were released at the sea surface. The initial locations of the simulated drifters followed those of the ARGOS drifts. Simulated circulation was represented by the pathway of the modeled surface drifters. The trajectories of the modeled drifters, which were deemed to be Lagrangian particles, were calculated by the fourth- order Runge–Kutta scheme [25]. The positions of the drifters were calculated every 3 computational hours until end of June 2021. Figure 5 suggests that the pattern of the modeled drifter trajectories successfully reproduced the observed ones of ARGOS surface drifters. The characteristics of the ARGOS trajectories were well-simulated by this experiment, once again showing that the present model can capture the realistic features of regional circulation.

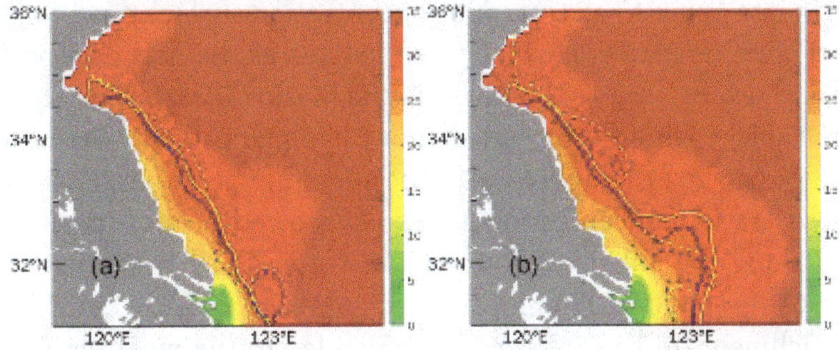

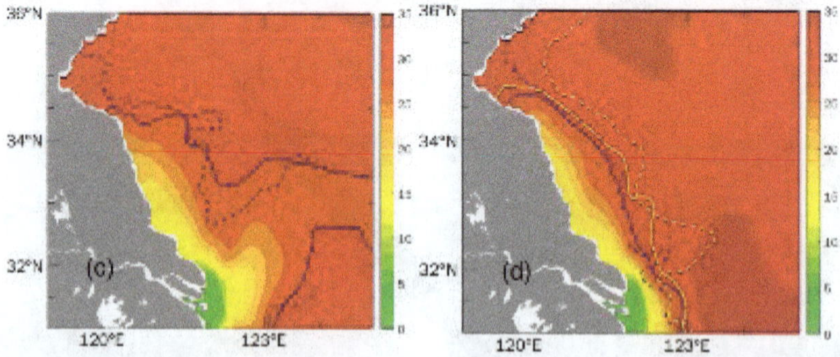

Figure 4. Simulated climatological salinity at 10 m of Exp.C in (a) winter, (b) spring, (c) summer, and (d) autumn (unit: PSU). The solid line is the simulated isohaline, and the dashed line is the observed isohaline digitalized from the work of Zhang et al. [19]. The blue and yellow lines represent the 30 and 31 PSU, respectively.

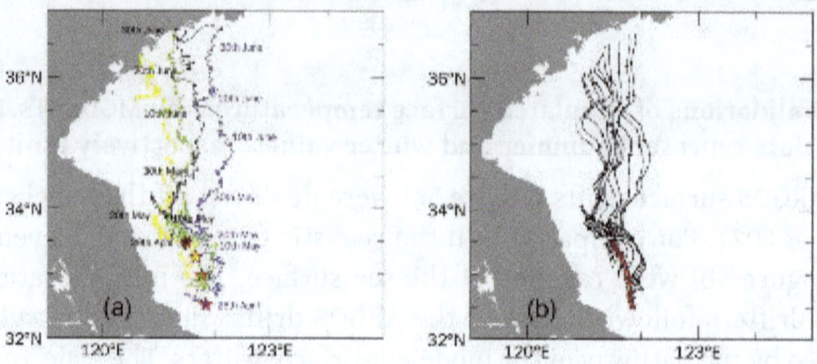

Figure 5. Observed (a) and simulated (b) trajectories of surface ARGOS drifters released in late April 2021 off the Subei coast. Star markers represent the released locations.

3.2 Hydrography in the Summer of 2021

The simulated tracer distributions at the surface until late April are shown in Figure 6. In Exp.C, a part of the Changjiang River water drifted northward, which could reach 33° N in late April. A relative high concentration surface patch was identified off the Subei coast (Figure 6a), which was reported as the first observed region of *Ulva prolifera* in several years (Bulletin of China Marine Disaster). In the meantime, as listed in Table 2, the discharges of the Changjiang River were significantly increased since late 2020 compared with the climatological ones. Correspondingly, the simulated tracer concentration in Exp.R showed a large positive anomaly distribution along the Subei coast. In other words, the larger Changjiang River input in the winter 2020–2021 and spring of 2021 significantly contributed to the nutrient enrichment off the Subei coast, which led to the most massive outbreak of the *Ulva prolifera* green tides in the south-western Yellow Sea the in summer of 2021.

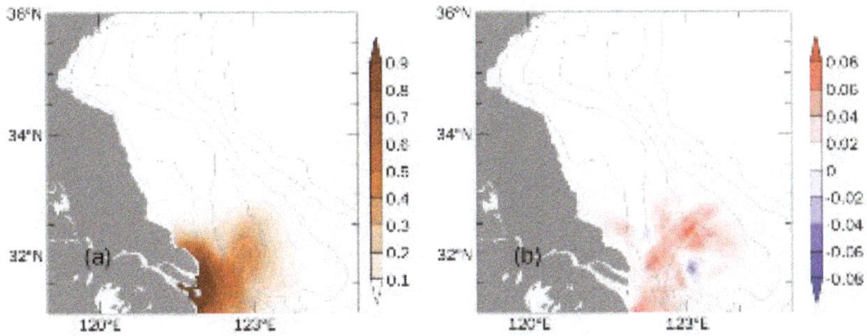

Figure 6. Simulated distribution of the passive tracer (a) in Exp.C and (b) anomaly distribution in Exp.R at the sea surface in late April with the bathymetry.

The tracer content results showed that in late April, T_C was found to be 1.20 × 10^{10} m^3 in Exp.C, which was almost 5% of the total Changjiang input, while in Exp.R, T_C approached 1.56 × 10^{10} m^3 in late April, which was more than 6% of the total Changjiang input during this month. The northward transportation proportion of Changjiang water also was increased. Thus, we checked the wind distribution in the winter of 2020–2021 (Figure 7). In the winter of 2020–2021, the southerly anomaly led to a much weaker north–south component of wind (Figure 7b). Thus, the weakened southward wind-driven current was at a disadvantage in competition with the northward tidal residual current [26,27]. The counter-wind transport caused by the northward tidal residual current was dominant during this winter. Thus, there was more remaining winter freshwater from Changjiang at the Subei coast in the spring 2021. Compared with the climatological status, it increased by about 30%.

Until June, the remaining tracer content from the Changjiang River was found to be significantly increased (Figure 8a) compared to that at the end of April, which provided necessary nutrient supplements for the population of *Ulva prolifera*. The passive tracer concentration in early summer of 2021 presented a stronger positive anomaly related to the long-term averaged case (Figure 8b). Without considering the biological absorption process, the passive tracer content results suggested that until late June, T_C was 4.79 × 10^{10} m^3 and 5.21 × 10^{10} m^3 for Exp.C and Exp.R, respectively.

The climatological wind during spring over the southwest Yellow Sea distributed northwestward (Figure 9a). It can be seen that there was a clear easterly wind anomaly during spring of 2021 (Figure 9b). The enhanced westward wind component contributed to the shoreward transport of *Ulva prolifera*, which led to its landing along the Lunan coast over long-distance migration.

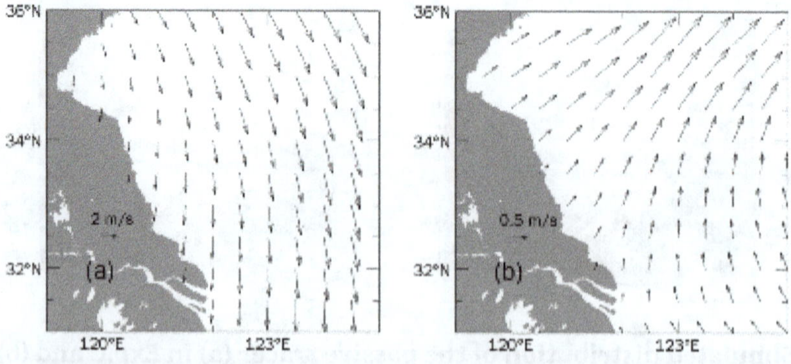

Figure 7. (a) Climatological winter wind distribution and (b) wind anomaly in the winter of 2020–2021 at 10 m above the south-western Yellow Sea.

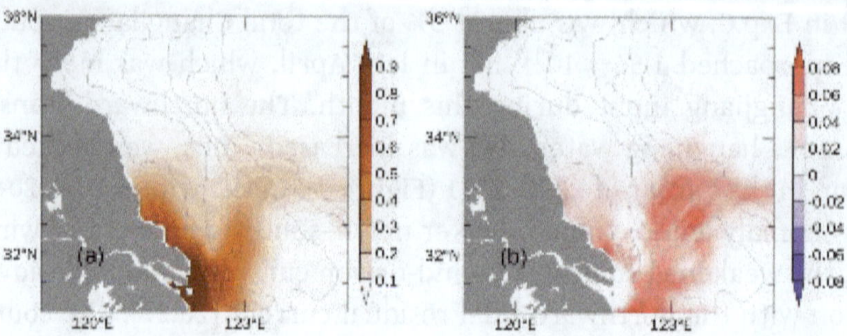

Figure 8. Simulated distribution of the passive tracer (a) in Exp.C and (b) anomaly distribution in Exp.R at the sea surface in late June with the bathymetry.

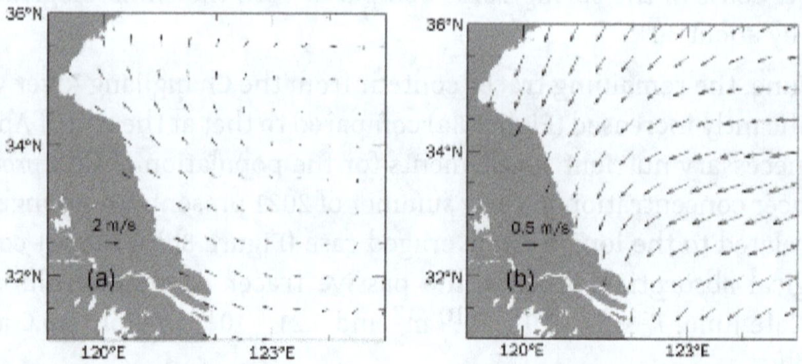

Figure 9. (a) Climatological spring wind distribution and (b) wind anomaly in the spring of 2021 at 10 m above the south-western Yellow Sea.

Altogether, these results suggest that the nutrient contents from the Changjiang River were increased in the winter 2020–2021 and spring 2021 due to the increase of discharges. Additionally, the southerly wind stress anomaly during winter 2020–2021 also favoured the upwind transport of Changjiang water. According to the tracer content, overall, the remaining freshwater coming from the Changjiang

River, which persisted in the upper layer of 32–35° N, 119–122° E region during spring 2021 (April, May, June), exceeded the climatological value by nearly 20%. At the same time, the stronger easterly wind component during spring of 2021 was conducive to the landing of *Ulva prolifera* along the Lunan coast. It should be pointed out that these experiments did not consider the nutrient concentration from Changjiang River in the individual years. Further quantitative investigations remain to be conducted in the future.

4. CONCLUDING REMARKS

The most massive outbreak on record of the *Ulva prolifera* green tides in the south-western Yellow Sea occurred in summer of 2021. The environmental factors were investigated based on observations and simulations.

The results suggested that the enhanced discharges of the Changjiang River and the southerly wind stress anomaly in winter 2020–2021 were crucial for the outbreak of the *Ulva prolifera* green tides in the south-western Yellow Sea, which could remarkably contribute to the nutrient enrichment off the Subei coast. The remaining freshwater coming from the Changjiang River, which persisted on the coast of Subei in the spring 2021, exceeded the climatological value by nearly 20%. We demonstrated that these large amounts of nutrient inputs, as an effective supplement, were the reason the green tides sharply emerged as an extensive outbreak in 2021. Subsequently, the strong easterly wind anomaly during spring of 2021 led to shoreward transport, which was helpful for the large number of *Ulva prolifera* to land off the Lunan coast.

REFERENCES

1. Yuan, D.; Zhu, J.; Li, C.; Hu, D. Cross-shelf circulation in the Yellow and East China Seas indicated by MODIS satellite observations. *J. Mar. Syst.* **2008**, *70*, 134–149. [CrossRef]
2. Hu, C.; He, M.X. Origin and offshore extent of floating algae in Olympic sailing area. *Eos Trans. AGU* **2008**, *89*, 302–303. [CrossRef]
3. Qiao, F.; Lü, X. Coastal upwelling in the South China Sea. In *Satellite Remote Sensing of South China Sea*; Liu, A.K., Ho, C.R., Liu, C.T., Eds.; Tingmao Publish Company: Taipei, Taiwan, 2008; pp. 135–158.
4. Liu, D.; Keesing, J.K.; Dong, Z.; Zhen, Y.; Di, B.; Shi, Y.; Fearns, P.; Shi, P. Recurrence of the world's largest green-tide in 2009 in Yellow Sea, China: Porphyra yezoensis aquaculture rafts confirmed as nursery for macroalgal blooms—Science Direct. *Mar. Pollut. Bull.* **2010**, *60*, 1423–1432. [CrossRef]
5. Taylor, R.; Fletcher, R.L.; Raven, J.A. Preliminary studies on the growth of selected 'green tide' algae in laboratory culture: Effects of irradiance

temperature, salinity and nutrients on growth rate. *Bot. Mar.* **2001**, *4*, 327–336. [CrossRef]

6. Zhang, X.; Wang, Z.; Li, R.; Li, Y.; Wang, X. Microscopic observation on population growth and reproduction of Eltrolnorphra prolifera under different temperature and salinity. *Adv. Mar. Sci.* **2012**, *30*, 276283.
7. Yuan, K.; Hong, C.; Ding, Y.; Song, X. Analysis of the seasonal and inter-annual changes of environmental factors in *Enteromorpha prolifera* green tide outbreak in the Yellow Sea. *J. Guangxi Acad. Sci.* **2018**, *34*, 204–209.
8. Wu, T.; Zhao, L.; Liu, H.; Wang, T.; Han, X.R.; Shi, X.Y. Preliminary study on the influence of Enteromorpha prolifera on nutrients. Mar. Environ. Sci. **2013**, 32, 347–352. (In Chinese)
9. Li, H.M.; Tang, H.J.; Shi, X.Y.; Zhang, C.S.; Wang, X.L. Increased nutrient loads from the Changjiang (Yangtze) River have led to increased harmful algal blooms. *Harmful Algae* **2014**, *39*, 92–101. [CrossRef]
10. Wang, J.; Yu, Z.; Wei, Q.; Yang, F.; Dong, M.; Li, D.; Gao, Z.; Yao, Q. Distributions of nutrients in the southwestern yellow sea in spring and summer of 2017 and their relationship with Ulva prolifera outbreaks. *Oceanol. Limnol. Sin.* **2018**, *49*, 1045–1053. (In Chinese)
11. Zhang, H.; Liu, K.; Su, R.; Shi, X.; Pei, S.; Wang, X.; Wang, G.; Wang, S. Study on the coupling relationship between the development of *Ulva prolifera* green tide and nutrients in the southern Yellow Sea in 2018. *Haiyang Xuebao* **2020**, *42*, 30–39. (In Chinese) [CrossRef]
12. Chen, Y.; Song, D.; Li, K.; Gu, L.; Wei, A.; Wang, X. Hydro-biogeochemical modeling of the early-stage outbreak of green tide (*Ulva prolifera*) driven by land-based nutrient loads in the Jiangsu coast. *Mar. Pollut. Bull.* **2020**, *153*, 111028. [CrossRef]
13. Wang, C.; Jiao, X.; Zhang, Y.; Zhang, L.; Xu, H. A light-limited growth model considering the nutrient effect for improved understanding and prevention of macroalgae bloom. *Environ. Sci. Pollut. Res.* **2020**, *27*, 1–9. [CrossRef]
14. Sun, K.; Ren, J.S.; Bai, T.; Zhang, J.; Liu, Q.; Wu, W.; Zhao, Y.; Liu, Y. A dynamic growth model of *Ulva prolifera*: Application in quantifying the biomass of green tides in the Yellow Sea, China. *Ecol. Model.* **2020**, *428*, 109072. [CrossRef]
15. Li, H.M.; Zhang, C.S.; Han, X.R.; Shi, X.Y. Changes in concentrations of oxygen, dissolved nitrogen, phosphate, and silicate in the southern yellow sea, 1980–2012: Sources and seaward gradients. *Estuar. Coast. Shelf Sci.* **2015**, *163*, 44–55. [CrossRef]

16. Zhang, Y.; He, P.; Li, H.; Li, G.; Liu, J.; Jiao, F.; Zhang, J.; Huo, Y.; Shi, X.; Su, R.; et al. *Ulva prolifera* green tide outbreaks and their environmental impact in the Yellow Sea, China. *Neurosurgery* **2019**, *4*, 825–838. [CrossRef]
17. Gong, N.; Shao, K.; Shen, K.; Gu, Y.; Ye, J.; Hu, C.; Shen, L.; Chen, Y.; Li, D.; Fan, J. Chemical control of overwintering green algae to mitigate green tide in the Yellow Sea. *Mar. Pollut. Bull.* **2021**, *168*, 112424. [CrossRef] [PubMed]
18. Luo, Q.J.; Yan, X.J.; Xu, S.L.; Xu, J.L.; Zhou, C.X.; Ma, B.; Yang, R.; Pei, L.Q. Method for Acid Treatment and Treatment of Green Algae and Infected Cells during Porphyra Haitanensis Cultivation. Chinese Patent CN 101822271B, 21 November 2012. (In Chinese)
19. Zhang, Z.X.; Guo, J.S.; Qiao, F.L.; Liu, Y.Y.; Guo, B.H. Whereabouts and freshwater origination of the Subei coastal water. *Oceano Logia Limnol. Sin.* **2016**, *47*, 527–532. (In Chinese)
20. Hirose, N.; Lee, H.C.; Yoon, J.H. Surface heat flux in the East China Sea and the Yellow Sea. *J. Phys. Oceanogr.* **1999**, *29*, 401–417. [CrossRef]
21. Matsumoto, K.; Takanezawa, T.; Ooe, M. Ocean tide models developed by assimilating TOPEX/POSEIDON altimeter data into hydrodynamical model: A global model and a regional model around Japan. *J. Oceanogr.* **2000**, *56*, 567–581. [CrossRef]
22. Kondo, J. Air-sea bulk transfer coefficients in diabetic conditions. *Bound. Layer Meteor.* **1975**, *9*, 91–112. [CrossRef]
23. Paulson, C.A.; Simpson, J.J. Irradiance measurements in the upper ocean. *J. Phys. Oceanogr.* **1977**, *7*, 952–956. [CrossRef]
24. Yang, H. A Research on Sustainable Development Strategic Option of Jiangsu Coast Beaches Resources. Ph.D. Thesis, China University of Geosciences, Beijing, China, 2012. (In Chinese)
25. Wang, B.; Hirose, N.; Moon, J.-H.; Yuan, D. Difference between the Lagrangian trajectories and Eulerian residual velocity fields in the southwestern Yellow Sea. *Ocean Dyn.* **2013**, *63*, 565–576. [CrossRef]
26. Wu, H.; Gu, J.; Zhu, P. Winter Counter-Wind Transport in the Inner Southwestern Yellow Sea. *J. Geophys. Res. Ocean.* **2018**, *123*, 411–436. [CrossRef]
27. Wei, Q.; Wang, B.; Fu, M.; Sun, J.; Yao, Q.; Xin, M.; Yu, Z. Spatiotemporal variability of physical-biogeochemical processes and intrinsic correlations in the semi-enclosed South Yellow Sea. *Acta Oceanol. Sin.* **2020**, *39*, 11–26. [CrossRef]

INDEX

A

Ability 82, 210, 235, 289

Abstract 93, 115, 276, 312, 314

Academic 330, 352, 354, 356, 359, 362-364, 367

Actors 18, 67, 83, 127, 152, 158, 226, 285-286, 290, 292, 307, 359, 361

Adults 41

Advertiser 142

Advisory 130, 152

Advocates 325

Agenda 21

Antitrust 225, 237, 242-243

Anxiety 204

Audits 261, 267, 335

Automation 108, 112, 117, 319

B

Backlash 47, 192

Bombard 71

Borders 104

Bottom-up 172

Brandeis 324-326

Breakdown 258

Breaking 18

Bricker 149

Britain 236

Broadcast 235-236, 239-240

Bulletin 211

C

Cations 284

Census 360, 365

Channels 162, 239, 286

Chilling 276

Ciampaglia 19, 27, 120

Citizens 4, 36, 40-41, 43-44, 46, 48-49, 109, 134, 200

Citron 75, 87, 308

Climate 191, 210, 220

Cognition 210-211, 213, 216, 218

Commission 131-132, 148, 152, 258, 260, 271, 279, 282, 335, 356, 368

Congress 134, 154

Consensus 64, 110, 123, 131, 227
Corporate 154, 325-327, 348-349, 364
Covid- 3
Crafting 305
Crawlers 142

D

Demand 15
Diffusion 18, 20, 22, 31
Digital 1, 4, 35, 37, 63, 67, 71, 109, 121, 125-126, 130, 135-138, 140-141, 144, 146-147, 151, 156, 158, 160-167, 170-174, 177-179, 181, 219, 233, 243, 247, 251, 280, 321-322, 327-328, 337, 339, 341, 344, 347-348, 356, 360, 367, 369
Domains 14
Domestic 73, 172
Dominance 240
Dominant 242
Donation 144
Dramatic 172
Drawbacks 141
Dynamics 83

E

Economics 55
Economists 237
Ecosystem 14-15
Elections 26, 29, 128, 133, 152, 156, 230, 260, 369
Electronic 120, 272
Elements 258
Emerged 321, 326
Emotions 204
Engines 167, 169-170, 179
Engstrom 264
Episodes 73
Evidence 21, 24-25, 29, 37, 40-41, 45, 51-52, 55, 72, 83, 114, 139, 145, 193, 197, 210, 248, 362

F

Footage 257
Forbes 92

G

Gender 64, 139, 171, 294
Genocide 76, 96
Graphic 332

H

Harassment 71, 165, 171
Harbor 356
Harmful 61-63, 67
Hypothesis 46

I

Ideology 200
Immunity 294-296, 301, 304, 310, 318
Informal 164
Interviews 12, 263
Investment 159

J

Journalism 107, 117-118, 157, 176, 179-180, 219, 314
Judgments 217
Justice 342

M

Machine 103, 112, 319, 343
Magazine 315

Management 92

Markets 167, 225, 236, 242

Mechanisms 51, 190, 224-225, 335, 337

Minority 39, 266

Mobile 85, 161

Models 194, 226, 244, 319

Moderate 304

Moderation 52, 79, 247, 254, 256, 263-264, 272, 276, 304, 330, 333, 348

Moderators 79, 185, 199, 262-263

Monitor 105, 251

N

Networking 42-43

Newsroom 57, 150, 161, 271-272, 345-349, 169-170, 177-179

O

Offensive 291

Omission 128

Ongoing 156

Online 2-3, 7, 9, 14-18, 21-22, 24-26, 28-29, 31-32, 39-43, 47, 53-56, 61-64, 66-67, 69-84, 86-91, 93-97, 100-105, 107, 109, 111, 113, 117-118, 120-123, 125-136, 138, 141, 144-149, 151-153, 163, 165-166, 168-173, 177, 179, 195, 247, 252, 265-266, 269, 271, 277-279, 283-295, 297-299, 301, 304, 308-309, 311, 317, 321, 346-347

Opinions 47, 131, 152

Opposition 111

P

Pandemic 3

Parliament 279, 281

Patient 209

Phenomenon 72, 362

Phones 153

Pluralism 226-229, 241

Policies 136, 241, 332-333, 368

Popularity 72-73

Porter 192, 217

Precursor 102

Primary 64, 184, 235, 256, 285

Principles 279

Promoting 228-229, 287

Promotion 269, 300

Propaganda 7, 9, 23, 25-28, 32, 99-101, 108-113, 116-117, 119, 123, 313-314, 317

Proposals 3, 132, 283, 297

Protection 233, 269, 355, 368

Psychology 57, 88, 209-211, 213-214, 218

Publicity 342

Publicly 335, 359

R

Radical 328, 346, 361

Reasoning 30, 185, 189, 201, 216, 294, 301

Recode 312

Regulatory 153, 236, 243, 285, 289, 302, 304

Reports 27, 41, 254-255, 258-262, 287, 334-335, 337

Revenues 161

Revolution 131

Rights 228, 251-253, 255, 261, 269, 273, 275, 278, 322, 331, 337, 348

Robust 300, 309

S

Scholars 23, 101, 126, 186, 191, 197, 202, 205, 207, 351, 353, 361, 366-367

Shaping 283

Silverman 31, 284

Singer 150

Software 101, 112

Solutions 113

Speech 61-70, 72-98, 224, 233, 239-240, 248, 250-252, 260, 265-266, 269, 271, 274, 279, 281-282, 306, 330, 332-333, 344, 346

Spending 125, 128, 134, 140-144

Sphere 73

Sponsor 131, 142-143

Sponsored 128, 130, 133

Sponsors 141

Spotlight 184

Stereotype 192

Subtyping 192

Sunstein 38

Supervisor 355

Supply 9, 11, 13

Surveys 74

Systems 108-109, 113, 157, 164, 172, 226, 247-248, 253, 264, 290, 304, 328-329, 332-333, 338, 341

T

Tactics 13, 300

Technology 2, 37, 94, 101, 113, 149-150, 158, 174, 231, 255, 264, 270, 274-275, 312, 314, 316, 321, 343, 351

Television 125, 128, 130, 138, 141, 145-146, 150, 161, 165-166, 172, 231, 235

Tensions 75

Tweets 12, 69, 72, 80

Twitter 12-13, 16-20, 29-32, 40, 50, 58, 64, 66, 68-73, 77-78, 82-83, 85-86, 92-94, 96-97, 100, 102-105, 110-111, 114, 116-122, 135, 254-255, 260, 273, 277, 279-281, 311-312, 331, 334, 341, 343, 349, 353, 369

U

Universe 137

Upticks 73

Useful 104, 337

V

Vaccine 216

Validity 30, 369

Valley 318

Variable 295

Variety 62, 144, 164, 233

Vector 65

Violations 255

Violence 35, 51, 61, 63-64, 73, 75-77, 79-80, 85, 87, 90, 94, 265, 332

Voters 22, 149

W

Wagner 335

Warnings 216

Warren 369, 371

Website 43, 142, 294

Welfare 10, 241

Whatsapp 23-24, 30, 365-366, 369

Y

Youtube 64, 68, 98, 145, 150, 234, 258-259, 272-273, 278, 282, 317, 320, 345

Z

Zealand 271

www.ingramcontent.com/pod-product-compliance
Lightning Source LLC
LaVergne TN
LVHW062225200926
848484LV00006B/1475